Spectroscopy

VOLUME TWO

SPECTROSCOPY

VOLUME ONE

Atomic spectra; Nuclear magnetic resonance spectroscopy; Nuclear quadrupole resonance spectroscopy; Electron spin resonance spectroscopy; Mössbauer spectroscopy.

VOLUME TWO

Molecular spectra; Symmetry and group theory; Microwave spectroscopy; Infra-red and Raman spectroscopy; Far-infrared spectroscopy; Force constants; Thermo-dynamic functions.

VOLUME THREE

Molecular quantum numbers; Electronic spectra of diatomic molecules; Dissociation energies of diatomic molecules; Electronic spectra of polyatomic molecules; Fluoresence and phosphorescence spectroscopy; Astrochemistry; Photoelectron spectroscopy.

Spectroscopy
VOLUME TWO

EDITED BY

B.P. STRAUGHAN Ph.D., F.R.I.C.

Department of Chemistry,
University of Newcastle upon Tyne,
England

AND

S. WALKER, M.A., D.PHIL., D.Sc.

Chairman, Department of Chemistry,
Lakehead University,
Ontario, Canada

LONDON
CHAPMAN AND HALL

A HALSTED PRESS BOOK
JOHN WILEY & SONS INC., NEW YORK

First published 1976
by Chapman and Hall Ltd.,
11 *New Fetter Lane, London EC4P 4EE*

Typeset by EWC Wilkins Ltd., London and Northampton
and printed in Great Britain by
Butler and Tanner Ltd., Frome and London

ISBN 0 412 13360 1 (*cased edition*)
ISBN 0 412 13370 9 (*paperbound edition*)

Distributed in the U.S.A. by Halsted Press,
a Division of John Wiley & Sons, Inc., New York

Library of Congress Cataloging in Publication Data
Main entry under title:

Spectroscopy.

Previous editions by S. Walker and H. Straw published in 1962 and 1967, entered under: Walker, Stanley.
Includes bibliographies.
1. Spectrum analysis. I. Straughan, B.P.
II. Walker, Stanley. Spectroscopy.
QC451.W33 1976   535'.84   75-45328
ISBN 0-470-15031-9 (v. 1)
0-470-15032-7 (v. 2)
0-470-15033-5 (v. 3)

Preface

It is fifteen years since Walker and Straw wrote the first edition of 'Spectroscopy' and considerable developments have taken place during that time in all fields of this expanding subject. In atomic spectroscopy, for example, where the principles required in a student text have been laid down for many years, there have been advances in optical pumping and double resonance which cannot be neglected at undergraduate level. In addition, nuclear quadrupole resonance (n.q.r.) and far infrared spectroscopy now merit separate chapters while addtional chapters dealing with Mössbauer spectroscopy, photoelectron spectroscopy and group theory are an essential requisite for any modern spectroscopy textbook.

When the idea for a new edition of Spectroscopy was first discussed it quickly became clear that the task of revision would be an impossible one for two authors working alone. Consequently it was decided that the new edition be planned and co-ordinated by two editors who were to invite specialists, each of whom had experience of presenting their subject at an undergraduate level, to contribute a new chapter or to revise extensively an existing chapter. In this manner a proper perspective of each topic has been provided without any sacrifice of the essential character and unity of the first edition.

The expansion of subject matter has necessitated the division of the complete work into three self contained volumes.

Volume 1 includes atomic, n.m.r., n.q.r., e.s.r. and Mössbauer spectroscopy.

Volume 2 contains chapters on molecular symmetry and group theory, microwave, infrared and Raman, far-infrared spectroscopy, force constants, evaluation of thermodynamic functions.

Volume 3 centres on the information which results when a valence electron(s) is excited or removed from the parent molecule. It includes electronic spectroscopy, quantum numbers, dissociation energies, fluorescence and phosphorescence spectroscopy, astrochemistry, photoelectron spectroscopy.

The complete work now provides a single source of reference for all the spectroscopy that a student of chemistry will normally encounter as an undergraduate. Furthermore, the depth of coverage should ensure the books' use on graduate courses and for those starting research work in one of the main branches of spectroscopy.

A continued source of confusion in the spectroscopic literature is the duplication of symbols and the use of the same symbol by different authors to represent different factors. The literature use of both SI and non SI units further complicates the picture. In this book we have tried to use SI units throughout or units such as the electron volt which are recognised for continued use in conjunction with SI units. The symbols and recognised values of physical constants are those published by the Symbols Committee of the Royal Society 1975.

B.P. Straughan
S. Walker

October, 1975

Acknowledgements

Although not involved in the production of this second edition, we would like to express our sincere thanks to Mr. H. Straw whose vital contribution to the first edition of Spectroscopy helped to ensure its widespread success and hence the demand for a new edition. One of us (S.W.) wishes to thank his wife, Kathleen, without whose help at many stages part of this work could not have gone forward.

Contributors to Volume Two

CHAPTER ONE AND CHAPTER FOUR

Dr. B.P. Straughan, University of Newcastle upon Tyne

CHAPTER TWO

Dr. A.V. Golton, Aston University, Birmingham

CHAPTER THREE AND CHAPTER FIVE

Professor S. Walker, Lakehead University, Ontario, Canada

CHAPTER SIX, CHAPTER SEVEN AND APPENDIX

Dr. J.K. Burdett, University of Newcastle upon Tyne

Contents

1 Introduction to molecular spectra

1.1 ABSORPTION AND EMISSION OF ELECTROMAGNETIC RADIATION

All spectra arise from transitions between energy states, and *molecular spectroscopy* is concerned with the change in internal energy when a molecule absorbs or emits electromagnetic radiation in discrete amounts or quanta.

Molecular energy is divided among several different motions within the molecule, and the measurement of the absorbed or emitted radition gives a value for the energy change involved. A *quantum of energy*, ΔE, is related to the wavelength λ of the radiation by the equation:

$$\Delta E = hc/\lambda$$

where h is Planck's constant and c is the velocity of the electromagnetic radiation in the same medium as the wavelength is measured. In addition, c and λ are related to the frequency, ν, of the electromagnetic radiation by the formula:

$$\lambda \times \nu = c$$

The values of λ and c are slightly dependent on whether the measurements are made in vacuum or in air. The corresponding frequency, however, is in each case given by the same type of ratio which is:

$$\frac{c_{\text{air}}}{\lambda_{\text{air}}} = \frac{c_{\text{vac}}}{\lambda_{\text{vac}}} = \nu$$

If the wavelength measurement is made in air (λ_{air}), it may be corrected to λ_{vac} by the addition to it of $(n - 1)\lambda_{\text{air}}$ where n is the refractive index of air at that particular wavelength.

It is generally more desirable to employ frequency (ν in Hz) rather than wavelength (λ in m) because (energy difference)/h between two molecular energy levels is numerically equal to the frequency of the radiation. This relationship is expressed by the *Bohr frequency rule*:

$$\Delta E/h = \nu \text{Hz} \quad \text{or} \quad \Delta E = E' - E'' = h\nu \text{ joules}$$

and it is the basis for all quantitative spectroscopy. Thus a molecule in an energy state E'' can only be excited into a *higher* energy state E' by the *absorption* of electromagnetic radiation of frequency $\Delta E/h$. Since E' and E'' are essentially precise energies, the absorbed radiation will be essentially monochromatic and all other frequencies will be undiminished in intensity. A spectrum arising from this sort of transition is called an *absorption spectrum.* Alternatively, a molecule may start off in an excited state E' and may undergo a transition to a lower energy state E''. The transition will result in the *emission* of monochromatic radiation of frequency $\Delta E/h$. This process gives rise to an *emission spectrum* which is clearly complementary to an absorption spectrum.

The majority of spectra discussed in these volumes are absorption spectra, i.e. the transitions originate from a ground state. The exceptions include atomic spectra and electronic spectra of diatomic molecules, which are more usually observed in emission, and Raman spectra which arise from a scattering process.

1.2 MOLECULAR ENERGY STATES

It was stated in the previous section that the energy levels of an isolated molecule are *essentially precise.* However, there is some uncertainty associated with the levels, and this uncertainty causes any molecular transition to have a natural line-width. Thus a spectrum does not consist of infinitely sharp lines but the absorptions and emissions appear as broad lines.

Readers will find that some authors in this book write about *line* spectra (which may be sharp, narrow, or broad) whereas others refer to *band* spectra. A distinction can be made if one accepts that a band represents the overall contours which arises when more than one transition has taken place, but the individual energy changes remain unresolved or only partially resolved. For example, if the rotational fine structure associated with a vibrational transition remains unresolved, then the spectrum exhibits a band. If the rotational structure can be fully resolved, then the spectrum consists of a series of rotational lines centred on the position of the vibrational transition. Thus, a line arises from a single transition.

Although this distinction is generally used, the terms tend to become interchangeable in imprecise discussions.

Three factors contribute to the natural line-width of a transition.

1.2.1. Heisenberg's Uncertainty Principle

If a molecule is isolated for a time, Δt seconds, in a particular energy state, then the energy of the state will be uncertain (blurred) to an extent ΔE where:

$$\Delta E \cdot \Delta t \approx \frac{h}{2\pi} \approx 10^{-34} \text{ Js}$$

h is Planck's constant and the relationship is known as Heisenberg's Uncertainty Principle. If we take into account also the Bohr frequency relationship and write it in terms of an uncertainty in the radiation frequency, $\Delta\nu$, then:

$$\Delta E = h\Delta\nu \approx \frac{h}{2\pi \cdot \Delta t}$$

and hence:

$$\Delta t \cdot \Delta\nu \approx \frac{1}{2\pi}$$

Thus we conclude that the *longer* a molecule remains in a particular energy level, the more precisely the energy will be defined.

For molecular systems it is usual to define the state of zero energy as the *ground state*, and it is conventional to label this state as E''. The latter becomes progressively occupied as the system in thermal equilibrium is cooled towards the absolute zero temperature. The ground state energy of the system will be *sharply* defined because Δt will be large.

In contrast, a molecule may occupy a state of higher energy, i.e. an excited state which is labelled E'. The lifetime of the excited state is usually much less than the lifetime of the ground state, and so the energy is not known precisely. For example, an excited singlet electronic state has a lifetime of 10^{-8} s and hence $\Delta E = 10^{-26}$ J. Assuming that the energy of the ground state is known precisely, a transition from the excited state to the ground state, $E' \rightarrow E''$ will have an uncertainty in the corresponding radiation frequency of:

$$\Delta\nu \approx \frac{1}{2\pi \cdot \Delta t} \approx 10^{8} \text{ Hz}$$

This uncertainty is small compared with the radiation frequency used to excite transitions between electronic energy levels (10^{14} – 10^{16} Hz; see Table 1.1), and so the *natural line-width* is said to be *small.* Compare that situation with an excited *electron spin state* with a lifetime Δt of $\sim 10^{-7}$ s. $\Delta\nu$ is now $\sim 10^{7}$ Hz for a transition to the ground state, and this is of the same order as the usual frequency of radiation used to excite such transitions (10^{8}–10^{9} Hz). This gives rise to a *broad line-width* situation which is a direct consequence of the uncertainty of the energy gap and the region of the electromagnetic spectrum used to excite the particular transition.

1.2.2 Collision broadening

Molecules in the gaseous and liquid phases are never stationary; they continually collide with each other even at temperatures approaching absolute zero. The collisions become more serious as the temperature increases and the buffeting perturbs the energies of the outer electrons as well as the rotational and vibrational energies of the molecules. Thus the collisions cause a blurring of these energy levels, and the corresponding electronic, vibrational, and rotational spectra exhibit *broad* rather than *sharp* lines.

The collisions in liquids are more severe than in gases at ambient temperature, and so the lines in the spectra of gaseous samples are usually sharper than the lines for liquid samples.

The spectra of solid samples are less subject to collision broadening because the random motions of molecules are severely curtailed in the condensed phase. However, the spectra of polycrystalline solids or powders still exhibit broad features caused by unresolved solid-state splitting effects (see Chapter 4, p. 206).

1.2.3 Doppler broadening

The random motions of molecules in the gaseous and liquid states also cause the absorption or emission frequencies to show a Doppler shift. This arises if there is a relative velocity between the instrument detector and the molecules in motion. Since the molecular motions are random, both positive and negative frequency shifts are encountered and a broad line spectrum is observed.

Since the effects of all three types of line broadening are present in a spectrum to some extent, the total line width can be expressed as the sum of all three.

Degeneracy

If two or more states have the same numerical values of energy, they are said to be *degenerate*. Double, triply, four-fold, etc. degenerate states are the names commonly used when 2, 3, 4, etc. energy states have the same numerical value. The degeneracy may be removed (split or lowered are other terms used in this context) by the effect of some external influence such as an electric or magnetic field.

1.3 CLASSIFICATION OF ENERGIES

Since molecules possess a very large number of different types of energy states (e.g. rotational, electronic, etc.), it is essential to use simplifications. The electromagnetic spectrum is usually broken down into various regions, and the regions are associated with a particular type of molecular energy (see Table 1.1). Although the boundaries between the regions are by no means precise, a radiation frequency in a particular region is of the right order of magnitude to bring about a transition between the appropriate molecular energy levels. For example, a frequency of 10^{13} Hz lies in the middle of the *infrared* region and would be associated with *vibrational* energy levels. It is important to remember, however, that Bohr's frequency rule must be obeyed when a *particular* transition is being studied, and that $\Delta E/h$ must be *precisely* equal to ν Hz. The exact frequencies which are required to accomplish the particular energy changes are selected by the molecules from the source of radiation. For absorption and emission processes in a molecule involving electronic, rotation-vibration, and pure rotational changes, it is very rare for only one frequency to be observed. Usually an appreciable number is involved. The set of frequencies at which the absorption occurs enables the molecule to be characterized and identified.

At frequencies above 3×10^{18} Hz (γ-ray region) the energy changes involve the rearrangement of nuclear particles.

In the 3×10^{18}–3×10^{16} Hz region (X-rays) the spectroscopic changes involve the inner electrons of molecules, and the transitions will involve energies of the order of 10^7 J mol^{-1} ($\sim$10 eV).

The visible and ultraviolet regions extend from approximately 3×10^{14} to 3×10^{16} Hz, and a transition involves the transfer of a valence electron from one molecular orbital to another. The study of valence electron transitions is called *electronic spectroscopy*, and, because the radiation expressed as a frequency (Hz) leads to very large numbers, it is common to find the frequency expressed as a wavenumber (cm^{-1}) or as a wavelength (nm). Thus 3×10^{15} Hz $\equiv 10^5$ cm^{-1} $\equiv$ 100 nm. The separations between the energy levels of the valence electrons are $\sim 10^5$ J mol^{-1}.

The infrared region is usually quoted as 3×10^{12}–3×10^{14} Hz, and because the energies are mainly associated with the vibrations of molecules, a study of this subject is termed *vibrational spectroscopy*. Again, the region expressed in Hz yields unwieldy numbers, and so the commonest unit is cm^{-1}, i.e. 100–10 000 cm^{-1}. Alternatively, the positions of vibrational bands can be quoted in wavelengths; they will lie in the range 100–1 μm (microns). The separations between vibrational levels are $\sim 10^3$–10^4 J mol^{-1}. The far-infrared region covers the approximate range 6×10^{12}–3×10^{11} Hz (200–10 cm^{-1}) and gives information about the vibrations of molecules containing heavy atoms as well as lattice vibrations.

The microwave region lies in the frequency range 3×10^{11}–10^9 Hz (30 cm to 10 cm^{-1}) and spectroscopy at these frequencies is concerned with transitions between rotational energy levels which are separated by hundreds of joules per mole.

Finally, we come to the radiofrequency region, 3×10^6–10^9 Hz (10 m to 30 cm) where the energy change involved arises from the reversal of spin of a nucleus or electron. The energy is here of the order of only 10^{-3}–10 joules per mole, and the techniques are called *nuclear magnetic resonance* (n.m.r.) and *electron spin resonance* (e.s.r.) *spectroscopy*.

One other spectroscopic technique which is not mentioned in Table 1.1 is

Table 1.1. Regions of the electromagnetic spectrum and associated molecular energies

Frequency (Hz)	(cm^{-1})	*Region*	*Wavelength* (m)	*Energy* (joules mol^{-1})	*Molecular energy*
10^{20}		γ-rays			Nuclear rearrangements (see Mössbauer)
	10^8		10^{-10}	10^9	
10^{18}					
	10^6	X-rays	10^{-8}	10^7	Electronic (inner shells)
10^{16}		Vacuum u.v.			
	10^4	Ultraviolet (u.v.) Visible	10^{-6}	10^5	Electronic (valence shells)
10^{14}					
		Infrared (i.r.)			
	10^2		10^{-4}	10^3	Vibrational
10^{12}		Far i.r.			
	1	Microwave	10^{-2}	10	
10^{10}				1	Rotational
			1	10^{-1}	
10^8					
		Radiofrequency	10^2	10^{-3}	Nuclear orientation
10^6					
10^4			10^4	10^{-5}	
		Audiofrequency	10^6	10^{-7}	
10^2					
1			10^8	10^{-9}	

Useful approximations: $1\ \text{cm}^{-1} \approx 10\ \text{J mol}^{-1}$

$$\approx 3 \times 10^{10}\ \text{Hz}$$

$$1\ \text{eV} \approx 8000\ \text{cm}^{-1} \approx 100\ \text{kJ mol}^{-1}$$

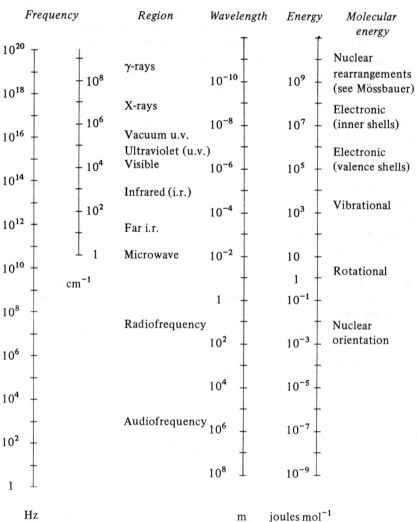

Raman spectroscopy. This is a scattering effect which provides information about vibrational and rotational energy levels. The information obtained is often complementary to the results obtained in the infrared and microwave regions but the the Raman scattered light occurs in the visible and ultraviolet regions of the spectrum.

1.4 THE INTENSITY OF SPECTRAL LINES

1.4.1 Transition probability

Whether an energy change can occur at all depends on the ability of a molecule to interact with the electromagnetic radiation. The likelihood of a molecule undergoing a transition from one energy level to another is known as the *transition probability*. The detailed calculation of *absolute transition probabilities* requires a knowledge of the numerical values of the quantum mechanical wavefunctions of the two energy states, and a detailed discussion is beyond the scope of this book. It is much easier, however, to decide whether the transition probability is zero or non-zero, i.e. to deduce the *selection rules* by a qualitative mathematical discussion (see individual chapters).

For example, if the absorption of electromagnetic radiation is to produce a change in vibrational energy, interaction may only take place provided that there is a change in the electric dipole moment of the vibrating unit during the vibration, while for a pure rotational energy change the molecule must possess either a permanent electric or magnetic dipole moment.

It is found in practice that transitions do not take place between all the possible energy levels; there are selection rules which limit the number of transitions. Such rules are occasionally defied, but if a transition occurs which is not permitted by the selection rule, then the intensity of that particular spectral transition is usually very low. The theoretical justification for employing such rules is given by wave mechanics, where in order to solve the equations it is necessary to introduce certain limitations. For example, when the wave-mechanical equation is formulated for the absorption or emission of rotational energy of a diatomic molecule, between a higher rotational energy state characterized by the rotational quantum number J' and a lower one characterized by J'', it is found necessary, in order to obtain an acceptable solution for the equation, that:

$$\Delta J = J' - J'' = \pm 1$$

where $(J' - J'')$ is represented by ΔJ. Thus, permissible pure rotational energy changes are limited by this selection rule to transitions between adjacent levels. Transitions with finite intensity are called *allowed* or *active* transitions. Transitions such as $\Delta J = \pm 2, \pm 3, \ldots$ have *zero intensity* and are not observed in the pure rotational spectrum of a diatomic molecule in either the far-infrared or microwave regions; they are termed *forbidden* or *inactive* transitions.

If the eigenfunctions are known for the two energy states between which transitions are being considered, then it is in theory possible to determine the selection rules, i.e. to decide which transitions are permissible. It should be noted that even if a transition is theoretically permissible, it does not follow that it will necessarily be experimentally detected.

In the analysis of all types of spectra, selection rules are found to be necessary, and examples will be found in individual chapters.

1.4.2 Population of energy states

In addition to the transition probability, the intensity of a transition depends upon the number of molecules which are in the state corresponding to the starting point of the transition. The number of molecules n_i in an upper state i relative to the number of molecules n_j in a lower state j is given by the Boltzmann law of energy distribution. For a system in thermal equilibrium:

$$\frac{n_i}{n_j} = \frac{g_i}{g_j} e^{-\Delta E/kT}$$

where ΔE is the energy difference between the two states, T is the temperature in K, g_i and g_j are the degeneracies or statistical weights of the two states, and $k = 1.38 \times 10^{-23}$ J K^{-1} is the Boltzmann constant. The reader can easily confirm that, if T is room temperature, then the number of molecules in an excited electronic state ($\Delta E \approx 10^6$ J mol^{-1}) is negligible whereas excited rotational energy levels ($\Delta E \approx 100$ J mol^{-1}) are well populated up to quite high J values (see p. 82). Thus, in an electronic absorption spectrum at room temperature, only those transitions which arise from the ground electronic state are observed. The rotational spectrum, however, normally contains lines whose transitions originate from excited rotational energy levels.

In conclusion, we have seen that a spectral transition has the important properties of position (i.e. frequency) and intensity. Both of these parameters will be developed further in the chapters dealing with individual techniques because they will usually form the basis of any spectroscopic discussion.

2 Molecular symmetry and group theory

2.1 INTRODUCTION

Molecules are able to possess shape because of the disparity in mass between nuclei and electrons. The nuclei are subject to forces arising largely from their mutual repulsion and an electric field provided by the much more rapidly moving electrons: a field which changes continuously in response to the nuclear configuration. In stable molecules these forces restrain the nuclei to the neighbourhood of an equilibrium configuration, corresponding to a well-defined potential energy minimum. For many sets of atoms several such minima exist. If these correspond to quite different configurations and are separated by high potential barriers they are referred to as isomers, they are chemically separable and are regarded as distinct compounds. On the other hand, relatively low barriers may separate one conformation from another. Such may be the case if one part of a molecule rotates with respect to the remainder, or if a non-planar molecule undergoes inversion. Not all such conformations will be distinct, but even those that are will not easily be separated. Nevertheless such configurations may still usefully be regarded as distinct entities, and throughout this chapter molecules will be considered as having one definite equilibrium shape.†

Because molecules possess shape, identical nuclei occupy equilibrium locations that are geometrically distinct but may usually be related to each other by a set of well-defined movements in space. Identical particles are inherently indistinguishable, and any hypothetical operation which interchanges them (with due regard for the conservation of spin) must leave the molecule unchanged. Such

† Readers are referred to Longuet-Higgins [2.9] for a treatment of molecules with several potential minima.

permutation operations include those which carry out movements on the molecule as a whole, bringing it into coincidence with itself. These latter movements are called *symmetry operations*, although this 'active' definition of their nature could be replaced by a 'passive' one in which the transformation is applied not to the molecule itself but to the coordinate system we use to describe it. Thus a forward rotation of a molecule, having the shape of an equilateral triangle, through an angle $+2\pi/3$, is equivalent to a backward rotation of the axes through $-2\pi/3$, as in Fig. 2.1.

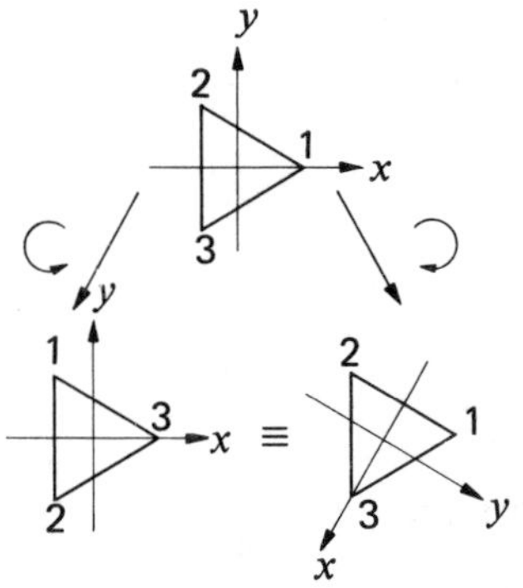

Fig. 2.1

This equivalence emphasizes the fact that a symmetry operation amounts to no more than a renumbering of identical particles and cannot therefore change any observable property of a molecule, such as its energy or the probability of its undergoing a transition from one state to another. We shall find this equivalence useful when applying symmetry operations to molecular systems but in order to give an unambiguous meaning to symmetry operation we shall take the 'active' *definition* because this is somewhat easier to visualize when describing the various kinds of symmetry operation and their relationships.

Although symmetry operations cannot affect observable quantities, they do generally change the molecular wavefunctions. Thus, in the simple case of a non-degenerate wavefunction, if ψ_i is a solution of the Schrödinger equation, $\mathbf{H}\psi_i = E_i\psi_i$, then $a\psi_i$ is also a solution corresponding to the same eigenvalue E_i, where a is any number, real or complex. However, if all solutions are normalized, the modulus of a can only be unity, and if a is real, $a = \pm 1$. Some symmetry operations therefore may leave ψ_i unchanged, but others may change it to $-\psi_i$: its pattern of behaviour with respect to the different symmetry operations characterizes what we shall later refer to as its *symmetry species*. The situation for degenerate wavefunctions is more complicated, but in all cases *the symmetry behaviour can be used to classify and label solutions of the Schrödinger equation.*

Observable quantities, such as those mentioned, are given, according to quantum mechanics, by integrals $\int \psi_n \Omega \psi_m \, d\tau$ involving two wavefunctions ψ_n and ψ_m and an operator Ω. For quantities like energy, which refer to a single molecular state, $n = m$, but for transition probabilities which relate two states together, n and m differ. From the symmetry species of ψ_n and ψ_m and of the operator $\mathbf{\Omega}$, it is possible to deduce whether such integrals *must be zero*. As a result, *spectroscopic selection rules can be obtained from a knowledge of molecular symmetry alone*, without any detailed information about the form of the wavefunctions, which are generally unknown because of the complexity of the Schrödinger equation. But symmetry arguments cannot provide the numerical magnitudes of the observable quantities, when these are not zero.

Finally, when the symmetry of a molecule is changed by an alteration of its constitution or environment there will be corresponding changes to the symmetry species of its wavefunctions and to the eigenvalues of their energies. *From the correlation of the molecular symmetries involved, the changes in the symmetry species of the wavefunctions may be deduced, with any consequent changes in the degeneracies of the states involved, and in the qualitative appearance of the molecular spectra.*

2.2 SYMMETRY OPERATIONS AND POINT GROUPS

2.2.1 Symmetry operations

We have already decided to regard symmetry operations as movements of the physical system which bring it into coincidence with itself, and we therefore define a symmetry operation as *a movement of a system which, when completed, brings every point either back to its original position or to a physically indistinguishable position within the system in its original orientation.* If it were carried out in our absence we should not be able to tell on our return whether it had actually been performed, because the position and orientation of the system would appear unchanged.

All such movements can be described with reference to a *symmetry element*, which is the point, line, or plane with respect to which the symmetry operation takes place. The same letters are used to denote both the elements and the operations, but we shall distinguish the symbols by printing those for the elements in light-face italic, but those for the operations in sans-serif type. Throughout this book we shall use only the Schönflies notation, favoured by spectroscopists. Crystallographers use the rather different Hermann–Mauguin system: see Buerger [2.10]. When we need general symbols for the operations we shall use the letters . . . **P**, **Q**, **R**, **S**, . . . and especially the letter **R**. Later on we shall use the same symbols to designate the *operators* which carry into effect the symmetry operations and bring about the consequent transformations of vectors and functions.

For an isolated molecule there are five kinds of symmetry operation:

1) a *rotation* about an axis, denoted by **C**;
2) a *reflection* in a mirror-plane, denoted by $\boldsymbol{\sigma}$;
3) an *inversion* through a central point, denoted by **i**;

4) a *rotary-reflection*, consisting of a rotation about an axis followed or preceded by a reflection in a plane perpendicular to the axis, and denoted by **S**;
5) the *identity*, corresponding to the return of all the points to their original positions, and denoted by **E**.

A rotation, **C**, and of course the trivial identity operation, **E** (which may be regarded as a complete revolution about any axis passing through the centre of mass), are both operations that may by physically carried out upon a ball-and-stick model of the molecule. They are referred to as *proper rotations*, and the axis concerned as a proper axis. The remaining operations cannot be so carried out and we refer to them collectively, for reasons that will soon be apparent, as *improper rotations* about improper or alternating axes.

The general symbol for a *proper rotation* is $\mathbf{C}_n$, where the subscript n denotes the *order* of the axis. By 'order' is meant the number of times the operation must be repeated to bring each point identically to its initial position. Thus a three-fold axis, C_3, corresponds to an operation of rotation about the axis by an angle of $2\pi/n = 2\pi/3 = 120°$. Three successive rotations by this angle will complete one revolution and return the system to its original position. An equilateral triangle possesses such an axis perpendicular to itself. In Fig. 2.2 we arbitrarily number the apices and thereby distinguish the successive rotations.

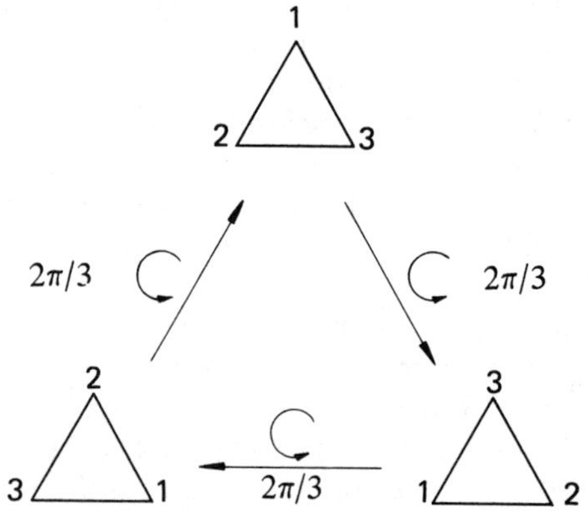

Fig. 2.2

A convenient symbol for the operation of rotation by $2 \times 2\pi/3$ is $\mathbf{C}_3^2$, while $\mathbf{C}_3^3 \equiv \mathbf{E}$ denotes the rotation by $3 \times 2\pi/3$, *where the equivalence sign implies that the result of carrying out the one operation is in every respect identical to the result of carrying out the other operation.* Clearly $\mathbf{C}_3^4 \equiv \mathbf{C}_3$, so the symbols $\mathbf{C}_3$ and $\mathbf{C}_3^2$, together with **E**, denote the full number of separate and distinct operations about the C_3 axis. In general n distinct rotations, $\mathbf{C}_n, \mathbf{C}_n^2, \mathbf{C}_n^3, \ldots, \mathbf{C}_n^n \equiv \mathbf{E}$, may be carried out about a C_n axis. We may note that the operation $\mathbf{C}_n^k$, corresponding to a rotation through an angle $2\pi k/n$, may be written in an alternative way if the fraction n/k is a whole number. Thus $\mathbf{C}_6^2 = \mathbf{C}_3$, $\mathbf{C}_6^3 = \mathbf{C}_2$, and $\mathbf{C}_6^4 = \mathbf{C}_3^2$, so a six-fold C_6 axis is seen to be coincident with a three-fold C_3 and a two-fold C_2 axis. For this reason the six distinct rotations $\mathbf{C}_6, \mathbf{C}_6^2, \mathbf{C}_6^3, \mathbf{C}_6^4, \mathbf{C}_6^5, \mathbf{C}_6^6$ are usually written as $\mathbf{C}_6, \mathbf{C}_3, \mathbf{C}_2, \mathbf{C}_3^2, \mathbf{C}_6^5, \mathbf{E}$.

It is also important to realize that our two rotated configuration of the equilateral triangle could equally well have been achieved by carrying out rotations in the opposite sense. We may extend our index notation by the use of a negative sign to indicate such a rotation. Clearly $\mathbf{C}_3^2 = \mathbf{C}_3^{-1}$ and $\mathbf{C}_3 = \mathbf{C}_3^{-2}$, or in general, $\mathbf{C}_n^{-k} = \mathbf{C}_n^{n-k}$.

The presence of one symmetry element places limitations upon the existence of others. Many molecules possess a unique *principal axis* C_n whose order is higher than that of any other axis. If this is so the only other compatible kind of axis is a two-fold axis perpendicular to the principal axis, since rotation about such an axis will merely send the principal axis into itself, without duplicating it in some other direction. And if there is *one* such C_2 axis there must clearly be n, since each $\mathbf{C}_n$ operation will generate another. It is customary to distinguish such axes, if n is *odd*, by a prime, as C_2'. If n is *even* there are still n such axes, but these generally occur in two equal sets, denoted by C_2' and C_2'', since the $\mathbf{C}_n$ operation will not convert members of one set into members of the other.

The *reflection* operation $\boldsymbol{\sigma}$ is carried out with respect to a mirror-plane which, since the movement merely carries the molecule into some physically indistinguishable configuration, must pass through the molecule rather than lie outside it. Each point is carried to an equivalent image point such that the line joining the two points is perpendicular to the mirror-plane and bisected by it.

Unlike the rotation axis C_n $(n > 2)$ about which more than one distinct operation may be carried out, the mirror-plane gives rise to only one operation and when this is repeated each point is restored to its original position: $\boldsymbol{\sigma}^2 = \mathbf{E}$.

Clearly any atoms in the molecule which lie outside the mirror-plane must occur in pairs, but any number of atoms may lie in the plane. If the molecule possesses only one atom of a given kind it must lie in every mirror-plane, on their point or line of intersection.

It is usual to distinguish mirror-planes and their reflection operations by the addition of subscripts. If there is a principal axis it must lie in every mirror-plane, along their line of intersection, except that it may be perpendicular to one mirror-plane. This axis is regarded as *vertical*, so those planes in which the axis lies are generally labelled σ_v, while the perpendicular plane is labelled σ_h. There must be n σ_v planes and, as in the case of the C_2' axes, when n is *even* there will usually be two separate sets which are distinguished by labelling the members of one set σ_d, where d stands for dihedral. Vertical planes constituting a single set are however also denoted σ_d when C_2' axes are present and the planes bisect the angles between them. When there is more than one C_3 or C_4 axis present, as in the case of a cube, the diagonal mirror-planes are also denoted σ_d.

The operation of *inversion*, $\mathbf{i}$, through a central point, or centre of symmetry, carries each point into an equivalent image point such that the line joining them has the centre as its mid-point.

Like the mirror-plane, only one distinct operation can be carried out with respect to the centre, and when this is repeated every point is returned to its original position, so that $\mathbf{i}^2 = \mathbf{E}$. It follows also that, apart from a possible unique atom located at the centre, every atom in the molecule must occur in pairs.

The *rotary-reflection* operation has the general symbol $\mathbf{S}_n$, where n denotes the order of the axis, which in this case is best defined by saying that the rotation involved is carried out through an angle $\theta = 2\pi/n$. This rotation is preceded or followed by a reflection in a plane *perpendicular* to the axis of rotation. This should be made clear by Fig. 2.3.

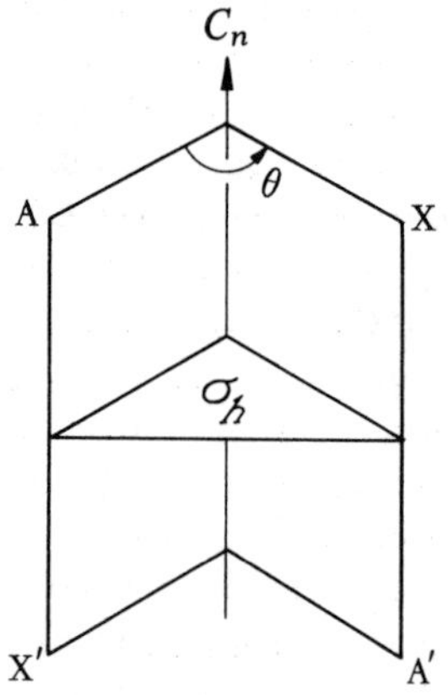

Fig. 2.3 The rotary-reflection operation, $\mathbf{S}_n$. A′ is the image-point of A, while either X or X′ is reached when only the first half of the operation has been carried out.

When n is *even*, $\mathbf{S}_n^n = \mathbf{E}$, since $\mathbf{C}_n^n = \mathbf{E}$ and $\boldsymbol{\sigma}^{\text{even } n} = \mathbf{E}$. Neither the axis C_n nor the plane σ_h need exist on its own, although if both do exist, then S_n must exist also. When n is *odd*, $\mathbf{S}_n^n = \boldsymbol{\sigma}_h$, since $\boldsymbol{\sigma}^{\text{odd } n} = \boldsymbol{\sigma}$, while $\mathbf{S}_n^{n+1} = \mathbf{C}_n^{n+1} = \mathbf{C}_n$, so both σ_h and C_n *must* exist independently. Then only $\mathbf{S}_n^{2n} = \mathbf{E}$.

The successive positions of a general point, for repetitions of $\mathbf{S}_3$ and $\mathbf{S}_4$, are shown in Figs. 2.4 and 2.5, from which it will also be seen why the axis is sometimes referred to as an *alternating axis*.

From Fig. 2.5 it will be appreciated that if n is even, the S_n axis must coincide with a proper axis of order $n/2$. We may now observe from Fig. 2.6 that $\mathbf{i}$ is equivalent to $\mathbf{S}_2$ about *any* axis through the centre, and from Fig. 2.7 that $\boldsymbol{\sigma}$ is equivalent to $\mathbf{S}_1$ about an axis *perpendicular* to the mirror-plane. Since these two operations may also be regarded as rotary-reflections, we may, when it is convenient, treat these collectively with the rotary-reflections as *improper rotations.*

The above relationships enable us to express the successive distinct rotary-reflections about an S_6 axis $\mathbf{S}_6, \mathbf{S}_6^2, \mathbf{S}_6^3, \mathbf{S}_6^4, \mathbf{S}_6^5, \mathbf{S}_6^6$ as $\mathbf{S}_6, \mathbf{C}_3, \mathbf{i}, \mathbf{C}_3^2, \mathbf{S}_6^5, \mathbf{E}$, and about an S_5 axis $\mathbf{S}_5$ to $\mathbf{S}_5^{10}$ as $\mathbf{S}_5, \mathbf{C}_5^2, \mathbf{S}_5^3, \mathbf{C}_5^4, \boldsymbol{\sigma}, \mathbf{C}_5, \mathbf{S}_5^7, \mathbf{C}_5^3, \mathbf{S}_5^9, \mathbf{E}$. We see that, for

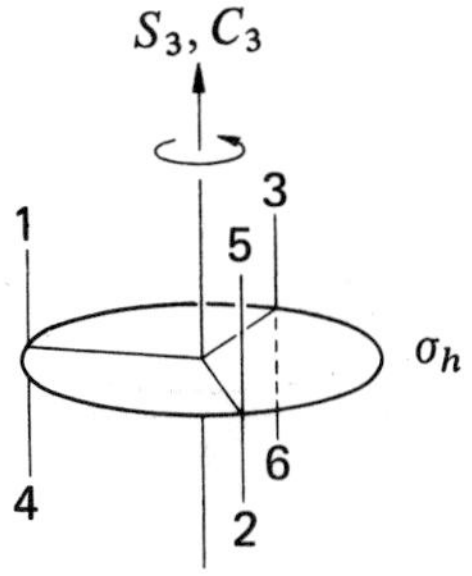

Fig. 2.4

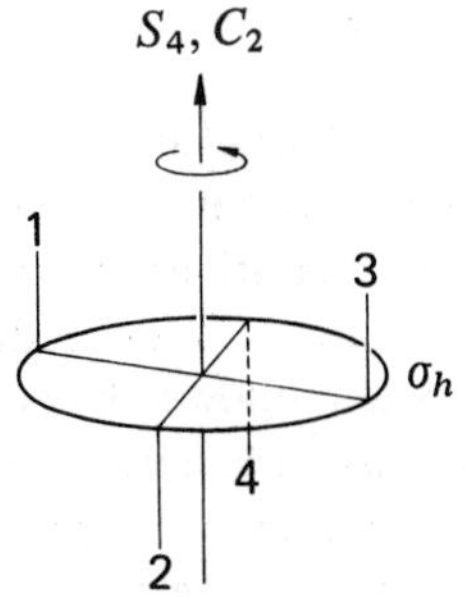

Fig. 2.5

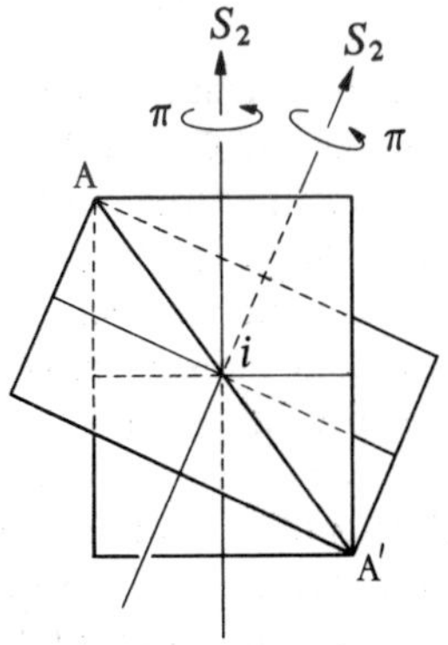

Fig. 2.6

$n > 2$, there are at least two $\mathbf{S}_n^k$ operations that cannot be expressed as equivalent to any one of the previous symmetry operations. We may also observe that $\mathbf{S}_6^5 = \mathbf{S}_6^{-1}$, $\mathbf{S}_5^3 = \mathbf{S}_5^{-7}$, or in general $\mathbf{S}_n^k = \mathbf{S}_n^{k-n}$ if n is even, or $\mathbf{S}_n^k = \mathbf{S}_n^{k-2n}$ if n is odd.

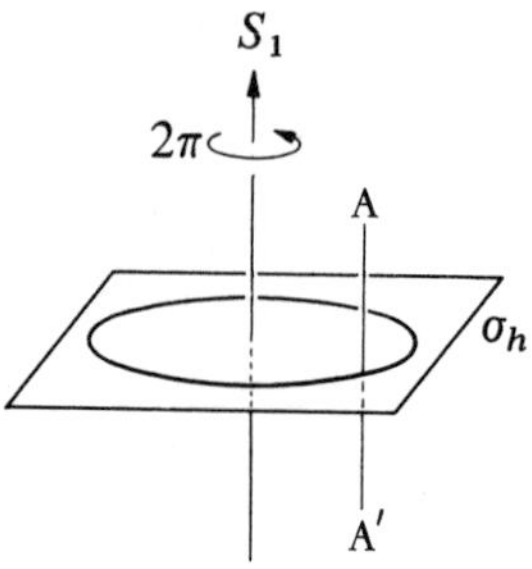

Fig. 2.7

2.2.2 Products of symmetry operations

We have already introduced equations of the type $\mathbf{P} = \mathbf{Q}$, expressing the equivalent effect of two symmetry operations $\mathbf{P}$ and $\mathbf{Q}$, and also the use of an index notation to represent successive repetitions of the same operation. The index notation follows the usual algebraic rules for products if we use this term to denote the effect of one operation followed by another. Thus the rotation $\mathbf{C}_n^k$ followed by $\mathbf{C}_n^p$ *about the same axis* is equivalent to the single rotation $\mathbf{C}_n^{k+p}$. If $p = -k$, this corresponds to a rotation of $+2\pi k/n$ followed by one of $-2\pi k/n$, so that all points are returned to their original position. This agrees with $\mathbf{C}_n^{k-k} = \mathbf{C}_n^0 = \mathbf{E}$. We now extend our 'product' notation to include *the effect of any operation*, $\mathbf{P}$, *followed by another*, $\mathbf{Q}$, by writing:

$$\mathbf{QP} = \mathbf{R} \tag{2.1}$$

The order in which we write the operations is important, since in general $\mathbf{QP} \neq \mathbf{PQ}$. If the order is immaterial and $\mathbf{QP} = \mathbf{PQ}$ we say that the two operations *commute*. Note that the operations are carried out on the molecule in the order in which they are written from *right to left* (in the same way that successive differential operators are applied to a function). Since each symmetry operation carries the molecule into a physically indistinguishable configuration it follows that the product $\mathbf{QP}$ must be equivalent to some single symmetry operation $\mathbf{R}$.

Two successive rotations (proper or improper) about the *same* axis will commute since their combined effect will be a rotation about the same axis through an angle which is the sum of the separate angles, together perhaps with a reflection in a σ_h plane. The result of two successive rotations about *different* axes depends however on the order in which they are performed, as seen in Fig. 2.8, so these operations do not commute. Clearly $\mathbf{E}$ and $\mathbf{i}$ will commute with all other operations. Thus:

$$\mathbf{RE} = \mathbf{ER} = \mathbf{R} \tag{2.2}$$

We can easily extend our definition of a product to include more than two factors. If $\mathbf{T}$ is another symmetry operation we can write $\mathbf{TR} = \mathbf{T(QP)}$ for the effect of carrying out the operations $\mathbf{P}$, $\mathbf{Q}$, and $\mathbf{T}$ successively, in that order.

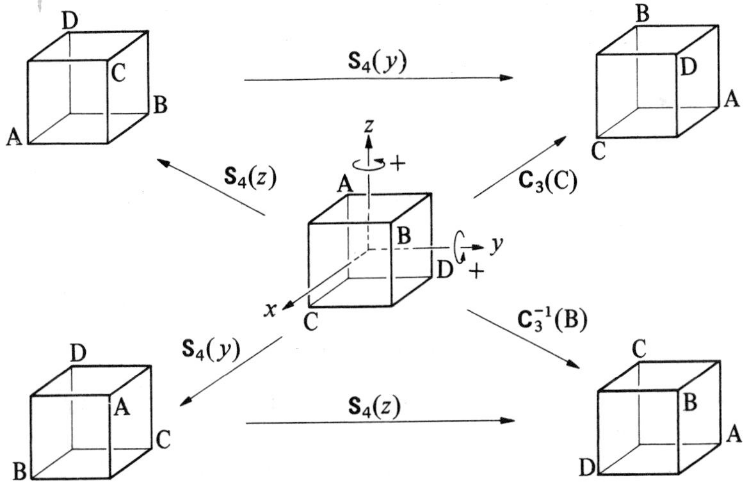

Fig. 2.8

Since any successive pair of these constitutes a symmetry operation we could equally well have written:

$$\mathbf{TR} = \mathbf{T(QP)} = \mathbf{(TQ)P} = \mathbf{TQP} \tag{2.3}$$

where the final expression implies that there is no ambiguity about the overall result if the brackets are omitted. We say that the product is *associative*.

We have already observed that a positive proper rotation followed by an equal negative rotation about the same axis restores every point to its original location, so that $\mathbf{C}_n^{-k}\mathbf{C}_n^{+k} = \mathbf{E}$. For every symmetry operation, $\mathbf{R}$, there is another, $\mathbf{S}$, that similarly reverses its effect, and this relationship must be symmetrical so that $\mathbf{R}$ reverses $\mathbf{S}$:

$$\mathbf{SR} = \mathbf{RS} = \mathbf{E} \tag{2.4}$$

We call $\mathbf{S}$ the *inverse operation* to $\mathbf{R}$ and write $\mathbf{S} = \mathbf{R}^{-1}$. Similarly $\mathbf{R} = \mathbf{S}^{-1}$. Thus $\mathbf{R}^{-1}\mathbf{R} = \mathbf{R}\mathbf{R}^{-1} = \mathbf{E}$ and we see that $\boldsymbol{\sigma}$ and $\mathbf{i}$ are their own inverses. Note that:

$$\mathbf{(PQ)(Q^{-1}P^{-1})} = \mathbf{P(QQ^{-1})P^{-1}} = \mathbf{PEP^{-1}} = \mathbf{PP^{-1}} = \mathbf{E}$$

from which it follows that:

$$\mathbf{(PQ)^{-1}} = \mathbf{Q^{-1}P^{-1}} \tag{2.5}$$

2.2.3 Groups and subgroups

The properties (2.1)–(2.4) of symmetry operations enable us to describe *the complete collection of distinct symmetry operations applicable to a molecule* as its symmetry *group*.

Mathematically any set of 'elements'† is said to form a group if it satisfies the following conditions:

1. We can define a combining rule between any two elements **Q** and **P** such that their 'product' **R** is also a member of the group:

$$\mathbf{QP} = \mathbf{R}$$

2. There is one element **E**, the identity, that commutes with all the others, leaving them unchanged:

$$\mathbf{RE} = \mathbf{ER} = \mathbf{R}$$

3. The associative law of multiplication holds between the elements of the group:

$$\mathbf{T(QP)} = \mathbf{(TQ)P}$$

   (But the commutative law **QP** = **PQ** need not apply. If it does, the group is said to be *Abelian*.)

4. Every element **R** has an inverse $\mathbf{R}^{-1}$ which is also a member of the group, such that:

$$\mathbf{R}^{-1}\mathbf{R} = \mathbf{R}\mathbf{R}^{-1} = \mathbf{E}.$$

The number of elements in a group is called the *order* of the group. This may be finite or infinite. Symmetry groups are of either kind, but most have only a limited number of symmetry operations. Those groups applicable to an individual molecule consist of operations that necessarily leave at least one point (the centre of mass) unshifted; for this reason they are referred to as symmetry *point groups*, to distinguish them from the groups applicable to arrays of molecules, which involve translational symmetry operations.

Before discussing the symmetry of actual molecules it is convenient to give the recommended rules for the choice of a reference set of Cartesian axes:

1. The origin is placed at the centre of mass.
2. A *principal axis* is taken as the z-axis. If there is no such unique axis the 2-fold axis is chosen which passes through the most atoms (or, failing that, cuts the most bonds), the x- and y-axes also being taken as 2-fold axes.
3. For *planar* molecules, when the z-axis lies in the plane, the x-axis is the normal; and when the z-axis is normal to the plane, the x-axis passes through the most atoms.
4. For *non-planar* molecules with a principal axis, the x-axis is taken as a C_2' axis, or as lying in a σ_v plane. Where both C_2' and C_2'' axes or both σ_v and σ_d planes occur the first named in each case passes through the most atoms.
5. The *positive* directions of the x-, y-, and z-axes are taken as corresponding respectively to those of the thumb, index, and middle fingers of the right hand when held mutually perpendicular.
6. The *positive, forward sense of rotation* about an axis advances a right-handed

† Not to be confused with *symmetry element* as defined in Section 2.2.1. In the present context these elements are the *symmetry operations*.

screw along the positive direction of the axis. A positive rotation about the z-axis therefore corresponds to a movement from the $+x$-axis towards the $+y$-axis.

We now examine the symmetry properties of two small molecules, water and ammonia. Readers are strongly advised to select other molecules, such as those referred to in Section 2.2.5, and to treat them in the same way, preferably with the help of models.

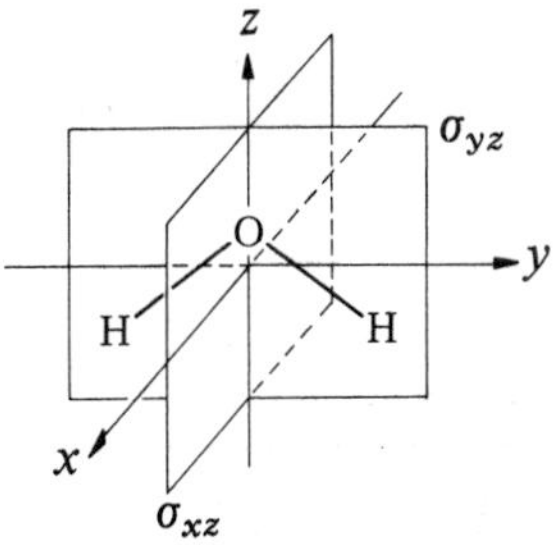

Symmetry operations

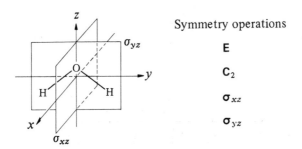

E

C_2

σ_{xz}

σ_{yz}

Fig. 2.9

Figure 2.9 shows the symmetry elements of water, and beside it is a list of the four distinct symmetry operations, which constitute the point group $\mathcal{C}_{2v}$. The effect of successive operations is conveniently summarized in Table 2.1 in a 'group multiplication table'.

Table 2.1

$\mathcal{C}_{2v}$	E	C_2	σ_{xz}	σ_{yz} ←Applied first
E	E	C_2	σ_{xz}	σ_{yz}
C_2	C_2	E	σ_{yz}	σ_{xz}
σ_{xz}	σ_{xz}	σ_{yz}	E	C_2
σ_{yz}	σ_{yz}	σ_{xz}	C_2	E

(Rows: Applied second)

Note that the product **QP** is placed at the intersection of column **P** and row **Q**. The corresponding diagrams and multiplication table for ammonia are given in Fig. 2.10, in which we have named the hydrogen atoms as A, B, and C, and in Table 2.2. The six distinct symmetry operations correspond to the point group $\mathcal{C}_{3v}$.

It must be emphasized that although we have labelled the mirror-planes σ_a, σ_b, σ_c after the *original* positions of the hydrogen atoms, these planes and their designations remain *unaltered*, fixed in relation to the coordinate system, while the molecule undergoes the symmetry operations, as in Fig. 2.11.

Both tables illustrate the general property that each row and column contains every element of the group once only. This follows from their uniqueness; were any two entries in the same row, **QP** and **QR**, equal it would follow that **P** = **R**.

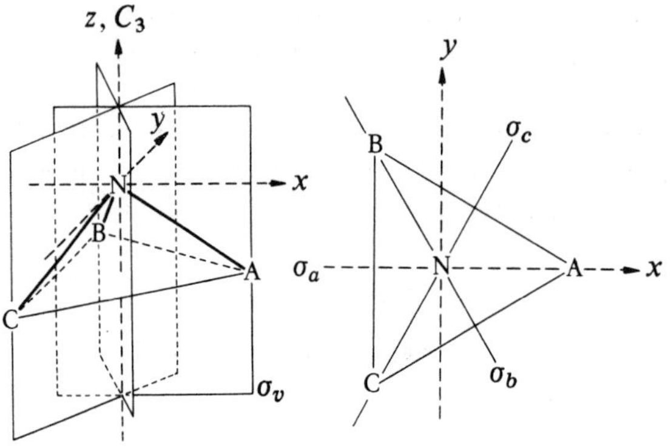

Symmetry operations:

E	σ_a
$\mathbf{C}_3$	σ_b
$\mathbf{C}_3^2 = \mathbf{C}_3^{-1}$	σ_c

Fig. 2.10

Table 2.2

$\mathcal{C}_{3v}$	E	C_3^+	C_3^-	σ_a	σ_b	σ_c
E	E	C_3^+	C_3^-	σ_a	σ_b	σ_c
C_3^+	C_3^+	C_3^-	E	σ_c	σ_a	σ_b
C_3^-	C_3^-	E	C_3^+	σ_b	σ_c	σ_a
σ_a	σ_a	σ_b	σ_c	E	C_3^+	C_3^-
σ_b	σ_b	σ_c	σ_a	C_3^-	E	C_3^+
σ_c	σ_c	σ_a	σ_b	C_3^+	C_3^-	E

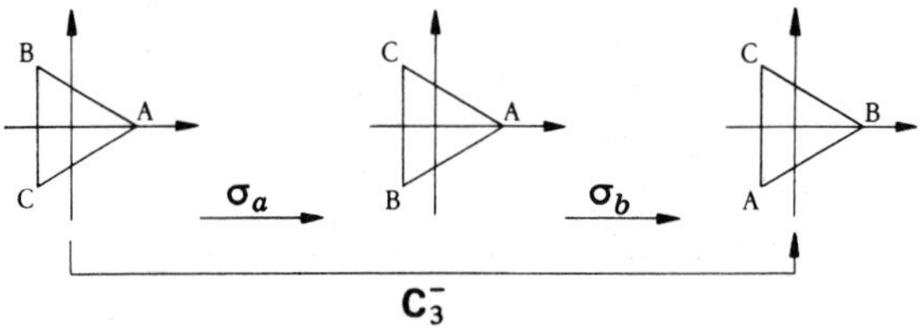

Fig. 2.11

The group multiplication table enables any product of several factors to be converted to an equivalent single operation. Conversely it enables us to express the members of the group in terms of the powers and products of a smaller number of *generating operations*. Thus in the case of $\mathcal{C}_{3v}$ it is sufficient to take one *typical rotation* and one *typical reflection*, say $\mathbf{C}_3$ and σ_a, since $\mathbf{E} = \mathbf{C}_3^3$, $\mathbf{C}_3^{-1} = \mathbf{C}_3^2$, $\sigma_b = \sigma_a \mathbf{C}_3$, $\sigma_c = \mathbf{C}_3 \sigma_a$, and these do, in fact, exhaust the number of distinct operations which can be formed. This concept proves useful in describing the various point groups in Section 2.2.5.

Any collection of the elements of a group, $\mathcal{G}$, which *by themselves* form a group, is called a *sub*group of $\mathcal{G}$. The multiplication table of $\mathcal{G}$ enables us to pick out the subgroups that it contains. Thus the sets $\{E, C_3^+, C_3^-\}$, $\{E, \sigma_a\}$, $\{E, \sigma_b\}$, $\{E, \sigma_c\}$ and of course $\{E\}$, form subgroups of $\mathcal{C}_{3v}$. It may be shown that the order of a subgroup is always a factor of the order of the group; these have orders 3, 2, and 1 only, while that of $\mathcal{C}_{3v}$ is 6. The importance of subgroups derives from situations in which the symmetry of a molecule is lowered. This is discussed in Section 2.6.3.

2.2.4 Conjugate operations and classes

The $\mathcal{C}_{3v}$ multiplication table is seen to be divided into blocks, corresponding to groupings of the operations. Earlier, in referring to the generating operations, we spoke of a *typical* rotation and a *typical* reflection. In doing so it was implied that in some way the distinctions between the rotations and between the reflections were arbitrary, depending only on our definition of positive or negative rotation or our lettering of the mirror-planes. An interconversion of these labels corresponds to a transformation of the coordinate system. Such transformations can be carried out by applying the symmetry operations of the group *to the coordinate system.* Thus the definitions of C_3^+ and C_3^- are interchanged by any of the σ_v reflections and those of the σ_v are permuted by the rotations. Let us examine the conversion of σ_a into σ_c. In sequence (Fig. 2.12) we apply (i) C_3^- to the coordinate system, (ii) σ_a *according to this transformed system* to the molecule, (iii) C_3^+ to the coordinate system to restore it to its original state. The overall result is seen to be equivalent to σ_c.

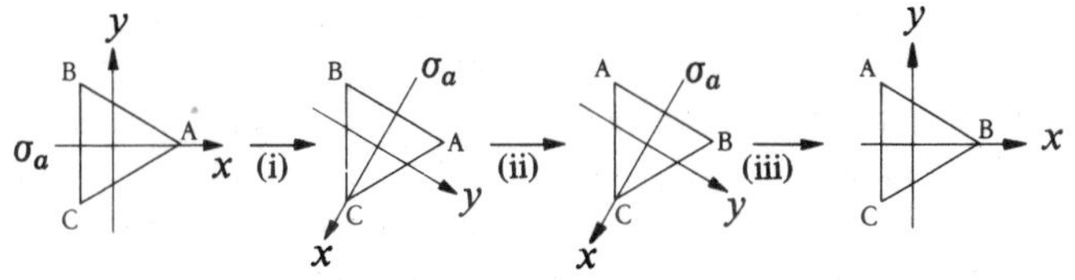

Fig. 2.12

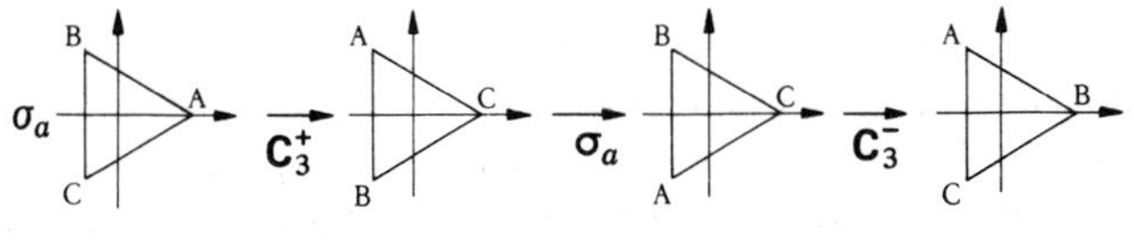

Fig. 2.13

Since operations on the coordinate system are equivalent to the *inverse* operation carried out on the molecule, we may replace the sequence in Fig. 2.12 by that in Fig. 2.13.

Two operations, **P** and **Q**, related in this way with respect to another, **R**, by the general expression:

$$\mathbf{Q} = \mathbf{R}^{-1}\mathbf{P}\mathbf{R} \tag{2.6}$$

are said to be *conjugate* to each other. The set of distinct operations, $\mathbf{Q}$, formed by taking each group operation, $\mathbf{R}$, in turn, is said to form a *class*. As a result the point group $\mathcal{C}_{3v}$ is found to consist of the following classes: $\{\mathbf{E}\}$, $\{\mathbf{C}_3^+, \mathbf{C}_3^-\}$, and $\{\sigma_a, \sigma_b, \sigma_c\}$. The identity always forms a class on its own, as does any element which commutes with all the others. This follows from Equation (2.6) since we may write:

$$\mathbf{RQ} = \mathbf{PR} = \mathbf{RP} \quad \text{or} \quad \mathbf{Q} = \mathbf{P} \text{ for all } \mathbf{R} \tag{2.7}$$

Hence in Abelian groups each operation is in a separate class. The subgroup of $\mathcal{C}_{3v}$ consisting of $\{\mathbf{E}, \mathbf{C}_3^+, \mathbf{C}_3^-\}$, called $\mathcal{C}_3$, is such a group; without the reflections σ_v there can be no interconversion of $\mathbf{C}_3^+$ and $\mathbf{C}_3^-$. The great convenience of division into classes will become apparent later.

2.2.5 The molecular point groups

We will now specify the point groups applicable to molecules, in terms of their generating operations. We have already indicated that severe limitations are necessarily imposed on the mutual existence of various symmetry elements, and it may be proved that the following list is a complete one. One or more molecular examples are named in most cases, and several of these are illustrated in Figs. 2.14 and 2.15.

1. *The pure rotation or cyclic groups, $\mathcal{C}_n$, for which the sole generator is* $\mathbf{C}_n$. All asymmetric molecules provide examples of the trivial $\mathcal{C}_1$ group; the molecules H_2O_2 and cyclohexene are said to 'belong' to the $\mathcal{C}_2$ point group, while triphenylarsine is probably an example of the $\mathcal{C}_3$ group.

2. *The groups, $\mathcal{C}_{nv}$, for which the generators are* $\mathbf{C}_n$ *and* σ_v. Water, dichloromethane, and pyridine have the symmetry of $\mathcal{C}_{2v}$, ammonia, methyl halides, and chloroform that of $\mathcal{C}_{3v}$, while $XeOF_4$ has the symmetry of $\mathcal{C}_{4v}$. Linear molecules such as HCl, HCN, and N_2O, which lack a centre of symmetry, belong to the infinite order group $\mathcal{C}_{\infty v}$.

3. *The groups, $\mathcal{C}_{nh}$, for which the generators are* $\mathbf{C}_n$ *and* σ_h. The non-linear molecule NOCl has the symmetry of $\mathcal{C}_{1h}$ which consists, apart from the identity, of a reflection operation σ only. This group, equivalent,to $\mathcal{C}_{1v}$, is generally denoted $\mathcal{C}_s$. *trans*-$(CHBr_2)_2$ and *trans*-$(CHBr)_2$ belong to $\mathcal{C}_{2h}$, while the planar configuration of H_3BO_3 and the most symmetrical configuration of $[C(NH_2)_3]^+$ are examples of $\mathcal{C}_{3h}$. A hexamethylbenzene molecule with its methyl groups similarly oriented, each with one H atom in the plane of the ring, has the symmetry of $\mathcal{C}_{6h}$. Note that the presence of both $\mathbf{C}_n$ and σ_h generates $\mathbf{S}_n$, and when n is even, $\mathbf{i}$.

4. *The groups, $\mathcal{S}_{2n}$, for which the sole generator is* $\mathbf{S}_{2n}$. The groups $\mathcal{S}_n$ with n *odd* are clearly equivalent to $\mathcal{C}_{nh}$, so we are left with $\mathcal{S}_2$, often denoted as $\mathcal{C}_i$, $\mathcal{S}_4$, $\mathcal{S}_6$, . . . $\mathcal{S}_6$ is sometimes denoted $\mathcal{C}_{3i}$. $\mathcal{C}_i$ has, apart from the identity, only the inversion operation, $\mathbf{i}$; the *trans*-conformation of $(CHBrCl)_2$ is an example

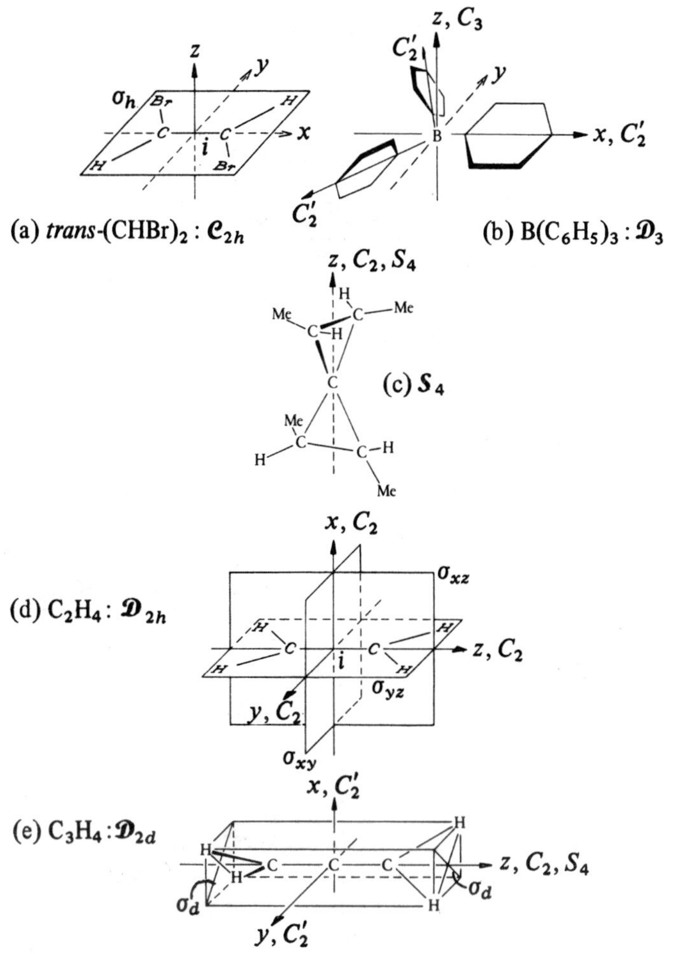

Fig. 2.14

of this symmetry. The molecule shown in Fig. 2.14(c) belongs to $\mathcal{S}_4$ if we treat the methyl groups as single atoms.

5. *The groups, $\mathcal{D}_n$, for which the generators are* $\mathbf{C}_n$ *and* $\mathbf{C}'_2$. Biphenyl with its rings neither coplanar nor perpendicular has the symmetry of the group $\mathcal{D}_2$, which is sometimes denoted $\mathcal{V}$, while triphenylboron with its phenyl groups similarly oriented has the symmetry of $\mathcal{D}_3$.

6. *The groups, $\mathcal{D}_{nh}$, for which the generators are* $\mathbf{C}_n$, $\mathbf{C}'_2$, *and* σ_h. These generators also produce $\mathbf{S}_n$ and $n\sigma_v$ $(= \mathbf{C}'_2\sigma_h)$. The group $\mathcal{D}_{2h}$, sometimes denoted $\mathcal{V}_h$, is a special case in that it has three mutually perpendicular σ planes and three C_2 axes along their lines of intersection, together with a centre of symmetry. It is an Abelian group with each operation in a separate class. Ethylene, naphthalene, and diborane have this symmetry. $\mathcal{D}_{3h}$ has many examples including gaseous

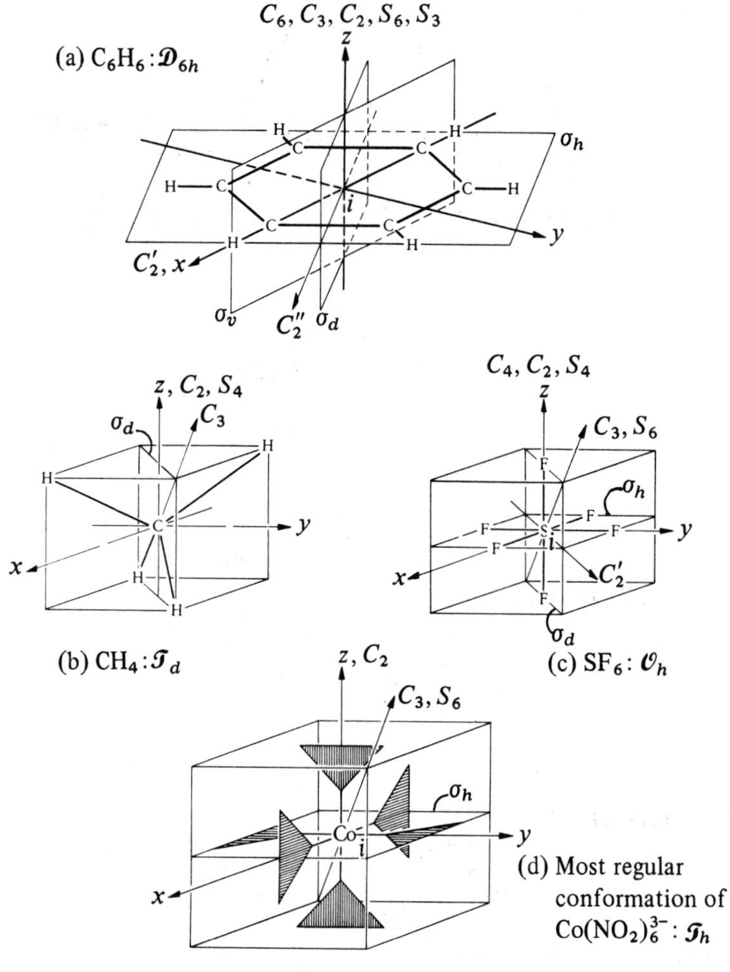

Fig. 2.15 One symmetry element of each kind is depicted.

PF_5, cyclopropane, 'eclipsed'-ethane, and the carbonate ion; $\mathcal{D}_{4h}$ has XeF_4 and $PtCl_4^{2-}$, and $\mathcal{D}_{6h}$ benzene as examples. Linear molecules such as N_2, acetylene, and CO_2, with centres of symmetry, belong to the infinite order group $\mathcal{D}_{\infty h}$.
7. *The groups, $\mathcal{D}_{nd}$, for which the generators are* $\mathbf{C}_n$, $\mathbf{C}_2'$, *and* $\boldsymbol{\sigma}_d$, *or more simply,* $\mathbf{S}_{2n}$ *and* $\mathbf{C}_2'$. Allene and gaseous B_2Cl_4 illustrate the symmetry of $\mathcal{D}_{2d}$, also written $\mathcal{V}_d$, while the 'staggered' conformation of ethane with the methyl groups rotated 60° from each other, and cyclohexane, both belong to $\mathcal{D}_{3d}$. S_8 has the symmetry of $\mathcal{D}_{4d}$, and ferrocene, with its staggered C_5H_5 groups, that of $\mathcal{D}_{5d}$.

This completes the list of groups with axial symmetry, the remainder corresponding to the symmetries of the five regular solid bodies, which have several axes with orders greater than 2 and no principal axis.

8. *The cubic point groups*. The regular octahedron belongs to the same group as the cube, while the regular tetrahedron, which can be inscribed in a cube, belongs to a subgroup.

The pure rotation group based on the tetrahedron is denoted $\mathcal{T}$ and has the generators $\mathbf{C}_2$ and $\mathbf{C}_3$; three C_2 axes join the mid-points of opposite faces of the cube, and four C_3 axes lie along the body diagonals. An added $\boldsymbol{\sigma}_d$ generator gives the full symmetry of the tetrahedron, the group $\mathcal{T}_d$; the six σ_d planes bisect the angles between the C_2 axes. This provides three $\mathbf{S}_4$ operations, since $\mathbf{S}_4 = \boldsymbol{\sigma}_d\mathbf{C}_2$. Also, since $\mathbf{S}_4^2 = \mathbf{C}_2$, we need only name *two* generators of the group $\mathcal{T}_d$: $\mathbf{S}_4$ and $\mathbf{C}_3$. $\mathcal{T}_d$ is a group of 24 operations altogether and is illustrated by the molecules CH_4, P_4, and $Ni(CO)_4$.

If instead of $\boldsymbol{\sigma}_d$ we add the generator $\mathbf{i}$ to those of $\mathcal{T}$ we obtain the group $\mathcal{T}_h$; this gives us, besides $\mathbf{C}_2$, $\mathbf{C}_3$, and $\mathbf{i}$, the reflections $\boldsymbol{\sigma}_h$ in planes perpendicular to the C_2 axes, and $\mathbf{S}_6$ about every C_3 axis. The ion $[Co(NO_2)_6]^{3-}$ would have this symmetry if opposite NO_2 groups were coplanar; in the solid Na salt they are in fact *gauche* and the symmetry is that of the subgroup $\mathcal{S}_6$.

The pure rotation group of the cube or octahedron, $\mathcal{O}$, may be obtained from $\mathcal{T}$ by replacing the generator $\mathbf{C}_2$ with $\mathbf{C}_4$ about the same axis. This also gives six C_2 axes joining the mid-points of opposite edges of the cube. These and the corresponding operations are usually distinguished by a prime. An added generator $\mathbf{i}$ gives the full symmetry of the cube or octahedron, the group $\mathcal{O}_h$, consisting of *all* the operations described above, 48 altogether. SF_6 belongs to this point group.

9. *The icosahedral point groups.* The regular icosahedron has 20 equilateral triangular faces, meeting at 12 vertices around C_5 axes. In the same way that the octahedron is related to the cube, being bounded by faces perpendicular to the axes through the cube vertices, so the pentagonal dodecahedron is related to the icosahedron. The corresponding pure rotation group, $\mathcal{I}$, has two generators: a $\mathbf{C}_5$, and a $\mathbf{C}_2$ about an axis joining the mid-points of opposite edges. There are 15 such axes and 10 C_3 axes joining mid-points of faces. The full symmetry group, $\mathcal{I}_h$, of the icosahedron requires the additional generator $\mathbf{i}$, which introduces $\mathbf{S}_6$, $\mathbf{S}_{10}$, and $\boldsymbol{\sigma}_v$ operations. The ion $B_{12}H_{12}{}^{2-}$ has been found to have this symmetry.

2.3 VECTORS AND TRANSFORMATIONS

2.3.1 Symmetry operations and position vectors

We have defined the symmetry operation as a certain kind of movement carried out on the molecule with respect to fixed symmetry elements, which we may regard as embedded in a fixed coordinate system. The movement brings each point in the molecule – that is, each point in space – into a physically equivalent point in the same space, which we may refer to as the *image* of the first point under that particular symmetry operation. There has been an unambiguous one-to-one *mapping* of each point onto its image point, so that the reverse mapping, the inverse symmetry operation, is always possible.

The description of the relationship between a point and its image is conveniently couched in the language of vectors. Each point in space can be associated with a *position vector*, corresponding to its range and direction from the origin.

A symmetry operation can now be described as a *reorientation* of all the position vectors to bring them into coincidence with the position vectors of the image points. It is a mapping of the space onto itself.

Later, in order to apply symmetry operations to vibrations and to wavefunctions, we shall extend the meaning of the words *vector* and *space*. These words have so far referred to real, physical space; we shall shortly use them to refer also to purely 'mathematical' spaces of more than three dimensions. We will therefore describe the appropriate mathematical language with this in mind. The reader who is unfamiliar with vectors and matrices is advised to consult one or other of the books listed at the end of this chapter. We shall however develop the subject so that it can be seen to apply first of all to real 3-dimensional space.

2.3.2 Vectors and coordinate systems

Vectors (and later, functions) will be denoted by lower-case letters, a, b, . . . in sans-serif type, to distinguish them from numbers, which we will refer to as scalars and denote, as before, by lightface italic letters: a, b, ...

We must next decide what is meant by saying that two position vectors are equal. If $\mathsf{a} = \mathsf{b}$ they are equal both in length and direction, and this will only be true if the two points, whose position vectors they are, coincide with each other.

Vector addition corresponds to the combination of two displacements of given extent and direction and is determined by the completion of the parallelogram whose sides are defined by the vectors, as in Fig. 2.16. The sum is expressed in magnitude and direction by the diagonal having a common origin with the two vectors, and algebraically by writing:

$$\mathsf{a} + \mathsf{b} = \mathsf{c} \tag{2.8}$$

This addition is *commutative*, that is, the order is immaterial:

$$\mathsf{a} + \mathsf{b} = \mathsf{b} + \mathsf{a} \tag{2.9}$$

The definition may be extended to any number of terms when, as is evident from Fig. 2.17, the overall displacement is independent of the route taken; that is, the addition is *associative*:

$$(\mathsf{c} + \mathsf{d}) + \mathsf{e} = \mathsf{c} + (\mathsf{d} + \mathsf{e}) = \mathsf{c} + \mathsf{d} + \mathsf{e} \tag{2.10}$$

The position vector of the origin has zero length and no direction. Denoted $\mathsf{0}$ it has the property of zero in ordinary algebra since $\mathsf{a} + \mathsf{0} = \mathsf{a}$, for all vectors a. We may define a *negative* vector, $-\mathsf{a}$, as being one which added to a gives the

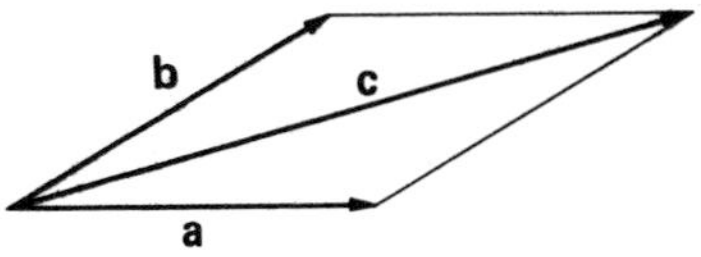

Fig. 2.16

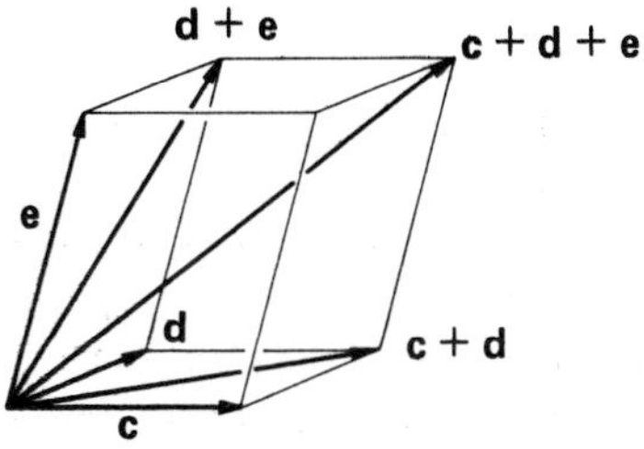

Fig. 2.17

zero vector:

$$(-\mathbf{a}) + \mathbf{a} = \mathbf{0} \tag{2.11}$$

from which it follows that $-\mathbf{a}$ is the vector whose length equals that of **a** and whose direction is exactly opposite to that of **a**, as in Fig. 2.18. Hence $-(-\mathbf{a}) = \mathbf{a}$, and if $\mathbf{a} + \mathbf{b} = \mathbf{c}$, then $\mathbf{b} = \mathbf{c} - \mathbf{a}$.

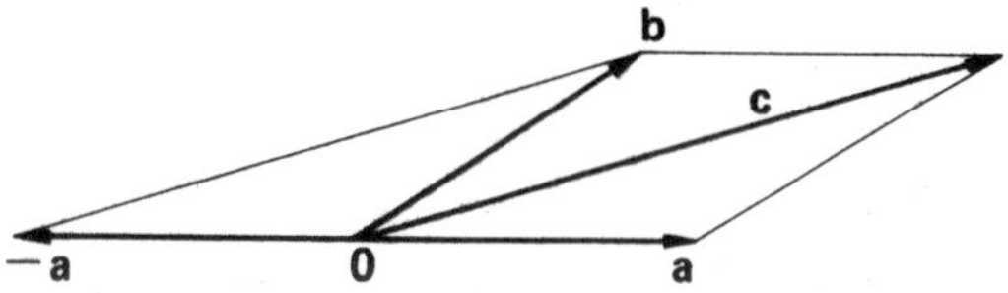

Fig. 2.18

If x is a positive real number and **a** is a position vector whose length is a, we define the product $x\mathbf{a} = \mathbf{a}x$ to be the position vector whose length is xa and whose direction is the same as that of **a**. We can extend this definition to negative values by saying, as before, that the negative sign reverses the direction. As a result we see that multiplication by a scalar is *associative* and *distributive*:

$$\begin{aligned} (\mathbf{a}x)y &= \mathbf{a}(xy) \\ \mathbf{a}(x + y) &= \mathbf{a}x + \mathbf{a}y \\ (\mathbf{a} + \mathbf{b})x &= \mathbf{a}x + \mathbf{b}x \end{aligned} \tag{2.12}$$

The meaning of the last equation may be understood from Fig. 2.19. In writing Equations (2.12) we have put the numerical factor *after* the symbol for the vector in order to be consistent with what we shall later find to be most convenient.

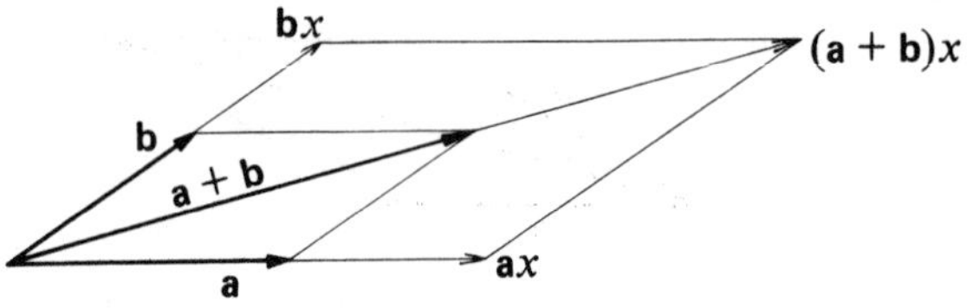

Fig. 2.19

Now let us suppose that $\mathbf{e}_1$ is the position vector of the point A (Fig. 2.20). Then the set of vectors $\mathbf{e}_1 x_1$, as x_1 varies from $-\infty$ to $+\infty$, defines all the points on the line OA; to each point there corresponds a unique value of x_1. Likewise if the point B lies off the line OA, and $\mathbf{e}_2$ is its position vector, the set of vectors $(\mathbf{e}_1 x_1 + \mathbf{e}_2 x_2)$ defines all points in the plane OAB. Finally, if the point C lies outside this plane and its position vector is $\mathbf{e}_3$, the vector $\mathbf{r} = \mathbf{e}_1 x_1 + \mathbf{e}_2 x_2 + \mathbf{e}_3 x_3$ defines the position of any point in space.

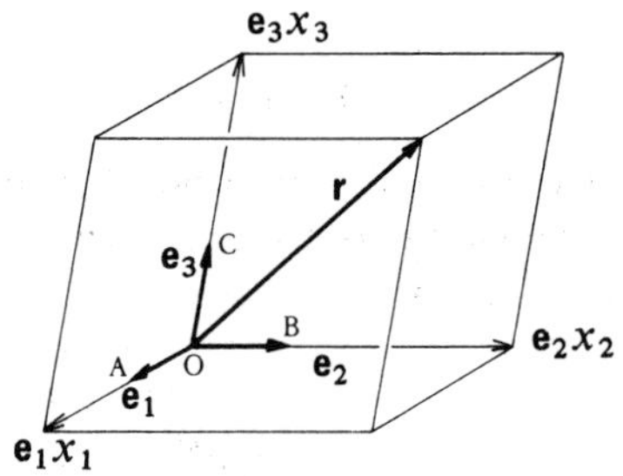

Fig. 2.20

In this way we have set up a *coordinate system.* The three non-coplanar vectors $\mathbf{e}_1$, $\mathbf{e}_2$, and $\mathbf{e}_3$ are said to be its *base vectors*. The unique set of three numbers x_1, x_2, and x_3 which characterize the position vector of each point in space are the *coordinates* of that point or the *components* of its position vector.

The condition that the base vectors are non-coplanar can be expressed algebraically by saying that the only solution of the equation:

$$\mathbf{e}_1 x_1 + \mathbf{e}_2 x_2 + \mathbf{e}_3 x_3 = \mathbf{0} \tag{2.13}$$

is $x_1 = x_2 = x_3 = 0$. For, if we suppose that it had a solution with, say $x_3 \neq 0$, dividing through by x_3 would give:

$$\mathbf{e}_3 = -\mathbf{e}_1(x_1/x_3) - \mathbf{e}_2(x_2/x_3)$$

which is impossible, since $\mathbf{e}_3$ does not lie in the plane of $\mathbf{e}_1$ and $\mathbf{e}_2$. We say that the base vectors are *linearly independent*. Clearly we can find a set of no more than three linearly independent base vectors, although such sets may be chosen in an infinite number of ways, and for this reason we speak of the physical space of the position vectors as having three dimensions. We also say that the base vectors *span* this space.

When we generalize the concept of position vector, the entities so named must continue to have the properties expressed in Equations (2.8)–(2.12), but Equation (2.13) must become:

$$\sum_{i=1}^{n} \mathbf{e}_i x_i = \mathbf{0} \text{ has no solution other than all } x_i = 0 \qquad (2.14)$$

there being then just n linearly independent base vectors. Just as the complete set of position vectors in physical space embraces every point in that space, so the complete set of any generalized vectors constitutes what we mean by the corresponding *n-dimensional vector space*.

The most convenient way of choosing the base vectors in ordinary space is to take three vectors of unit length which are mutually perpendicular to each other. This defines a Cartesian coordinate system in which a general position vector

$$\mathbf{r} = \mathbf{e}_1 x_1 + \mathbf{e}_2 x_2 + \mathbf{e}_3 x_3$$

has components x_1, x_2, x_3 which are normally labelled x, y, and z, since the vectors $\mathbf{e}_1$, $\mathbf{e}_2$, and $\mathbf{e}_3$ respectively then define the x-, y-, and z-axes.

To express algebraically the geometrical notion of two vectors being perpendicular we make use of the *scalar product*. For two vectors, **a** and **b**, of lengths a and b, this is defined as the scalar quantity $ab\cos\theta$, where θ is the angle between the vectors, measured from **a** to **b**. It is written by placing a dot between the vectors:

$$\mathbf{a}.\mathbf{b} = |\mathbf{a}|\,|\mathbf{b}| \cos(\mathbf{a}, \mathbf{b}) = ab\cos\theta \qquad (2.15)$$

It follows geometrically that scalar products have the following properties:

$$\begin{aligned} \mathbf{a}.\mathbf{b} &= \mathbf{b}.\mathbf{a} \\ \mathbf{a}.(\mathbf{b} + \mathbf{c}) &= \mathbf{a}.\mathbf{b} + \mathbf{a}.\mathbf{c} \\ \mathbf{a}.(\mathbf{b}x) &= (\mathbf{a}.\mathbf{b})x \\ \mathbf{a}.\mathbf{a} &> 0, \text{ for all } \mathbf{a} \neq \mathbf{0} \end{aligned} \qquad (2.16)$$

If two vectors are at right angles, $\cos\theta = 0$, and the scalar product is zero. The converse is true provided that neither **a** nor **b** is a zero vector. Such vectors are said to be *orthogonal*. If $\mathbf{a} = \mathbf{b}$, $\cos\theta = 1$ and $\mathbf{a}.\mathbf{a} = a^2$. A unit vector, for which this is unity, is said to be *normalized*. The Cartesian base vectors are said to be *orthonormal*, and this property may be summarized by writing:

$$\mathbf{e}_i.\mathbf{e}_j = \delta_{ij} \qquad (2.17)$$

where δ_{ij} is the *Kronecker delta*, defined to be zero unless $i = j$, when it is equal to unity.

If we express two vectors in terms of orthonormal base vectors:

$$\mathbf{a} = \mathbf{e}_1 a_1 + \mathbf{e}_2 a_2 + \mathbf{e}_3 a_3$$

$$\mathbf{b} = \mathbf{e}_1 b_1 + \mathbf{e}_2 b_2 + \mathbf{e}_3 b_3$$

we obtain a useful expression for the scalar product, since:

$$\begin{aligned} \mathbf{a.b} &= (\mathbf{e}_1 a_1 + \mathbf{e}_2 a_2 + \mathbf{e}_3 a_3).(\mathbf{e}_1 b_1 + \mathbf{e}_2 b_2 + \mathbf{e}_3 b_3) \\ &= a_1 b_1 + a_2 b_2 + a_3 b_3 \end{aligned} \quad (2.18)$$

2.3.3 Matrix algebra

It is convenient at this point to introduce the symbolism and conventions of matrix algebra.

A *matrix* is a rectangular array – in horizontal rows and vertical columns – of entities which are usually scalars (real or complex) but may also be vectors. A matrix of n rows and m columns is designated an $n \times m$ matrix. The entity located in the i-th row and the j-th column is called the i,j-th element of the matrix, and will be denoted in lightface italic, with subscripts: a_{ij}. The entire matrix will be denoted $[a_{ij}]$ or by a single letter in boldface: in lower case, **a**, if $1 \times m$ (a single row) or $n \times 1$ (a single column), or in upper case, **A**, otherwise. Throughout this chapter, after this section, such matrices will always be square ($n \times n$). If $\mathbf{A} = \mathbf{B}$, it is implied that all corresponding elements are identical.

The most characteristic operation involving matrices is that of matrix multiplication, which we shall use repeatedly. *The element in the i-th row and j-th column of the product matrix* **AB** *is obtained by multiplying all the elements in the i-th row of* **A** *into the corresponding elements of the j-th column of* **B**, *and summing the products formed in this way:*

$$c_{ij} = \sum_{k=1}^{m} a_{ik} b_{kj} \quad (2.19)$$

It is important to note that multiplication is possible only if the number of columns in **A** is equal to the number of rows in **B**.

Column: k $\quad j \quad j$

$$\text{Row:} \quad i: \begin{bmatrix} & \vdots & \\ a_{i1} \ldots & a_{ik} & \ldots a_{im} \\ & \vdots & \end{bmatrix} \begin{bmatrix} & b_{1j} & \\ & \vdots & \\ \ldots & b_{kj} & \ldots \\ & \vdots & \\ & b_{mj} & \end{bmatrix} :k \longrightarrow \begin{bmatrix} & \vdots & \\ \ldots & c_{ij} & \ldots \\ & \vdots & \end{bmatrix} :i$$

$$\underset{(n \times m)}{\mathbf{A}} \qquad \underset{(m \times p)}{\mathbf{B}} \qquad \underset{(n \times p)}{\mathbf{C}}$$

Matrix multiplication is generally *non-commutative*, that is, $\mathbf{AB} \neq \mathbf{BA}$, so the order of the factors must not be confused. (If **A** and **B** are rectangular, these products are also of different sizes!) Matrix multiplication is however *associative*:

$$\mathbf{A}(\mathbf{BC}) = (\mathbf{AB})\mathbf{C} = \mathbf{ABC}$$

We now show how matrix algebra serves as a convenient shorthand, anticipating the use to which we will shortly put it.

Suppose two sets of variables x_i and y_j, which may be vector components, are related by a set of linear equations:

$$\begin{aligned} y_1 &= p_{11}x_1 + p_{12}x_2 + \ldots + p_{1m}x_m \\ y_2 &= p_{21}x_1 + p_{22}x_2 + \ldots + p_{2m}x_m \\ &\vdots \\ y_n &= p_{n1}x_1 + p_{n2}x_2 + \ldots + p_{nm}x_m \end{aligned}$$

This relationship may be expressed conveniently by a matrix equation:

$$\begin{bmatrix} y_1 \\ y_2 \\ \vdots \\ y_n \end{bmatrix} = \begin{bmatrix} p_{11} & p_{12} & \cdots & p_{1m} \\ p_{21} & p_{22} & \cdots & p_{2m} \\ \vdots & \vdots & & \vdots \\ p_{n1} & p_{n2} & & p_{nm} \end{bmatrix} \begin{bmatrix} x_1 \\ x_2 \\ \vdots \\ x_m \end{bmatrix} \quad \text{or} \quad \mathbf{y} = \mathbf{Px}$$

where $\mathbf{y}$ is the $(n \times 1)$ column matrix of the y_j variables, $\mathbf{x}$ is the $(m \times 1)$ column matrix of the x_i variables, and $\mathbf{P}$ is the $(n \times m)$ matrix $[p_{ij}]$.

Such a relationship is often spoken of as a *linear transformation* from an m-dimensional space into an n-dimensional one, and the matrix $\mathbf{P}$ is regarded as representing a *linear operator* P which gives effect to the transformation, relating each point $(x_1, x_2, \ldots, x_m)$ with a point $(y_1, y_2, \ldots, y_n)$. This is called a *mapping* of the first point onto the second.

We need to define a few more properties of matrices. The *principal diagonal* of a matrix consists of those elements for which $i = j$: $a_{11}, a_{22}, a_{33}, \ldots$ The sum of these diagonal elements is called the *trace*:

$$\operatorname{Tr} \mathbf{A} = \sum_{i=1}^{n} a_{ii} \tag{2.20}$$

An important property concerns the trace of a product of square matrices:

$$\operatorname{Tr} \mathbf{AB} = \sum_i^n (\mathbf{AB})_{ii} = \sum_i^n\sum_k^n a_{ik}b_{ki} = \sum_k^n\sum_i^n b_{ki}a_{ik} = \sum_k^n (\mathbf{BA})_{kk} = \operatorname{Tr} \mathbf{BA} \tag{2.21}$$

A *diagonal matrix* is a square matrix with non-zero elements only on the principal diagonal. Two such matrices of the same order, $\mathbf{A}$ and $\mathbf{B}$, always commute since the product is a diagonal matrix whose diagonal elements are $a_{11}b_{11}$, $a_{22}b_{22}, \ldots, a_{nn}b_{nn}$ in which the order of the factors is immaterial.

The *unit matrix* $\mathbf{1}$ (often denoted $\mathbf{E}$) is a diagonal matrix whose diagonal elements are all unity: $\mathbf{1} = [\delta_{ij}]$. It plays the part of unity in ordinary algebra since $\mathbf{1A} = \mathbf{A1} = \mathbf{A}$.

Transposition of a matrix involves the interchange of its rows and columns; a square matrix is 'rotated' about its diagonal. We shall denote the *transpose* of **A** as $\mathbf{A}^T$. Since $a_{ij}^T = a_{ji}$, the important result follows that:

$$(\mathbf{AB})^T = \left[\sum_k a_{jk} b_{ki}\right] = \left[\sum_k b_{ik}^T a_{kj}^T\right] = \mathbf{B}^T\mathbf{A}^T \tag{2.22}$$

If $\mathbf{A} = \mathbf{A}^T$, the matrix is said to be *symmetrical*.

A square matrix **A** has an *inverse* $\mathbf{A}^{-1}$, defined by:

$$\mathbf{AA}^{-1} = \mathbf{A}^{-1}\mathbf{A} = \mathbf{1} \tag{2.23}$$

provided that the determinant of **A** *is not zero*. It follows that:

$$(\mathbf{AB})^{-1} = \mathbf{B}^{-1}\mathbf{A}^{-1} \tag{2.24}$$

If $\mathbf{A}^T = \mathbf{A}^{-1}$, the matrices **A** and $\mathbf{A}^T$ are said to be *orthogonal*. The reason for this name will appear shortly.

2.3.4 Vectors and operators represented by matrices

The general expression for a vector in terms of a set of base vectors:

$$\mathsf{a} = \mathsf{e}_1 a_1 + \mathsf{e}_2 a_2 + \ldots + \mathsf{e}_n a_n = \sum_{i=1}^{n} \mathsf{e}_i a_i \tag{2.25}$$

may now be written:

$$\mathsf{a} = [\mathsf{e}_1 \quad \mathsf{e}_2 \quad . \quad . \quad \mathsf{e}_n] \begin{bmatrix} a_1 \\ a_2 \\ \vdots \\ a_n \end{bmatrix} = \mathsf{ea} \tag{2.26}$$

and we shall continue to follow the rule that *vector components appear in columns, sets of vectors in rows.*

Before we can speak of orthogonal *n*-dimensional vectors we must reconsider our expression for the scalar product. For the present we assume this will continue to have the properties (2.16) and write:

$$\begin{aligned}\mathsf{a.b} &= (\mathsf{e}_1 a_1 + \mathsf{e}_2 a_2 + \ldots + \mathsf{e}_n a_n).(\mathsf{e}_1 b_2 + \mathsf{e}_2 b_2 + \ldots + \mathsf{e}_n b_n)\\ &= (\mathsf{e}_1.\mathsf{e}_1) a_1 b_1 + (\mathsf{e}_1.\mathsf{e}_2) a_1 b_2 + \ldots + (\mathsf{e}_1.\mathsf{e}_n) a_1 b_n + \\ &\quad (\mathsf{e}_2.\mathsf{e}_1) a_2 b_1 + (\mathsf{e}_2.\mathsf{e}_2) a_2 b_2 + \ldots + (\mathsf{e}_2.\mathsf{e}_n) a_2 b_n + \\ &\quad .\ . \qquad .\ . \qquad .\ . \\ &\quad (\mathsf{e}_n.\mathsf{e}_1) a_n b_1 + (\mathsf{e}_n.\mathsf{e}_2) a_n b_2 + \ldots + (\mathsf{e}_n.\mathsf{e}_n) a_n b_n .\end{aligned}$$

If we write the matrix $[\mathsf{e}_i.\mathsf{e}_j]$ as $\mathsf{e}^T\mathsf{e} = \mathbf{M}$ (it is called the metrical matrix) we may extend the notion of scalar product:

$$\mathbf{a.b} = (\mathrm{ea})^{\mathrm{T}}(\mathrm{eb}) = \mathrm{a}^{\mathrm{T}}\mathrm{e}^{\mathrm{T}}\mathrm{eb} = \mathrm{a}^{\mathrm{T}}\mathbf{M}\mathrm{b} \tag{2.27}$$

When $\mathbf{M} = \mathbf{1}$ we have the same requirement as (2.17) and we may speak of the n-dimensional base vectors as orthonormal. We then have the general result that:

$$\mathbf{a.b} = \mathrm{a}^{\mathrm{T}}\mathrm{b} \tag{2.28}$$

which we may compare to Equation (2.18). If the base vectors are orthonormal and $\mathrm{a}^{\mathrm{T}}\mathrm{b} = 0$, then $\mathbf{a}$ and $\mathbf{b}$ are orthogonal (provided that neither is zero). From the definition of an orthogonal matrix, that $\mathbf{A}^{\mathrm{T}} = \mathbf{A}^{-1}$ or $\mathbf{A}^{\mathrm{T}}\mathbf{A} = \mathbf{1}$, it is seen that its columns (or rows) consist of the normalized components of mutually orthogonal vectors.

Before we may express the effect of a symmetry operation upon any general vector we need to investigate *how a change in basis alters the components* of a vector.

Consider, as well as an original set of base vectors, $\mathbf{e}_1, \mathbf{e}_2, \ldots, \mathbf{e}_n$, a new set $\bar{\mathbf{e}}_1, \bar{\mathbf{e}}_2, \ldots, \bar{\mathbf{e}}_n$, of which there will be the same number of vectors to span the same space. Then:

$$\bar{\mathbf{e}}_j = \mathbf{e}_1 r_{1j} + \mathbf{e}_2 r_{2j} + \ldots + \mathbf{e}_n r_{nj} = \mathbf{e}\mathbf{r}_j \tag{2.29}$$

where the coefficients r_{ij} are the components of $\bar{\mathbf{e}}_j$ along the axes defined by the basis $\mathbf{e}$, and form the elements of the column matrix $\mathbf{r}_j$. We now assemble the n $\bar{\mathbf{e}}_j$ into a single row matrix, and the successive column matrices $\mathbf{r}_j$ into a square $n \times n$ matrix $\mathbf{R}$ whose elements are the coefficients r_{ij}:

$$\bar{\mathbf{e}} = \mathbf{e}\mathbf{R} \tag{2.30}$$

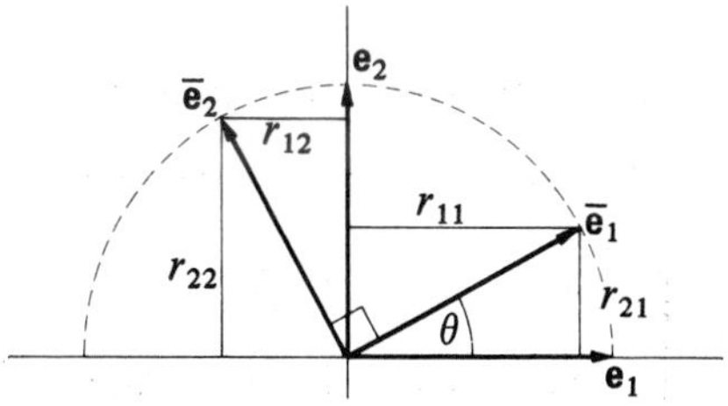

Fig. 2.21 The change of basis from $\mathbf{e}_1, \mathbf{e}_2$ to $\bar{\mathbf{e}}_1, \bar{\mathbf{e}}_2$, both being orthonormal: $r_{11} = r_{22} = \cos\theta$ and $-r_{12} = r_{21} = \sin\theta$.

Figure 2.21 exemplifies this relationship in 2-dimensional space with a pair of orthonormal bases. Here Equation (2.30) becomes:

$$[\mathbf{e}_1 \quad \mathbf{e}_2] \begin{bmatrix} \cos\theta & -\sin\theta \\ \sin\theta & \cos\theta \end{bmatrix} = [\bar{\mathbf{e}}_1 \quad \bar{\mathbf{e}}_2] \tag{2.31}$$

Since $\bar{\mathbf{e}}$, like $\mathbf{e}$, is a linearly independent set of vectors, the converse relationship will exist:

$$\mathbf{e} = \bar{\mathbf{e}}\mathbf{S} \tag{2.32}$$

Multiplying both sides of this equation from the right by $\mathbf{R}$ we obtain:

$$\mathbf{eR} = \bar{\mathbf{e}}\mathbf{SR} = \bar{\mathbf{e}}.$$

Therefore

$$\mathbf{S} = \mathbf{R}^{-1}$$

Having established the relationship between the two sets of base vectors, now consider a general vector $\mathbf{a}$ which on the old basis can be written $\mathbf{ea}$ (2.26) and on the new $\bar{\mathbf{e}}\bar{\mathbf{a}}$:

$$\mathbf{a} = \mathbf{ea} = \bar{\mathbf{e}}\mathbf{R}^{-1}\mathbf{a} = \bar{\mathbf{e}}\bar{\mathbf{a}}$$

Therefore

$$\bar{\mathbf{a}} = \mathbf{R}^{-1}\mathbf{a} \tag{2.33}$$

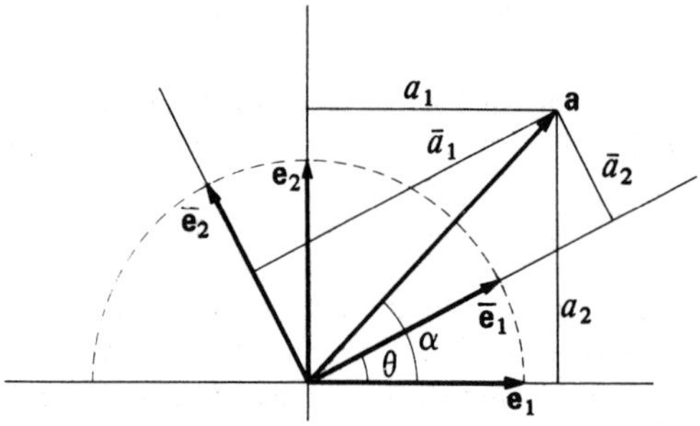

Fig. 2.22

Figure 2.22 shows the components of an arbitrary vector $\mathbf{a}$ in terms of the same bases as in Fig. 2.21:

$$a_1 = a\cos\alpha \quad \bar{a}_1 = a\cos(\alpha-\theta) = a(\cos\alpha\cos\theta + \sin\alpha\sin\theta)$$

$$a_2 = a\sin\alpha \quad \bar{a}_2 = a\sin(\alpha-\theta) = a(\sin\alpha\cos\theta - \cos\alpha\sin\theta)$$

Hence

$$\begin{bmatrix} \cos\theta & \sin\theta \\ -\sin\theta & \cos\theta \end{bmatrix}\begin{bmatrix} a_1 \\ a_2 \end{bmatrix} = \begin{bmatrix} \bar{a}_1 \\ \bar{a}_2 \end{bmatrix}$$

Also

$$\begin{bmatrix} \cos\theta & \sin\theta \\ -\sin\theta & \cos\theta \end{bmatrix}\begin{bmatrix} \cos\theta & -\sin\theta \\ \sin\theta & \cos\theta \end{bmatrix} = \begin{bmatrix} 1 & 0 \\ 0 & 1 \end{bmatrix}$$

Equation (2.33) is therefore verified for this example.

We may contrast Equations (2.30) and (2.33) which express the ways in which the base vectors and the components of an arbitrary fixed vector transform under a change of basis. This difference is called *contragredience*. The invariance of the fixed vector itself may be brought out by writing:

$$\mathbf{a} = \bar{\mathbf{e}}\bar{\mathbf{a}} = \mathbf{eRR}^{-1}\mathbf{a} = \mathbf{ea} = \mathbf{a}$$

We are now in a position to consider a vector $\mathbf{a}'$ *which has the same components in the new basis as* **a** *had in the old.* We may regard it as the vector which results when **a** is no longer fixed but is transformed along with the basis **e**, when this is converted into $\bar{\mathbf{e}}$; it is called the image of **a** under this transformation or mapping. Then

$$\mathbf{a}' = \bar{\mathbf{e}}\bar{\mathrm{a}}' = \bar{\mathbf{e}}\mathrm{a} = \mathbf{eRa} = \mathbf{e}\mathrm{a}'$$

which expresses $\mathbf{a}'$ *relative to the original basis*, so that its components can be written:

$$\mathrm{a}' = \mathbf{R}\mathrm{a} \tag{2.34}$$

Note the distinction between Equations (2.33) and (2.34); (2.33) refers to the components of a fixed vector **a** under a *transformation of axes*, and (2.34) to the components of a *transformed vector* $\mathbf{a}'$ while the axes remain unaltered.

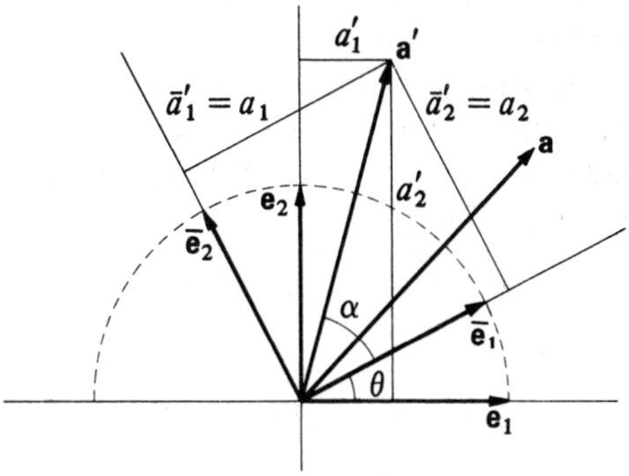

Fig. 2.23 The vector $\mathbf{a}'$ has the same components in $\bar{\mathbf{e}}_1$, $\bar{\mathbf{e}}_2$ as **a** has in $\mathbf{e}_1$, $\mathbf{e}_2$. It is the image of **a** under the mapping $\mathbf{e} \to \bar{\mathbf{e}}$, that is under the rotation through the angle θ, which sends $\mathbf{e}_1$, $\mathbf{e}_2$ into $\bar{\mathbf{e}}_1$, $\bar{\mathbf{e}}_2$.

Figure 2.23 refers to our 2-dimensional example and shows the components of $\mathbf{a}'$ in terms of the same bases as in Figs. 2.21 and 2.22, $\mathbf{a}'$ being the image of **a** under the mapping $\mathbf{e} \to \bar{\mathbf{e}}$:

$$a'_1 = a\cos(\alpha + \theta) = a(\cos\alpha\cos\theta - \sin\alpha\sin\theta)$$

$$a'_2 = a\sin(\alpha + \theta) = a(\sin\alpha\cos\theta + \cos\alpha\sin\theta)$$

Hence

$$\begin{bmatrix} \cos\theta & -\sin\theta \\ \sin\theta & \cos\theta \end{bmatrix} \begin{bmatrix} a_1 \\ a_2 \end{bmatrix} = \begin{bmatrix} a'_1 \\ a'_2 \end{bmatrix}$$

and Equation (2.34) is also verified for this example.

We may consider the transformation of **a** into $\mathbf{a}'$ as involving an operator **R**. Then

$$\mathbf{a}' = \mathbf{Ra} = \mathbf{Rea} = \bar{\mathbf{e}}\mathrm{a} = \mathbf{eRa} \tag{2.35}$$

Writing out these expressions we find they imply the following distributive behaviour for **R**:

$$\mathbf{R}(\mathbf{e}_1 a_1 + \mathbf{e}_2 a_2 + \ldots) = \mathbf{R}\mathbf{e}_1 a_1 + \mathbf{R}\mathbf{e}_2 a_2 + \ldots = \bar{\mathbf{e}}_1 a_1 + \bar{\mathbf{e}}_2 a_2 + \ldots \tag{2.36}$$

This equation (2.36) expresses the characteristic property of *linear operators*. Hence, on the basis of Equation (2.35) we may write:

$$\mathbf{R}\mathbf{e} = \mathbf{e}\mathbf{R} \tag{2.37}$$

which is the important result we require. The effect of $\mathbf{R}$ upon *any* vector $\mathbf{a}$ in an n-dimensional space is completely determined as soon as we know its effect upon n base vectors spanning that space. The matrix $\mathbf{R}$ results from carrying out the operation $\mathbf{R}$ upon the base vectors, in accordance with Equation (2.37). *Every operation* $\mathbf{R}$ *has its own matrix* $\mathbf{R}$ *to represent its effect*, and the form of that matrix depends on the choice of base vectors.

This result applies whether or not the base vectors are orthonormal, and includes transformations, such as changes of scale, which are not symmetry operations. Two important consequences may be deduced.

In the first place, we may seek the relationship between the two matrices $\mathbf{P}$ and $\bar{\mathbf{P}}$, which represent the same operation $\mathbf{P}$ applied respectively to the two different bases, $\mathbf{e}$ and $\bar{\mathbf{e}}$, which are themselves related by Equation (2.30): $\bar{\mathbf{e}} = \mathbf{e}\mathbf{R}$. Making use of Equation (2.37):

$$\mathbf{P}\mathbf{e} = \mathbf{e}\mathbf{P} \quad \text{and} \quad \mathbf{P}\bar{\mathbf{e}} = \bar{\mathbf{e}}\bar{\mathbf{P}}$$

from which it follows that:

$$\mathbf{P}\bar{\mathbf{e}} = \mathbf{P}\mathbf{e}\mathbf{R} = \mathbf{e}\mathbf{P}\mathbf{R} = \bar{\mathbf{e}}\mathbf{R}^{-1}\mathbf{P}\mathbf{R}$$

and hence

$$\bar{\mathbf{P}} = \mathbf{R}^{-1}\mathbf{P}\mathbf{R} \tag{2.38}$$

The representative matrices $\bar{\mathbf{P}}$ and $\mathbf{P}$ are said to be related by a *similarity transformation*, and are described as *equivalent*. We may note, using Equation (2.21), that:

$$\mathrm{Tr}\,\bar{\mathbf{P}} = \mathrm{Tr}(\mathbf{R}^{-1}\mathbf{P}\mathbf{R}) = \mathrm{Tr}(\mathbf{P}\mathbf{R}\mathbf{R}^{-1}) = \mathrm{Tr}\,\mathbf{P} \tag{2.39}$$

The trace of a matrix is unaltered by a similarity transformation.

In the second place we may ask what kind of matrices represent symmetry operations. These necessarily leave molecules undistorted and therefore correspond to transformations in which the scalar products of pairs of vectors remain unchanged. Hence

$$\mathbf{a}.\mathbf{b} = \mathbf{a}^{\mathrm{T}}\mathbf{M}\mathbf{b} = (\mathbf{R}\mathbf{a})^{\mathrm{T}}\mathbf{M}(\mathbf{R}\mathbf{b}) = \mathbf{a}^{\mathrm{T}}\mathbf{R}^{\mathrm{T}}\mathbf{M}\mathbf{R}\mathbf{b}$$

and

$$\mathbf{R}^{\mathrm{T}}\mathbf{M} = \mathbf{M}\mathbf{R}^{-1}$$

We may always choose an orthonormal basis, for which $\mathbf{M} = \mathbf{1}$, when it follows that:

$$\mathbf{R}^{\mathrm{T}} = \mathbf{R}^{-1} \tag{2.40}$$

Symmetry operations may always be represented by orthogonal matrices. The 2×2 matrices in the illustrated 2-dimensional examples above are clearly matrices of this kind, representing rotations through an angle θ about an axis perpendicular to the plane of the paper.

2.3.5 The representation of a group

We now apply the results of the previous section to vectors associated with an actual molecule. For this purpose we take ammonia, whose symmetry ($\mathcal{C}_{3v}$) we have already determined. To describe the effect of the symmetry operations upon any general point, we obtain the matrices that result from applying the operations to a suitable set of base vectors. Let us consider two such sets: (1) unit vectors defining the Cartesian axes, and (2) unit vectors directed along the NH bonds, as shown and lettered in Fig. 2.24.

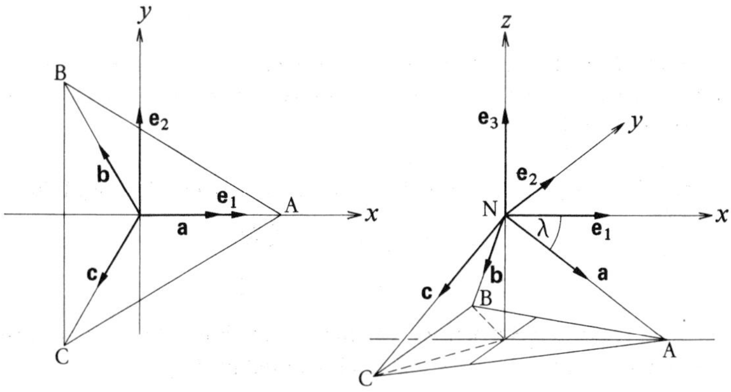

Fig. 2.24

(1) Cartesian unit vectors

The behaviour of $\mathbf{e}_3$ is clearly distinct from that of $\mathbf{e}_1$ and $\mathbf{e}_2$, since it lies in all the symmetry elements and is left unchanged by every symmetry operation:

$$\mathbf{R}\mathbf{e}_3 = \mathbf{e}_3(+1) \tag{2.41}$$

while $\mathbf{e}_1$ and $\mathbf{e}_2$ are transformed into other vectors lying in the same plane. It follows that the corresponding matrices will all have the form

$$[\bar{\mathbf{e}}_1 \quad \bar{\mathbf{e}}_2 \quad \bar{\mathbf{e}}_3] = [\mathbf{e}_1 \quad \mathbf{e}_2 \quad \mathbf{e}_3]\begin{bmatrix} & & 0 \\ & & 0 \\ 0 & 0 & 1 \end{bmatrix} \tag{2.42}$$

The 2 x 2 submatrices describing $\bar{\mathbf{e}}_1$ and $\bar{\mathbf{e}}_2$ may be derived by inspection of, for example, Figs. 2.25 and 2.26, and the use of Equation (2.37).

$$\begin{aligned} \mathbf{C}_3^+\!: \ [\bar{\mathbf{e}}_1 \quad \bar{\mathbf{e}}_2] &= [\mathbf{e}_1(-1/2) + \mathbf{e}_2(\sqrt{3}/2) \quad \mathbf{e}_1(-\sqrt{3}/2) + \mathbf{e}_2(-1/2)] \\ &= [\mathbf{e}_1 \quad \mathbf{e}_2]\begin{bmatrix} -1/2 & -\sqrt{3}/2 \\ \sqrt{3}/2 & -1/2 \end{bmatrix} \end{aligned}$$

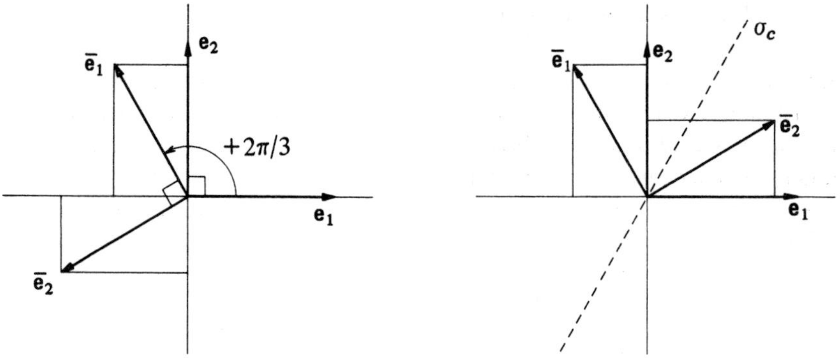

Figs. 2.25 and 2.26 The application of the operations $\mathbf{C}_3^+$ and σ_c, respectively, to the base vectors $\mathbf{e}_1$, $\mathbf{e}_2$. The transformed base vectors $\bar{\mathbf{e}}_1$, $\bar{\mathbf{e}}_2$ can be expressed in terms of $\mathbf{e}_1$, $\mathbf{e}_2$, cos 30°, and sin 30°.

$$\sigma_c\colon\ [\bar{\mathbf{e}}_1 \quad \bar{\mathbf{e}}_2] = [\mathbf{e}_1(-1/2) + \mathbf{e}_2(\sqrt{3}/2) \quad \mathbf{e}_1(\sqrt{3}/2) + \mathbf{e}_2(1/2)]$$
$$= [\mathbf{e}_1 \quad \mathbf{e}_2]\begin{bmatrix} -1/2 & \sqrt{3}/2 \\ \sqrt{3}/2 & 1/2 \end{bmatrix}$$

The complete set of 3 x 3 matrices, obtained by applying the group operations to $\mathbf{e}_1$, $\mathbf{e}_2$, and $\mathbf{e}_3$, is given in Table 2.3, and denoted by Γ_{xyz}.

(2) NH bond unit vectors

In this case all three vectors are related in a similar way to the symmetry elements, and the set of 3 x 3 matrices, denoted by Γ_{abc} in Table 2.3, expresses their permutation under the various symmetry operations. Thus for $\mathbf{C}_3^+$:

$$[\bar{\mathbf{a}} \quad \bar{\mathbf{b}} \quad \bar{\mathbf{c}}] = [\mathbf{b} \quad \mathbf{c} \quad \mathbf{a}] = [\mathbf{a} \quad \mathbf{b} \quad \mathbf{c}]\begin{bmatrix} 0 & 0 & 1 \\ 1 & 0 & 0 \\ 0 & 1 & 0 \end{bmatrix}$$

The matrices of Γ_{xyz} or Γ_{abc} may be used in Equation (2.34) to describe how, in the chosen coordinate system, the components of any general vector transform under the group operations or, to use an equivalent expression, how the 3-dimensional space of the molecule maps onto itself. This reminds us that we first described the symmetry operations by their effect upon the NH_3 molecule, and in that way obtained the group multiplication Table 2.2. It follows that the matrices, multiplied together, should give the same table.

In this way, using matrices of Γ_{abc}, we check the following relationships:

$$\mathbf{C}_3^+\sigma_b = \sigma_a \qquad \begin{bmatrix} 0 & 0 & 1 \\ 1 & 0 & 0 \\ 0 & 1 & 0 \end{bmatrix}\begin{bmatrix} 0 & 0 & 1 \\ 0 & 1 & 0 \\ 1 & 0 & 0 \end{bmatrix} = \begin{bmatrix} 1 & 0 & 0 \\ 0 & 0 & 1 \\ 0 & 1 & 0 \end{bmatrix}$$

Table 2.3

$\mathcal{C}_{3v}$	$\mathbf{E}$	$\mathbf{C}_3^+$	$\mathbf{C}_3^-$	σ_a	σ_b	σ_c
Γ_{xyz}	$\begin{bmatrix} 1 & 0 & 0 \\ 0 & 1 & 0 \\ 0 & 0 & 1 \end{bmatrix}$	$\begin{bmatrix} -1/2 & -\sqrt{3}/2 & 0 \\ \sqrt{3}/2 & -1/2 & 0 \\ 0 & 0 & 1 \end{bmatrix}$	$\begin{bmatrix} -1/2 & \sqrt{3}/2 & 0 \\ -\sqrt{3}/2 & -1/2 & 0 \\ 0 & 0 & 1 \end{bmatrix}$	$\begin{bmatrix} 1 & 0 & 0 \\ 0 & -1 & 0 \\ 0 & 0 & 1 \end{bmatrix}$	$\begin{bmatrix} -1/2 & -\sqrt{3}/2 & 0 \\ -\sqrt{3}/2 & 1/2 & 0 \\ 0 & 0 & 1 \end{bmatrix}$	$\begin{bmatrix} -1/2 & \sqrt{3}/2 & 0 \\ \sqrt{3}/2 & 1/2 & 0 \\ 0 & 0 & 1 \end{bmatrix}$
Γ_{abc}	$\begin{bmatrix} 1 & 0 & 0 \\ 0 & 1 & 0 \\ 0 & 0 & 1 \end{bmatrix}$	$\begin{bmatrix} 0 & 0 & 1 \\ 1 & 0 & 0 \\ 0 & 1 & 0 \end{bmatrix}$	$\begin{bmatrix} 0 & 1 & 0 \\ 0 & 0 & 1 \\ 1 & 0 & 0 \end{bmatrix}$	$\begin{bmatrix} 1 & 0 & 0 \\ 0 & 0 & 1 \\ 0 & 1 & 0 \end{bmatrix}$	$\begin{bmatrix} 0 & 0 & 1 \\ 0 & 1 & 0 \\ 1 & 0 & 0 \end{bmatrix}$	$\begin{bmatrix} 0 & 1 & 0 \\ 1 & 0 & 0 \\ 0 & 0 & 1 \end{bmatrix}$
χ	3	0	0	1	1	1

$$\sigma_a\sigma_c = C_3^- \qquad \begin{bmatrix} 1 & 0 & 0 \\ 0 & 0 & 1 \\ 0 & 1 & 0 \end{bmatrix} \begin{bmatrix} 0 & 1 & 0 \\ 1 & 0 & 0 \\ 0 & 0 & 1 \end{bmatrix} = \begin{bmatrix} 0 & 1 & 0 \\ 0 & 0 & 1 \\ 1 & 0 & 0 \end{bmatrix}$$

The reader should check other parts of the multiplication table using the corresponding matrices of Γ_{xyz} or Γ_{abc}. It is clear that, in obtaining Γ_{xyz} or Γ_{abc}, it would have been sufficient to establish the matrices corresponding to the generators, C_3^+ and σ_a; the others could have been obtained from them by matrix multiplication.

Since both Γ_{xyz} and Γ_{abc} reproduce the behaviour of the symmetry operations we call them *representations of the $\mathcal{C}_{3v}$ point group* and speak of each as being *carried* or *generated* by the corresponding set of base vectors. We shall find it convenient to have a notation $\boldsymbol{\Gamma}_i(\mathbf{R})$ to refer to the matrix $\mathbf{R}$ belonging to a particular representation Γ_i. (Many use $\mathbf{D}_i(\mathbf{R})$ instead.)

Let us look more closely at this concept of *representation*. Formally we could have expressed the relationship between three group operations Q, P, and R and their corresponding matrices $\mathbf{Q}$, $\mathbf{P}$, and $\mathbf{R}$ as follows. If $\mathsf{QP} = \mathsf{R}$, then according to Equation (2.37):

$$\mathsf{QP}\mathbf{a} = \mathsf{QP}\mathbf{ea} = \mathsf{Q}\mathbf{ePa} = \mathbf{eQPa}$$

and

$$\mathsf{QP}\mathbf{a} = \mathsf{R}\mathbf{a} = \mathsf{R}\mathbf{ea} = \mathbf{eRa}$$

Hence $\mathbf{QP} = \mathbf{R}$, since $\mathbf{a}$ is arbitrary and $\mathbf{e}$ is chosen to be linearly independent.

For an arbitrary vector $\mathbf{a}$ in 3-dimensional space there will be a distinct mapping for each symmetry operation R, and there will be one-to-one relationships between these mappings, the operations R and the corresponding 3×3 matrices $\mathbf{R}$. We say that the set of mappings and the set of matrices $\mathbf{R}$ each form a group *isomorphic* with the point group, sharing the same multiplication table. Note especially that the unit matrix $\mathbf{1}$ represents the identity operation E and that the inverse matrix, $\mathbf{R}^{-1}$, represents the inverse operation R^{-1}.

Our experience with the Cartesian base vectors allows us to consider the situation in a space of fewer than 3 dimensions. We saw that $\mathbf{e}_3$ had only *one* mapping, to which there corresponded a *single* 1×1 matrix, namely the number $+1$, since there is always a one-to-one correspondence between mappings and matrices. Now, although this may seem trivial, the set of $+1$'s – one for each group operation – is also a representation of the group, since multiplied together they also must agree with the group multiplication table. Where there is such a one-to-many correspondence between the matrices $\mathbf{R}$ and the operations R, the relationship is said to be only *homomorphic*.

Another example of this exists for the $\mathcal{C}_{3v}$ point group. The following association:

$$\underbrace{\mathsf{E} \quad C_3^+ \quad C_3^-}_{+1} \qquad \underbrace{\sigma_a \quad \sigma_b \quad \sigma_c}_{-1} \tag{2.43}$$

corresponds to another 1-dimensional representation, as may be seen on replacing the operations in the multiplication Table 2.2 by the appropriate +1 or −1. This representation is not carried by any vector in ordinary space, but it is carried by the 'pseudovector' representing an infinitesimal *rotation* about the z-axis (R_z). Unaffected by the rotations $\mathbf{C}_3$, the sense of R_z is reversed by a σ_v reflection.

The 2 × 2 matrices carried by $\mathbf{e}_1$ and $\mathbf{e}_2$ also form a representation, as the reader should check. Since each of these matrices is distinct, there is in this case a one-to-one correspondence, or isomorphism, with the point group.

It should be appreciated that, in the sense of Section 2.3.4, the Γ_{xyz} and Γ_{abc} representations are *equivalent*, since they both convey the same, complete information about mappings in 3-dimensional space. A similarity transformation must relate corresponding matrices, so we may write Equation (2.38) as:

$$\boldsymbol{\Gamma}_{abc}(\mathsf{R}) = \mathrm{T}^{-1}\boldsymbol{\Gamma}_{xyz}(\mathsf{R})\mathrm{T} \quad \text{or} \quad \mathrm{T}\boldsymbol{\Gamma}_{abc}(\mathsf{R}) = \boldsymbol{\Gamma}_{xyz}(\mathsf{R})\mathrm{T} \tag{2.44}$$

The matrix **T** may be obtained from the relationship between the two sets of base vectors. From Fig. 2.24 we see that:

$$[\mathbf{a} \quad \mathbf{b} \quad \mathbf{c}] = [\mathbf{e}_1 \quad \mathbf{e}_2 \quad \mathbf{e}_3]\begin{bmatrix} c & -c/2 & -c/2 \\ 0 & c\sqrt{3}/2 & -c\sqrt{3}/2 \\ -s & -s & -s \end{bmatrix} \tag{2.45}$$

where $c = \cos\lambda$, $s = \sin\lambda$. We may use this matrix **T** to check Equation (2.44).

However, it is not necessary that we should carry out this transformation; the evidence of their equivalence lies in the common values of the traces of their corresponding matrices, in agreement with Equation (2.39). Because of this equivalence the traces of representative matrices are called their *characters* and denoted χ. The character system of Γ_{xyz} and Γ_{abc} is shown in Table 2.3. Note also that, because symmetry operations belonging to the same class are related by Equation (2.6), $\mathsf{Q} = \mathsf{R}^{-1}\mathsf{P}\mathsf{R}$, their representative matrices are related by similarity transformations and they all have the same character. We shall shortly see that it is sufficient for our purposes to use characters instead of the representations themselves. It will therefore only be necessary to consider a typical operation of each class.

The characters of the representation carried by the Cartesian vectors can be rapidly evaluated for any point group if it is appreciated that the general form of $\boldsymbol{\Gamma}_{xyz}(\mathsf{R})$ is given, on the basis of Equations (2.31), (2.41) and its counterpart for S_n, and (2.42), by:

$$\boldsymbol{\Gamma}_{xyz}(\mathsf{R}_z) = \begin{bmatrix} \cos\theta & -\sin\theta & 0 \\ \sin\theta & \cos\theta & 0 \\ 0 & 0 & \pm 1 \end{bmatrix} \tag{2.46}$$

the choice of sign depending on whether the symmetry operation is a proper rotation (+1) or an improper rotation (−1). If the axis of rotation is not the

z-axis, a suitable orthogonal transformation is required, but the character remains unchanged:

$$\chi_{xyz}(\mathbf{R}) = 2\cos\theta \pm 1 \tag{2.47}$$

Note that $\theta = 0$ for $\mathbf{E}\ (= \mathbf{C}_1)$ and $\boldsymbol{\sigma}(= \mathbf{S}_1)$ and $\theta = \pi$ for $\mathbf{C}_2$ and $\mathbf{i}\ (= \mathbf{S}_2)$.

Readers should confirm that Equation (2.47) gives the character system of Table 2.3. A further example of its application is provided in Table 2.4 for the $\mathcal{O}_h$ point group. [See SF_6 in Fig. 2.15(c).]

Table 2.4

$\mathcal{O}_h$	$\mathbf{E}$	$\mathbf{C}_3$	$\mathbf{C}_2$	$\mathbf{C}_4$	$\mathbf{C}_2'$	$\mathbf{i}$	$\mathbf{S}_6$	σ_h	$\mathbf{S}_4$	σ_d	
θ	0	$2\pi/3$	π	$\pi/2$	π	π	$\pi/3$	0	$\pi/2$	0	
$\cos\theta$	1	$-1/2$	-1	0	-1	-1	$1/2$	1	0	1	
$2\cos\theta \pm 1$	3	0	-1	1	-1	-3	0	1	-1	1	$= \chi_{xyz}$

2.4 IRREDUCIBLE REPRESENTATIONS

2.4.1 The meaning of reduction

When we were obtaining Γ_{xyz} from the action of the symmetry operations $\mathbf{R}$ upon $\mathbf{e}$, we saw that every matrix would take a form with square blocks along the diagonal, and zeros elsewhere:

$$\boldsymbol{\Gamma}_{xyz}(\mathbf{R}) = \begin{bmatrix} r_{11} & r_{12} & 0 \\ r_{21} & r_{22} & 0 \\ 0 & 0 & r_{33} \end{bmatrix} = \left[\begin{array}{c|c} \boldsymbol{\Gamma}_1(\mathbf{R}) & 0 \\ \hline 0 & \boldsymbol{\Gamma}_2(\mathbf{R}) \end{array}\right] \tag{2.48}$$

This came about because of the differing behaviour of $\mathbf{e}_3$ and of $\mathbf{e}_1$ and $\mathbf{e}_2$ with respect to each of the operations. As a result, we may speak of the matrices as having been *decomposed*, their products with other matrices being divided into parts involving either $\boldsymbol{\Gamma}_1(\mathbf{R})$ or $\boldsymbol{\Gamma}_2(\mathbf{R})$, but not both:

$$\begin{bmatrix} r_{11} & r_{12} & 0 \\ r_{21} & r_{22} & 0 \\ 0 & 0 & r_{33} \end{bmatrix}\begin{bmatrix} a_1 \\ a_2 \\ a_3 \end{bmatrix} = \begin{bmatrix} r_{11}a_1 + r_{12}a_2 \\ r_{21}a_1 + r_{22}a_2 \\ r_{33}a_3 \end{bmatrix}$$

Equally, *since every* $\boldsymbol{\Gamma}_{xyz}(\mathbf{R})$ *is 'blocked out' in the same way*, we may speak of the *representation* as having been decomposed into parts which act independently. We have already seen that these parts are, by themselves, representations of the point group. That consisting of 2 x 2 matrices, Γ_1, is referred to as the E representation; the other, Γ_2, consisting of the numbers, +1, as the A_1 representation. The other 1-dimensional representation (2.43) is called A_2. The combination of two (or more) representations by placing their respective matrices along the principal diagonals of larger matrices, whose other elements are all zeros, is referred to as their *direct sum*, and written, in this case, as:

$$\Gamma_{xyz} = E + A_1 \tag{2.49}$$

Γ_{xyz} is said to *contain* E and A_1, or to be *reducible* into E and A_1. In general we may write:

$$\Gamma_\rho = n_1\Gamma_1 + n_2\Gamma_2 + \ldots \tag{2.50}$$

where Γ_ρ will refer to a reducible representation, and $n_1, n_2, \ldots$ are integers, allowing for the possibility of particular representations occurring more than once in Γ_ρ.

Moreover, since Γ_{xyz} and Γ_{abc} are equivalent, we may also write:

$$\Gamma_{abc} = E + A_1$$

it being understood that the equality may involve a similarity transformation. The same applies to the representation E, since base vectors other than $\mathbf{e}_1$ and $\mathbf{e}_2$ might have been chosen; in Section 2.5.3 we shall meet such a choice.

However, no choice of base vectors to replace $\mathbf{e}_1$ and $\mathbf{e}_2$ will allow the E representation to be decomposed into two 1-dimensional representations. For, if this were possible, the resultant diagonal matrices would commute and form only a homomorphic and not an isomorphic representation of the point group. But E was found to be isomorphic, so this decomposition or reduction is impossible.

The A_1, A_2, and E representations are referred to as *irreducible representations*, and are, as it happens, the only ones possible for the $\mathcal{C}_{3v}$ point group. If the decomposition described by Equation (2.50) is carried as far as irreducible representations, it is called *complete reduction*.

Algebraically complete reduction implies that each $\mathbf{\Gamma}_\rho(\mathsf{R})$ matrix can be transformed into a block matrix with a corresponding pattern of blocks on its diagonal:

$$\mathbf{\Gamma}_\rho(\mathsf{R}) = \begin{bmatrix} \boxed{\mathbf{\Gamma}_1(\mathsf{R})} & & & 0 \\ & \boxed{\mathbf{\Gamma}_2(\mathsf{R})} & & \\ & & \boxed{\mathbf{\Gamma}_3(\mathsf{R})} & \\ & 0 & & \ddots \end{bmatrix} \tag{2.51}$$

Geometrically it implies a new choice of base vectors, spanning the same n-dimensional vector space as those carrying Γ_ρ, but, like the Cartesian base vectors under the operations of $\mathcal{C}_{3v}$, divided into subsets, the members of each of which *only transform among themselves*, under the group operations. We say that each subset defines an *invariant subspace*, and that such a choice of base vectors is *symmetry-adapted*. From this property we see that any two such *base vectors, corresponding to different irreducible representations, must necessarily be orthogonal*.

Examination of Fig. 2.24 or the calculation of $\mathbf{T}^{-1}$ [compare Equation (2.45)] shows how the base vectors **a**, **b**, and **c** may be combined together to give symmetry-adapted vectors which transform like $\mathbf{e}_1$, $\mathbf{e}_2$, and $\mathbf{e}_3$:

$$\begin{aligned} A_1:&\quad \mathbf{s}_1 = N_1(\mathbf{a}+\mathbf{b}+\mathbf{c}) = -\mathbf{e}_3 \quad \text{when} \quad N_1 = 1/3 \sin\lambda \\ E:&\quad \begin{cases} \mathbf{s}_2 = N_2(2\mathbf{a}-\mathbf{b}-\mathbf{c}) = \mathbf{e}_1 \quad \text{when} \quad N_2 = 1/3\cos\lambda \\ \mathbf{s}_3 = N_3(\mathbf{b}-\mathbf{c}) = \mathbf{e}_2 \quad \text{when} \quad N_3 = 1/\sqrt{3}\cos\lambda \end{cases} \end{aligned} \tag{2.52}$$

2.4.2 The characters as vector components

Our ability to effect a reduction depends on the theorem (McWeeny [2.7] p.126) that, for any point group, *there are only as many distinct, inequivalent, irreducible representations as there are classes of operations*. Equation (2.50) can be completed as:

$$\Gamma_\rho = n_1\Gamma_1 + n_2\Gamma_2 + \ldots + n_i\Gamma_i + \ldots + n_m\Gamma_m \tag{2.53}$$

where the Γ_i are irreducible representations and m is the number of classes.

As a result, a representation of a finite group carried by a set of base vectors associated with a molecule can be described in terms of relatively few irreducible representations. The set of symmetry-adapted vectors into which the original set may be transformed can then be classified according to these irreducible representations whose pattern of symmetry behaviour they follow. For this reason the irreducible representations are also called *symmetry species*, to which the symmetry-adapted vectors are said to belong.

Since Equation (2.53) relates to a suitable similarity transformation, under which the traces of the representative matrices are invariant, it follows that:

$$\chi_\rho = n_1\chi_1 + n_2\chi_2 + \ldots + n_i\chi_i + \ldots + n_m\chi_m \tag{2.54}$$

Tables listing $\chi_i(\mathbf{R}_k)$, the characters under the typical operations $\mathbf{R}_k$ for each irreducible representation Γ_i, are available for every point group. That for $\mathcal{C}_{3v}$ is given in Table 2.5, where the number preceding each $\mathbf{R}_k$ refers to h_k the order of the k-th class.

Table 2.5

$\mathcal{C}_{3v}$	E	$2C_3$	$3\sigma_v$
A_1	1	1	1
A_2	1	1	−1
E	2	−1	0

For the simple reducible representations, Γ_{xyz} and Γ_{abc}, considered so far, we could have deduced the integers n_i by inspection as 1, 0, and 1. However, for a Γ_ρ of more dimensions, a general method is desirable. This will involve the

the solution of a set of m simultaneous linear equations, each of the form (2.54), in the m unknowns n_i. Now, as we have seen (p.31), such a set of equations can be conveniently expressed in matrix form as

$$\boldsymbol{\chi}_\rho = \mathbf{X}\mathbf{n} \tag{2.55}$$

in which the column matrix $\boldsymbol{\chi}_\rho$ consists of the $\chi_\rho(\mathbf{R}_k)$; so these may be regarded as *components* of an m-dimensional vector $\boldsymbol{\chi}_\rho$. Similarly the n_i form the elements of the column matrix $\mathbf{n}$. The square $m \times m$ matrix $\mathbf{X}$ is the *transpose* of the square array in the character Table 2.5, so its k,j-th element may be written $(\mathbf{X})_{kj} = \chi_j(\mathbf{R}_k)$. Thus, for our simple example, Equation (2.55) becomes:

$$\begin{bmatrix} 3 \\ 0 \\ 1 \end{bmatrix} = \begin{bmatrix} 1 & 1 & 2 \\ 1 & 1 & -1 \\ 1 & -1 & 0 \end{bmatrix} \begin{bmatrix} 1 \\ 0 \\ 1 \end{bmatrix}$$

To solve Equation (2.55) we pre-multiply both sides by the inverse matrix $\mathbf{X}^{-1}$:

$$\mathbf{X}^{-1}\boldsymbol{\chi}_\rho = \mathbf{X}^{-1}\mathbf{X}\mathbf{n} = \mathbf{n} \tag{2.56}$$

which, for our example, gives:

$$\begin{bmatrix} 1/6 & 1/3 & 1/2 \\ 1/6 & 1/3 & -1/2 \\ 1/3 & -1/3 & 0 \end{bmatrix} \begin{bmatrix} 3 \\ 0 \\ 1 \end{bmatrix} = \begin{bmatrix} 1 \\ 0 \\ 1 \end{bmatrix}$$

The elements of the inverse matrix could have been obtained by standard methods (Anderson [2.1] p.68) but, in fact, the general i,k-th element is given directly by the remarkable relationship:

$$(\mathbf{X}^{-1})_{ik} = \frac{h_k}{g}\chi_i(\mathbf{R}_k) \tag{2.57}$$

where g is the order of the point group, the number of its symmetry operations.

Equations (2.56) and (2.57) may be conveniently combined:

$$n_i = \frac{1}{g}\sum_k h_k \chi_i(\mathbf{R}_k)\chi_\rho(\mathbf{R}_k) \tag{2.58}$$

For $\mathcal{C}_{3v}$ this becomes:

$$n_i = \tfrac{1}{6}[1 \times \chi_i(\mathbf{E}) \times \chi_\rho(\mathbf{E}) + 2 \times \chi_i(\mathbf{C}_3) \times \chi_\rho(\mathbf{C}_3) + 3 \times \chi_i(\sigma_v) \times \chi_\rho(\sigma_v)]$$

and for our example this may be set out as follows:

$$n_{A_1} = \tfrac{1}{6}[(1 \times 1 \times 3) + (2 \times 1 \times 0) + (3 \times 1 \times 1)] = 1$$

$$n_{A_2} = \tfrac{1}{6}[(1 \times 1 \times 3) + (2 \times 1 \times 0) + (3 \times -1 \times 1)] = 0$$

$$n_E = \tfrac{1}{6}[(1 \times 2 \times 3) + (2 \times -1 \times 0) + (3 \times 0 \times 1)] = 1$$

The relationship (2.57) implies that the $\chi_i(\mathbf{R})$ possess a number of specific properties. Since $\mathbf{X}^{-1}\mathbf{X} = \mathbf{1}$, it follows from Equations (2.19) and (2.57), that:

$$\frac{1}{g}\sum_k h_k \chi_i(\mathbf{R}_k)\chi_j(\mathbf{R}_k) = \frac{1}{g}\sum_{\mathbf{R}} \chi_i(\mathbf{R})\chi_j(\mathbf{R}) = \delta_{ij} \qquad (2.59)$$

where the first summation is taken over every class and the second over every distinct symmetry operation. It can be seen therefore that *the* $\chi_i(\mathbf{R})$ *may be regarded as components, in a g-dimensional space, of vectors* $\boldsymbol{\chi}_i$ *which are orthogonal and all of length* $\sqrt{g}$. It is a sufficient criterion of irreducibility that:

$$\sum_{\mathbf{R}} [\chi(\mathbf{R})]^2 = g \qquad (2.60)$$

Thus for χ_E of $\mathcal{C}_{3v}$ we have $1(2)^2 + 2(-1)^2 + 3(0)^2 = 6$.

Equation (2.59) is a particular consequence of more general orthogonality relationships between vectors whose components are corresponding elements of the matrices of the irreducible representations (McWeeny [2.7] pp.116–123). Equation (2.53) is a further consequence of these relationships, while another is that the sum of the squares of the dimensions of the irreducible representations of a group is equal to g, its order. Since the dimensions of the matrices are given by the $\chi_i(\mathbf{E})$, we may express this as:

$$\sum_i [\chi_i(\mathbf{E})]^2 = g \qquad (2.61)$$

Thus for $\mathcal{C}_{3v}$: $1^2 + 1^2 + 2^2 = 6$. It is on the basis of these various relationships that all the character tables have been obtained.

2.4.3 The symmetry species of molecular vibrations

In the theory of molecular vibrations it may be shown that any distortion of a molecule from its equilibrium conformation, provided that distortion is only a very small one, gives rise to vibrations of the N nuclei about their equilibrium positions which may be expressed as the linear sum, or superposition, of $(3N-6)$† independent *normal modes of vibration* in each of which the nuclei oscillate in phase together at the same frequency. The course of these oscillations may be described in terms of *normal coordinates*, q_i, which are components that multiply $(3N-6)$ orthonormal base vectors $\mathbf{l}_i$. These latter describe the *relative* nuclear displacements from equilibrium in the i-th mode, which stay in the same ratio throughout the oscillation. They are illustrated for ammonia in Fig. 2.27.

If the q_i are mass-weighted, that is, if every spatial displacement x_j is multiplied by $\sqrt{m_j}$, where j refers to the j-th nucleus, then the potential energy may be expressed by:

† $(3N-5)$ throughout, for *linear* molecules.

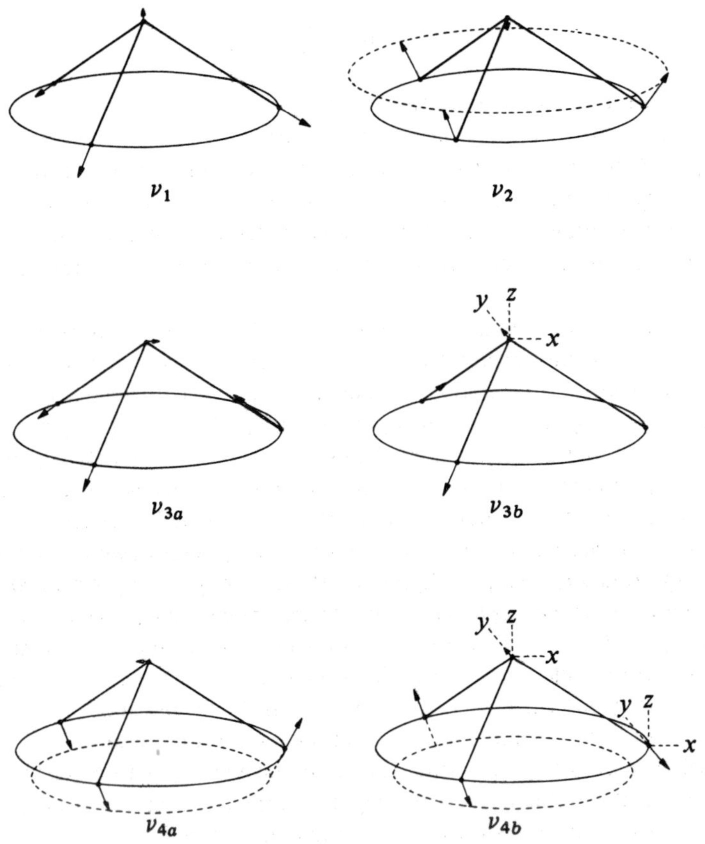

Fig. 2.27 The normal vibrational modes of NH_3. In each diagram the nuclear displacements shown (schematically and not strictly to scale) are proportional to the elements of the base vector $\mathbf{l}_i$. The displacements remain in the same ratio throughout the oscillation. The ν_1 and ν_2 modes are non-degenerate (a_1), while ν_{3a}, ν_{3b} and ν_{4a}, ν_{4b} are doubly degenerate pairs (e). (After Fig. 6.4 of McWeeny [2.7].)

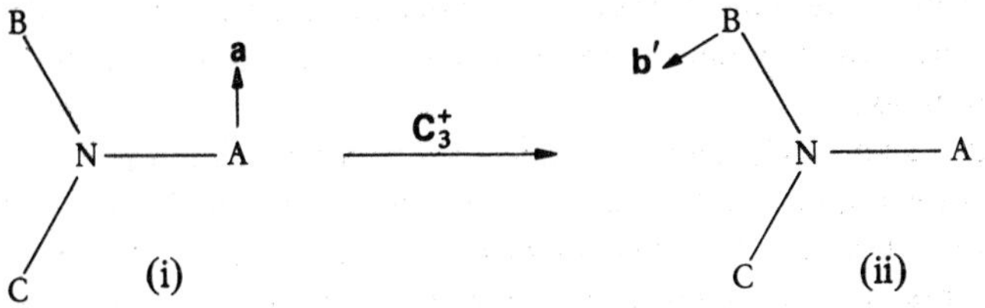

Fig. 2.28 Two arbitrary distortions of the NH_3 molecule, related by a $\mathbf{C}_3^+$ symmetry operation applied to the nuclear displacement vectors. The displacement vectors have been named according to the nucleus displaced.

$$2V = 4\pi^2 c^2 \sum_i \omega_i^2 q_i^2$$

where ω_i is the wavenumber of the i-th mode.

If we now consider two arbitrary distortions of a molecule, such as (i) and (ii) in Fig. 2.28, which are related by a symmetry operation carried out upon the nuclear displacement vectors, in which these are transferred to an equivalent (or the same) nucleus, we see that they must have identical potential energies. It follows that V is quite generally invariant to the symmetry operations of the point group.

If a particular ω_i is unique, the invariance of V implies the invariance of q_i^2, so that q_i is only multiplied by ± 1 and has to transform according to a 1-dimensional irreducible representation. On the other hand, if two or three ω_i are equal, say $\omega_i = \omega_j = \omega_k$, we require the invariance of $(q_i^2 + q_j^2 + q_k^2)$. Clearly q_i, q_j, and q_k must only transform among themselves under the symmetry operations, $\mathbf{l}_i$, $\mathbf{l}_j$, and $\mathbf{l}_k$ acting as a basis of a representation which must be irreducible if ω_i, ω_j, and ω_k are not accidentally coincident. The wavenumber ω_i and the coordinates q_i, q_j, q_k are said to be 3-fold degenerate. In general *a wavenumber (or frequency) is said to be n-fold degenerate if the corresponding normal coordinates transform according to an n-dimensional irreducible representation.*

Clearly the $(3N - 6)$ base vectors $\mathbf{l}_i$ constitute a set of symmetry-adapted vectors. Their calculation requires a knowledge of the atomic masses, the molecular geometry and the forces acting between the nuclei or, since the latter are not usually known, an experimental determination of the ω_i, although these, in themselves, may not be sufficient. This would take us beyond the scope of this chapter, but the problem of discovering the symmetry species of the $(3N - 6)$ normal modes provides us with an application of Section 2.4.2.

We commence with an arbitrarily chosen set of vectors spanning the same space as the $\mathbf{l}_i$, find the reducible representation they carry, and then reduce it to its constituent irreducible representations, which we may then identify as those of the normal modes. We therefore choose base vectors which could describe any arbitrary distortion of the molecule. The most convenient choice consists of 3 unit vectors attached to each nucleus, parallel to the Cartesian axes, $3N$ base vectors in all, so that our reducible representation will include the symmetry species of the translations and rotations of the undistorted molecule, which will have to be obtained separately and subtracted to leave those of the genuine vibrations.

The invariance of V under the symmetry operations requires us to consider that their effect upon the displacement vectors is, as in Fig. 2.28, to rotate the vectors but to *leave the molecule unmoved*, so that we regard the vectors as displacements of the nuclei to which they are transferred. Each symmetry operation produces a distinct mapping of any arbitrary displacement so that the corresponding matrices from a group isomorphic to the point group, and therefore a representation of it.

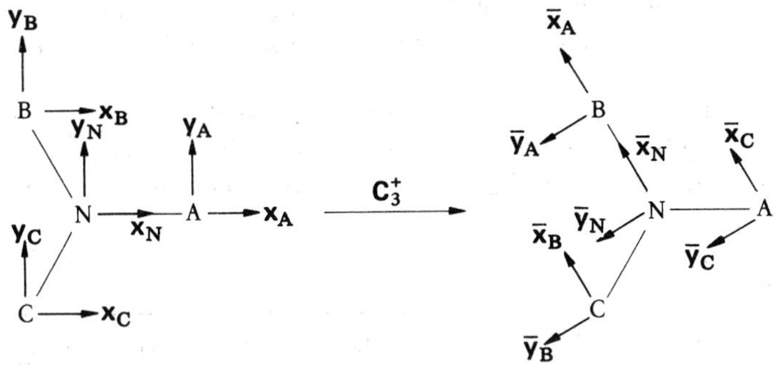

Fig. 2.29 The symmetry operation $\mathbf{C}_3^+$ applied to the Cartesian unit vectors set up at each nucleus. For clarity the $\mathbf{z}$ vectors are omitted. In the transformed basis the vectors are named according to the nucleus to which they were originally attached.

To obtain these matrices we now carry out the symmetry operations upon the $3N$ base vectors, as in Fig. 2.29 where we continue to take ammonia as our example. Note that, unlike the displacement vectors themselves, the base vectors are *not* renamed according to the nucleus to which they find themselves attached. Hence for $\mathbf{C}_3^+$ we have:

$$[\bar{\mathbf{x}}_A\ \bar{\mathbf{y}}_A\ \bar{\mathbf{z}}_A\ \bar{\mathbf{x}}_B\ \bar{\mathbf{y}}_B\ \bar{\mathbf{z}}_B\ \bar{\mathbf{x}}_C\ \bar{\mathbf{y}}_C\ \bar{\mathbf{z}}_C\ \bar{\mathbf{x}}_N\ \bar{\mathbf{y}}_N\ \bar{\mathbf{z}}_N]$$

$$= [\mathbf{x}_A\ \mathbf{y}_A\ \mathbf{z}_A\ \mathbf{x}_B\ \mathbf{y}_B\ \mathbf{z}_B\ \mathbf{x}_C\ \mathbf{y}_C\ \mathbf{z}_C\ \mathbf{x}_N\ \mathbf{y}_N\ \mathbf{z}_N] \left[\begin{array}{c|c|c|c} 0 & 0 & \mathbf{\Gamma}_{xyz}(\mathbf{C}_3^+) & 0 \\ \hline \mathbf{\Gamma}_{xyz}(\mathbf{C}_3^+) & 0 & 0 & 0 \\ \hline 0 & \mathbf{\Gamma}_{xyz}(\mathbf{C}_3^+) & 0 & 0 \\ \hline 0 & 0 & 0 & \mathbf{\Gamma}_{xyz}(\mathbf{C}_3^+) \end{array}\right] \quad (2.62)$$

where $\mathbf{\Gamma}_{xyz}(\mathbf{C}_3^+)$ has its previous meaning, as in Table 2.3.

The corresponding matrices for the other symmetry operations are soon written down but it is, of course, unnecessary that we should do so; all we require are the $\chi_{3N}(\mathbf{R})$ in order to find the irreducible representations contained in Γ_{3N}. We can see from the above example that the only block contributing to $\chi_{3N}(\mathbf{C}_3)$ corresponds to the displacement vectors of the N nucleus; those of the H nuclei are permuted among themselves and give rise to blocks that are off the diagonal. *Contributions to $\chi_{3N}(\mathbf{R})$ can only be made by displacement vectors that remain attached to the same atom throughout, and for Cartesian base vectors these contributions will be $\chi_{xyz}(\mathbf{R})$ in each case.* We may therefore write:

$$\chi_{3N}(\mathbf{R}) = U_{\mathbf{R}}\chi_{xyz}(\mathbf{R}) = U_{\mathbf{R}}(2\cos\theta \pm 1) \quad (2.63)$$

in which the value of χ_{xyz} obtained in Equation (2.47) has been used and $U_{\mathbf{R}}$ denotes the number of atoms *unshifted* when the operation $\mathbf{R}$ is carried out on the *molecule*. Thus, in the case of NH_3, $U_E = 4$; $U_{C_3} = 1$ since only the N atom lies on the C_3 axis; and $U_{\sigma_v} = 2$ since the N atom and one H atom lie in each σ_v plane.

Note that U_{S_n} corresponds to *the number of atoms that lie both on the S_n axis and in the plane perpendicular to this axis.* Thus, if our example had been the molecule SF_6 ($\mathcal{O}_h$), while $U_{C_4} = 3$, $U_{S_4} = 1$ only. Similarly, $U_{C_2} = 3$, but $U_i = 1$.

Table 2.6

NH_3:	$\mathcal{C}_{3v}$	E	$2C_3$	$3\sigma_v$	n_{3N}	n_{trans}	n_{rot}	n_{vib}
	A_1	1	1	1	3	1	0	2
	A_2	1	1	−1	1	0	1	0
	E	2	−1	0	4	1	1	2
	θ	0	$2\pi/3$	0				
	$\cos\theta$	1	−1/2	1				
$\chi_{xyz} = 2\cos\theta \pm 1$		3	0	1				
$\chi_{rot} = 1 \pm 2\cos\theta$		3	0	−1				
	U_R	4	1	2				
$\chi_{3N} = U_R\chi_{xyz}$		12	0	2				

Equation (2.63) is applied to ammonia in Table 2.6 and to methane in Table 2.13. These tables also set out the numbers of irreducible representations into which Γ_{3N} may be decomposed. Thus, using Equation (2.58), we may deduce that, in the case of ammonia:

$$n_{A_1} = \tfrac{1}{6}[(1 \times 1 \times 12) + (2 \times 1 \times 0) + (3 \times 1 \times 2)] = 3$$

$$n_{A_2} = \tfrac{1}{6}[(1 \times 1 \times 12) + (2 \times 1 \times 0) + (3 \times -1 \times 2)] = 1$$

$$n_E = \tfrac{1}{6}[(1 \times 2 \times 12) + (2 \times -1 \times 0) + (3 \times 0 \times 2)] = 4$$

or

$$\Gamma_{3N} = 3A_1 + A_2 + 4E \tag{2.64}$$

These are the symmetry species of all the possible displacements of the nuclei, not only of the vibrations but of the molecular translations and rotations as well.

A translation of the entire molecule corresponds to a *common displacement* of all the atoms, which may therefore be expressed in terms of 3 base vectors, conveniently chosen as unit vectors along the Cartesian axes. The common displacement may then be regarded as the position vector of the molecular centre of mass. Hence:

$$\chi_{trans}(R) = \chi_{xyz}(R) = 2\cos\theta \pm 1$$

from Equation (2.47), and for a molecule belonging to the $\mathcal{C}_{3v}$ point group:

$$\Gamma_{trans} = \Gamma_{xyz} = A_1 + E$$

according to Equation (2.49).

A rigid rotation of the molecule corresponds to a *common angular displacement* of the atoms. Finite rotations are not vector quantities; we have already observed their non-commutative, tensor character. However, we are only concerned with very small displacements, and very small rotations have some of the properties of vectors, such as vector-addition, and are therefore called pseudo-vectors. They can be represented by axial vectors: directed line segments along the axis of rotation, which define the sense of rotation by pointing along the line of advance of a right-handed screw.

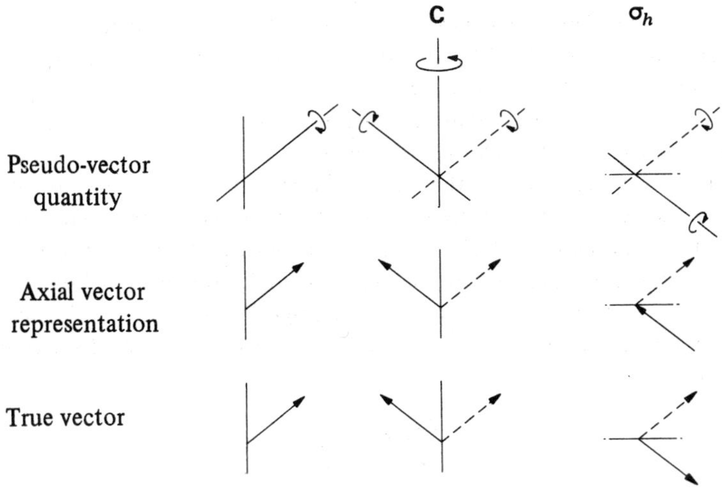

Fig. 2.30

Under a proper rotation an axial vector behaves like an ordinary vector; see Fig. 2.30. But an improper rotation consists of a proper rotation followed by a reflection in a mirror-plane, and the latter reverses the sense of the angular displacement and hence of the associated axial vector. Thus an axial vector behaves like a true vector, except that under an improper rotation it is multiplied by an extra factor of -1. Hence:

$$\chi_{rot}(\mathbf{R}) = \pm(2\cos\theta \pm 1) = 1 \pm 2\cos\theta \tag{2.65}$$

which for $\mathcal{C}_{3v}$ leads to:

$$\Gamma_{rot} = A_2 + E \tag{2.66}$$

(For linear molecules molecular rotation is confined to axes lying in the xy-plane, so the 1-dimensional representation corresponding to R_z must be rejected.) The decompositions of Γ_{xyz} and Γ_{rot} are usually recorded with the character tables, in additional column to the right, as in Tables 2.7 and 2.8.

Subtracting from (2.64) the symmetry species of the translations (2.49) and the rotations (2.66), we are left with those corresponding to the normal modes

Table 2.7

$\mathscr{C}_{3v}$	E	$2C_3$	$3\sigma_v$		
A_1	1	1	1	z	$x^2+y^2; z^2$
A_2	1	1	-1	R_z	
E	2	-1	0	$(x,y); (R_x, R_y)$	$(x^2-y^2, xy); (xz, yz)$

of vibration, in agreement with Fig. 2.27:

$$\Gamma_{\text{vib}} = 2A_1 + 2E \tag{2.67}$$

The character Table 2.8 for the point group $\mathscr{C}_h$ is given so that the reader may verify for himself that SF_6, for which χ_{xyz} and some values of U_R have been given in Table 2.4 and on p.50, has $\Gamma_{\text{vib}} = A_{1g} + E_g + F_{2g} + 2F_{1u} + F_{2u}$.

Instead of using $3N$ Cartesian base vectors we may use sets of equivalent internal coordinates, such as changes in bond-lengths, Δr_i, and interbond angles, $\Delta\beta_{ij}$. In this way we eliminate the translations and rotations. In the case of ammonia, shown in Fig. 2.31, there are just 6 internal coordinates and 6 normal modes. However methyl chloride, for example, has 10 internal coordinates but only 9 normal modes, and we often have the complication provided by such redundancies. Nevertheless, it is usually more significant to express the normal modes in terms of internal coordinates than in terms of Cartesian displacements.

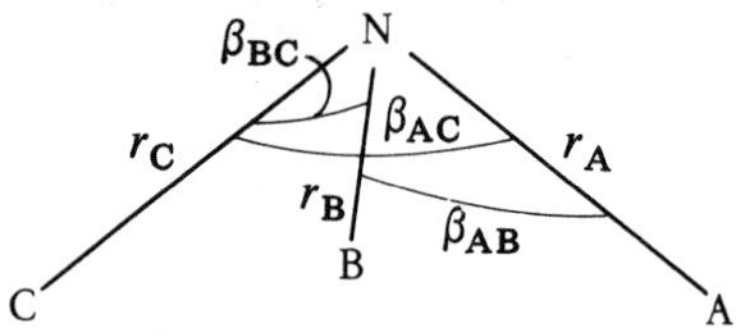

Fig. 2.31 The bond-lengths r_i and interbond angles β_{ij} whose *changes*, Δr_i and $\Delta\beta_{ij}$ respectively, constitute the internal coordinates of NH_3.

Since no symmetry operation mixes the two sets of equivalent internal coordinates of ammonia, we may consider separately their participation in the normal modes. The characters of the reducible representations carried by each set may be obtained as follows. Those coordinates that are permuted will not contribute, while those not shifted contribute +1. For each set $\chi(\mathbf{E}) = 3$, $\chi(\mathbf{C}_3) = 0$, $\chi(\sigma_v) = 1$, so that $\Gamma_{\Delta r} = \Gamma_{\Delta\beta} = A_1 + E$, giving us, as expected, the same overall result as before:

$$\Gamma_{\text{vib}} = \Gamma_{\Delta r} + \Gamma_{\Delta\beta} = 2A_1 + 2E$$

Physically this implies that none of the normal modes consists entirely of either bond-stretchings or bond-bendings, but *each* is some mixture of the two. As a first step to a description of the normal modes in terms of the internal coordinates it is usual to set up symmetry coordinates. Each calculation then

Table 2.8

$\mathcal{O}_h$	**E**	$8\mathbf{C}_3$	$3\mathbf{C}_2$	$6\mathbf{C}_4$	$6\mathbf{C}_2'$	**i**	$8\mathbf{S}_6$	$3\sigma_h$	$6\mathbf{S}_4$	$6\sigma_d$		
A_{1g}	1	1	1	1	1	1	1	1	1	1		$x^2+y^2+z^2$
A_{2g}	1	1	1	−1	−1	1	1	1	−1	−1		
E_g	2	−1	2	0	0	2	−1	2	0	0		$(2z^2-x^2-y^2, x^2-y^2)$
F_{1g}	3	0	−1	1	−1	3	0	−1	1	−1	(R_x, R_y, R_z)	
F_{2g}	3	0	−1	−1	1	3	0	−1	−1	1		(xy, xz, yz)
A_{1u}	1	1	1	1	1	−1	−1	−1	−1	−1		
A_{2u}	1	1	1	−1	−1	−1	−1	−1	1	1		
E_u	2	−1	2	0	0	−2	1	−2	0	0		
F_{1u}	3	0	−1	1	−1	−3	0	1	−1	1	(x, y, z)	
F_{2u}	3	0	−1	−1	1	−3	0	1	1	−1		

involves only the modes of one symmetry species at a time. With the help of Equations (2.52) the symmetry coordinates can be immediately given as:

$$A_1 \begin{cases} S_1 = (\Delta r_A + \Delta r_B + \Delta r_C)/\sqrt{3} \\ S_2 = (\Delta\beta_{AB} + \Delta\beta_{AC} + \Delta\beta_{BC})/\sqrt{3} \end{cases}$$

$$E \begin{cases} S_{3a} = (2\Delta r_A - \Delta r_B - \Delta r_C)/\sqrt{6} \\ S_{4a} = (2\Delta\beta_{BC} - \Delta\beta_{AC} - \Delta\beta_{AB})/\sqrt{6} \end{cases} \quad \text{and} \quad \begin{matrix} S_{3b} = (\Delta r_B - \Delta r_C)/\sqrt{2} \\ S_{4b} = (\Delta\beta_{AC} - \Delta\beta_{AB})/\sqrt{2} \end{matrix}$$

in which we have followed the customary convention of normalizing them on the basis that the unnamed vectors associated with the internal coordinates are treated as orthonormal.

2.4.4 The nomenclature of the irreducible representations

The irreducible representations or symmetry species are designated by letters, further distinguished by suffixes if more than one of a kind occur in the same group.

A, B: These both denote 1-dimensional representations. A is symmetrical under a rotation about the principal axis, $\chi(\mathbf{C}_n) = +1$; B is antisymmetrical, $\chi(\mathbf{C}_n) = -1$. Alternatively we say that $\mathbf{C}_n$ has *even and odd parity*, respectively.

E: A 2-dimensional representation, so $\chi(\mathbf{E}) = 2$.

F or T: A 3-dimensional representation. Most spectroscopists favour F, but T is used increasingly, especially by inorganic chemists, to avoid confusion with F ($L = 3$) atomic states.

G, H: 4- and 5-dimensional representations. ($\mathcal{I}, \mathcal{I}_h$ groups only.)

Σ^+, Σ^-: 1-dimensional representations of the linear groups $\mathcal{C}_{\infty v}$ and $\mathcal{D}_{\infty h}$. The superscripts denote that $\chi(\sigma_v) = +1$ or -1, respectively.

$\Pi, \Delta, \Phi, ..$: 2-dimensional representations of $\mathcal{C}_{\infty v}$ and $\mathcal{D}_{\infty h}$.

A superscript prime, A' or E', denotes that σ_h has even parity; a superscript double prime, A'' or E'', denotes that σ_h has odd parity.

A subscript g or u denotes symmetrical or antisymmetrical behaviour respectively with respect to inversion at the centre.

A subscript 1 or 2 after A or B denotes that $\mathbf{C}_2'$ (or in its absence, σ_v) has even or odd parity, respectively.

Upper case letters: $A, B, \ldots, \Sigma, \Pi, \ldots$ are used to denote molecular *states*, but lower case letters: $a, b, \ldots, \sigma, \pi, \ldots$ are generally used to denote *vibrational modes* and molecular *orbitals*.

2.5 SYMMETRY OPERATIONS AND FUNCTIONS

2.5.1 Functions and function spaces

As discussed in the Introduction, we wish to describe the effect of symmetry operations upon the molecular wavefunctions. Hitherto we have interpreted symmetry operations in terms of vectors and their associated vector spaces. Now it is helpful to find that a consistent analogy may be drawn between functions and vectors, and that function spaces can be described which are the analogues of vector spaces.

The concept of function involves three things: a region of the number scale, say from p to q, called the *interval* in which the function is being considered; a *variable*, say x, which can independently take values in the interval; and thirdly, a prescribed *rule* that for any value of x there exists a definite value of $\mathbf{f}$. We then say that $\mathbf{f}$ is a function of x in the range $p \leqslant x \leqslant q$, or $\mathbf{f} = f(x)$. This definition may be extended to several variables, each with its appropriate interval.

A *set of functions* is a collection of functions of the same variable(s), defined in the same interval(s), and characterized by a connected set of rules. If such a set is to provide a *function space*, analogous to a vector space, there must be analogues to Equations (2.8)–(2.12), (2.14), and (2.27).

Suppose that $\mathbf{f}, \mathbf{g}, \mathbf{h}, \ldots$ are a set of functions, with the variable x. Then function addition, $\mathbf{h} = \mathbf{f} + \mathbf{g}$, applies when this equation is satisfied for every value of x. A zero function, $\mathbf{0}$, which takes the value zero for every value of x may be defined and used to give meaning to $-\mathbf{f}$. If a is an arbitrary number, $a\mathbf{f}$ is a function of x which, for every value of x, has the value $f(x)$ multiplied by a. This set of functions forms a function space if, for any two members of the set, $\mathbf{f}$ and $\mathbf{g}$, the function $(\mathbf{f} + \mathbf{g})$ is also a member, and if, for any member $\mathbf{f}$ and number a, the function $a\mathbf{f}$ is also a member.

We are concerned particularly with sets of eigenfunctions which are solutions to the Schrödinger equation:

$$\mathbf{H}\psi = E\psi \tag{2.68}$$

where $\mathbf{H}$ is the Hamiltonian operator. This belongs to the class of linear operators, which have the property given in Equation (2.36). Thus if $\psi_1, \psi_2, \ldots$ are solutions of (2.68), corresponding to the *same* eigenvalue E_0, then *the set of functions* $\psi_1, \psi_2, \ldots$ *constitutes a function space*, since $a\psi_i$ and $(\psi_i + \psi_j)$ are also solutions:

$$\mathbf{H}(a\psi_1) = a\mathbf{H}\psi_1 = aE_0\psi_1 = E_0(a\psi_1) \tag{2.69}$$

and

$$\mathbf{H}(\psi_1 + \psi_2) = \mathbf{H}\psi_1 + \mathbf{H}\psi_2 = E_0\psi_2 + E_0\psi_2 = E_0(\psi_1 + \psi_2) \tag{2.70}$$

As another example we may take the operator $\mathbf{d}^2/\mathbf{d}\phi^2$, which is also a linear operator, and consider the solutions of the eigenvalue equation $\mathbf{d}^2\mathbf{f}/\mathbf{d}\phi^2 = -\mathbf{f}$. This has a general solution:

$$\mathbf{f} = f(\phi) = a_1 \cos\phi + a_2 \sin\phi$$

where a_1 and a_2 are arbitrary constants, either of which may be zero. Thus $\mathbf{f}_1 = \cos\phi$ and $\mathbf{f}_2 = \sin\phi$ are also solutions. These are linearly independent and therefore act as base vectors in what is evidently a 2-dimensional space, a_1 and a_2 being the components of the general vector $\mathbf{f}$.

Before we can describe $\mathbf{f}_1$ and $\mathbf{f}_2$ as orthogonal we must define a suitable scalar product. This number must be associated with the two functions as a whole and not depend on a particular choice of the variable(s) of which they are functions. The appropriate choice, which is also seen to be a logical extension of Equation (2.28), is an integration over the defined interval of the two functions:

$$\mathbf{f}.\mathbf{g} = \int_p^q f(x)\,g(x)\,\mathrm{d}x \qquad (2.71)^{\ddagger}$$

For functions of several variables the appropriate multiple integral is required. The importance of stating the interval is now apparent, since the value of this definite integral depends upon it. In our example, if the interval is from $-\pi$ to $+\pi$, $\mathbf{f}_1$ and $\mathbf{f}_2$ are orthogonal since $\mathbf{f}_1.\mathbf{f}_2 = 0$. They are not however normalized, since $\mathbf{f}_1.\mathbf{f}_1 = \mathbf{f}_2.\mathbf{f}_2 = \pi$.

2.5.2 Complex functions, representations and characters§

In one important respect functions differ from the vectors with which, so far, we have compared them. Functions may take complex values whereas our vectors, even when n-dimensional, have been defined in terms appropriate to real, Euclidean space.

Because functions may be complex, the effects of operations on them will, in general, be represented by matrices that have complex elements and traces. Such a matrix $\mathbf{A}$ with elements a_{ij} has a complex conjugate $\mathbf{A}^*$ with elements a_{ij}^*. The *Hermitian transpose*, or *adjoint*, of a matrix $\mathbf{A}$ is denoted by $\mathbf{A}^\dagger$ and formed by transposing the complex conjugate of $\mathbf{A}$: $\mathbf{A}^\dagger = (\mathbf{A}^*)^{\mathrm{T}}$. It follows that $(\mathbf{AB})^\dagger = \mathbf{B}^\dagger\mathbf{A}^\dagger$. A matrix is said to be *Hermitian* if $\mathbf{A} = \mathbf{A}^\dagger$, and *unitary* if $\mathbf{A}^{-1} = \mathbf{A}^\dagger$.

Many functions of real variables are functions of position in physical space or an n-dimensional Euclidean space; in the same way complex functions may be regarded as functions of position in an abstract non-Euclidean, so-called Hermitian, space. The analogy between functions and vectors can be usefully extended by describing vectors in such a space in terms of complex base vectors and components. The properties (2.8)–(2.12) and (2.14) still apply except that

‡ This quantity, referred to as the *inner product* of the two functions, is usually denoted by $\langle f\,|\,g\rangle$, but we continue to use the symbolism appropriate to vectors in order to emphasize the analogy.

§ Readers may omit this section. It is included because wavefunctions are, in general, complex and because several point groups have complex character systems.

for every vector:

$$\mathbf{a} = \mathbf{e}_1 a_1 + \mathbf{e}_2 a_2 + \ldots + \mathbf{e}_n a_n = \mathbf{ea}$$

we may define another, called its *dual*:

$$\mathbf{a}^* = \mathbf{e}_1^* a_1^* + \mathbf{e}_2^* a_2^* + \ldots + \mathbf{e}_n^* a_n^* = \mathbf{e}^* \mathbf{a}^*$$

The scalar product of a Euclidean vector with itself is a positive real quantity – its $(\text{length})^2$ – and we retain this property by defining the Hermitian scalar product of two vectors **a** and **b** as:

$$\mathbf{a}^*.\mathbf{b} = (\mathbf{ea})^\dagger (\mathbf{eb}) = \mathbf{a}^\dagger \mathbf{e}^\dagger \mathbf{eb} = \mathbf{a}^\dagger \mathbf{Mb} \qquad (2.72)$$

This is the analogue of (2.27), with $\mathbf{M} = \mathbf{e}^\dagger \mathbf{e}$. When $\mathbf{M} = \mathbf{1}$ the base vectors are said to be unitary or just orthonormal. Then:

$$\mathbf{a}^*.\mathbf{b} = \mathbf{a}^\dagger \mathbf{b} = a_1^* b_1 + a_2^* b_2 + \ldots + a_n^* b_n = \sum_i a_i^* b_i \qquad (2.73)$$

Note that now $\mathbf{a}^*.\mathbf{b} = (\mathbf{b}^*.\mathbf{a})^*$, so the Hermitian scalar product is a *complex number* associated with an *ordered pair* of vectors. The corresponding function scalar product or *inner product* (2.71) must now be redefined as:

$$\mathbf{f}^*.\mathbf{g} = \int_p^q f(x)^* \, g(x) \, dx \qquad (2.74)$$

We found that scalar products were invariant under an orthogonal transformation in real Euclidean space. The corresponding invariance in Hermitian space is achieved under a *unitary transformation*:

$$\mathbf{a}^*.\mathbf{b} = \mathbf{a}^\dagger \mathbf{Mb} = (\mathbf{Ua})^\dagger \mathbf{M}(\mathbf{Ub}) = \mathbf{a}^\dagger \mathbf{U}^\dagger \mathbf{MUb}$$

Hence $\mathbf{M} = \mathbf{U}^\dagger \mathbf{MU}$, and if $\mathbf{M} = \mathbf{1}$, then $\mathbf{U}^\dagger \mathbf{U} = \mathbf{1}$, which defines a unitary matrix.

Since symmetry operations may, more generally, be represented therefore by unitary matrices, we anticipate that some irreducible representations will consist of matrices with complex elements.

This applies to several Abelian groups, particularly the cyclic groups, $\mathcal{C}_n$, for in these all operations commute with each other, and therefore, by (2.7), belong to separate classes. Since there are as many irreducible representations as there are classes, it follows from (2.61) that these are all 1-dimensional. The $\mathcal{C}_n$ groups have the one generator $\mathbf{C}_n$ and therefore the numbers representing the operations, which will also be the characters, $\chi(\mathbf{R})$, must be such that:

$$[\chi^*(\mathbf{R})]\,[\chi(\mathbf{R})] = 1 \quad \text{and} \quad [\chi(\mathbf{R})]^n = 1$$

This is fulfilled if $\chi(\mathbf{R}) = \exp \pm 2\pi i/n$, a number which is referred to as the complex n-th root of unity.

Because of such characters, Equations (2.57)–(2.61) should properly be written as:

$$(\mathbf{X}^{-1})_{ik} = (h_k/g)\chi_i^*(\mathbf{R}_k) \qquad (2.75)$$

$$n_i = (1/g) \sum_k h_k \chi_i^*(\mathbf{R}_k) \chi_\rho(\mathbf{R}_k) \qquad (2.76)$$

$$(1/g)\sum_k h_k \chi_i^*(\mathbf{R}_k)\chi_j(\mathbf{R}_k) = (1/g)\sum_{\mathbf{R}} \chi_i^*(\mathbf{R})\chi_j(\mathbf{R}) = \delta_{ij} \quad (2.77)$$

$$\sum_{\mathbf{R}} \chi^*(\mathbf{R})\chi(\mathbf{R}) = g \quad (2.78)$$

$$\sum_i \chi_i^*(\mathbf{E})\chi_i(\mathbf{E}) = g \quad (2.79)$$

Finally, we may note that the *real* vectors $\mathbf{e}_1$ and $\mathbf{e}_2$ along the x- and y-axes carry 2-dimensional representations of most Abelian groups of order 3 or more because these directions are physically equivalent. The mutually complex conjugate irreducible representations to which these real representations may be reduced are therefore bracketed together as an E representation.

In a similar way, pairs of normal mode vectors occur, corresponding to a single molecular vibration frequency. In such cases we add together the corresponding characters of the bracketed representations, in order to work with a real 2-dimensional one.

2.5.3 Transformations of functions

We wish to know the change induced by a symmetry operation in a function which behaves as a vector. We continue to use the symbol $\mathbf{R}$ to denote the linear operators which effect this change:

$$\mathbf{Rf} = \mathbf{f}' \quad (2.80)$$

where $\mathbf{f}'$ represents the transformed function, always another function in the same space as $\mathbf{f}$, so that $\mathbf{R}$ corresponds to a 'mapping', as before. For the symmetry operations $\mathbf{R} = \mathbf{QP}$, we require that:

$$\mathbf{Rf} = \mathbf{Q}(\mathbf{Pf}) \quad (2.81)$$

so that the mappings will be at least homomorphic with the point group. Using a set of base functions it will then be possible to obtain a set of representative matrices **R**, in accordance with Equation (2.37).

We are able to fulfil the condition (2.81) because the functions with which we are concerned are continuous functions of position in physical space or, more generally, in an n-dimensional space, whose transformations under the symmetry operations we can already express in this manner. Such functions are also called *fields*, and Fig. 2.32 represents in two dimensions the rotation of such a field.

Let $\mathbf{f} = f(x)$, where x now stands for all the independent variables. These variables are the components of a general field point P with respect to some set of axes, which must now be defined. Using similar terminology to that in Section 2.3.4 we will denote the components with respect to a fixed set of axes as x and with respect to a set of axes embedded in the system as $\bar{x}$. Suppose we now carry out the rotation $\mathbf{R}$ upon the system and denote the image of P by P′ with coordinates x' or $\bar{x}'$.

Since $\bar{x}' = x$, it follows that $f(\bar{x}') = f(x)$, which expresses the fact that while

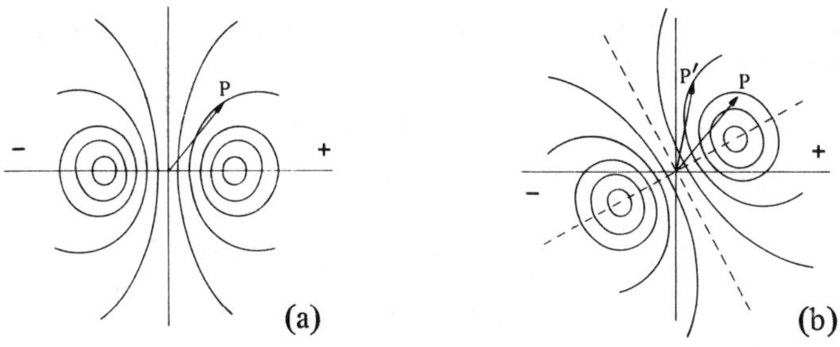

Fig. 2.32 The rotation of a field. The value of the transformed function at P′ in (b), which is the image of P after rotation, is the same as that of the original function at P in (a). (After Fig. 7.3 of McWeeny [2.7].)

system and axes are transformed together there is no change in the function. However, we wish to describe the change resulting from a rotation of the system but not of the axes. Now *the value of the transformed function at the point* P′ *is the same as that of the original function at the point* P. That is:

$$\mathbf{R}f(x') = f'(x') = f(x) \tag{2.82}$$

Since $x' = \mathbf{R}x$ [compare with (2.34)] and therefore $x = \mathbf{R}^{-1}x'$, it follows that:

$$\mathbf{R}f(x') = f'(x') = f(\mathbf{R}^{-1}x') \tag{2.83}$$

As P′ is as much a general point as P, we may drop the primes from each side, and write:

$$\mathbf{R}f(x) = f'(x) = f(\mathbf{R}^{-1}x) \tag{2.84}$$

which implies that the transformed function corresponds to the use of the co-ordinates of a *backwards-rotated* field point. Alternatively, since $\bar{x} = \mathbf{R}^{-1}x$ [compare with (2.33)], we may regard the new function as resulting from coordinates related to axes that have been rotated forwards.

We now apply Equation (2.84) to the base functions $\mathbf{f}_1$ and $\mathbf{f}_2$ of our previous example, identifying the parameter ϕ with an angle in the xy-plane (Fig. 2.33).

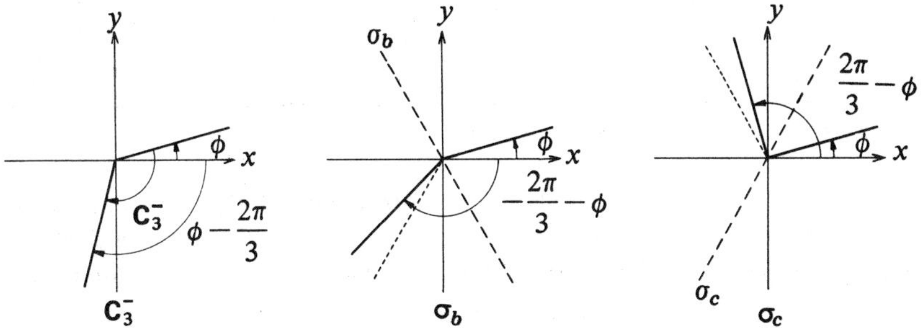

Fig. 2.33 The effects upon ϕ of carrying out the symmetry operations $\mathbf{C}_3^-$, σ_b, and σ_c, defined by Fig. 2.10, on an arbitrary radial line in the xy-plane. The parameter ϕ is defined as the angle between the x-axis and the radial line.

From Fig. 2.33 it may be seen that the rotation $\mathbf{C}_3^{-1}$ changes ϕ to $(\phi - 2\pi/3)$, from which it follows that:

$$\mathbf{C}_3^+\mathbf{f}_1 = \cos(\phi - 2\pi/3) = \cos\phi\cos 2\pi/3 + \sin\phi\sin 2\pi/3$$

$$\mathbf{C}_3^+\mathbf{f}_2 = \sin(\phi - 2\pi/3) = \sin\phi\cos 2\pi/3 - \cos\phi\sin 2\pi/3$$

or

$$[\mathbf{f}_1' \quad \mathbf{f}_2'] = [\mathbf{f}_1 \quad \mathbf{f}_2]\begin{bmatrix}\cos 2\pi/3 & -\sin 2\pi/3\\ \sin 2\pi/3 & \cos 2\pi/3\end{bmatrix}$$

Clearly, $\mathbf{C}_3^+$ changes ϕ to $(\phi + 2\pi/3)$, and σ_a changes ϕ to $-\phi$. Figure 2.33 shows that σ_b changes ϕ to $(-\phi - 2\pi/3)$ and σ_c changes it to $(2\pi/3 - \phi)$. So, in a similar way we may obtain representative matrices for each of the symmetry operations of $\mathcal{C}_{3v}$. It is soon found that these are identical to the E representation contained in Γ_{xyz} (Table 2.3).

For an example of a set of complex base functions, we may combine $\mathbf{f}_1$ and $\mathbf{f}_2$ as follows:

$$\mathbf{f}_+ = \mathbf{f}_1 + i\mathbf{f}_2 = e^{+i\phi}$$

$$\mathbf{f}_- = \mathbf{f}_1 - i\mathbf{f}_2 = e^{-i\phi}$$

Therefore

$$\mathbf{C}_3^+\mathbf{f}_+ = e^{+i(\phi-2\pi/3)} = \mathbf{f}_+ e^{-2\pi i/3}$$

$$\mathbf{C}_3^+\mathbf{f}_- = e^{-i(\phi-2\pi/3)} = \mathbf{f}_- e^{+2\pi i/3}$$

or

$$[\mathbf{f}_+' \quad \mathbf{f}_-'] = [\mathbf{f}_+ \quad \mathbf{f}_-]\begin{bmatrix}e^{-2\pi i/3} & 0\\ 0 & e^{+2\pi i/3}\end{bmatrix}$$

The reader may soon verify that $\mathbf{f}_+$, $\mathbf{f}_-$ carry the representation of $\mathcal{C}_{3v}$ given in Table 2.9 which is, of course, also of symmetry species E.

Table 2.9

$\mathcal{C}_{3v}$	E	C_3^+	C_3^-	σ_a	σ_b	σ_c
Γ	$\begin{bmatrix}1 & 0\\ 0 & 1\end{bmatrix}$	$\begin{bmatrix}\epsilon^* & 0\\ 0 & \epsilon\end{bmatrix}$	$\begin{bmatrix}\epsilon & 0\\ 0 & \epsilon^*\end{bmatrix}$	$\begin{bmatrix}0 & 1\\ 1 & 0\end{bmatrix}$	$\begin{bmatrix}0 & \epsilon\\ \epsilon^* & 0\end{bmatrix}$	$\begin{bmatrix}0 & \epsilon^*\\ \epsilon & 0\end{bmatrix}$
χ	2	-1	-1	0	0	0

where $\epsilon = \exp + 2\pi i/3 = \cos 2\pi/3 + i \sin 2\pi/3$

These real and complex base functions correspond closely to **p** atomic orbital wavefunctions. The solutions to the Schrödinger equation for a single electron moving in a central field may be written, for values of the quantum numbers $l = 1, m = +1$ and -1, respectively, as:

$$\mathbf{p}_{+1} = F(r).\sin\theta.e^{+i\phi} = (1/\sqrt{2})(\mathbf{p}_x + i\mathbf{p}_y)$$

$$\mathbf{p}_{-1} = F(r).\sin\theta.e^{-i\phi} = (1/\sqrt{2})(\mathbf{p}_x - i\mathbf{p}_y)$$

where $F(r).\sin\theta$ is a function of the radial distance r and the polar angle θ (Fig. 2.34) which remains invariant under the symmetry operations of $\mathcal{C}_{3v}$ if we assume that the atom is located, like the N atom in NH_3, on all the $\mathcal{C}_{3v}$ symmetry elements. The real forms of the **p** atomic orbitals, $\mathbf{p}_x$ and $\mathbf{p}_y$, correspond to our $\mathbf{f}_1$ and $\mathbf{f}_2$, and are so designated because $\cos\phi = x/r$ and $\sin\phi = y/r$. As we have seen, they transform under the symmetry operations in the same way that vectors do, directed along x and y, respectively.

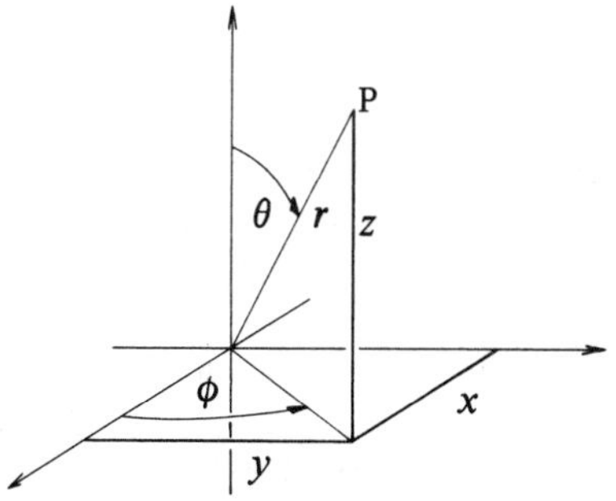

Fig. 2.34

2.5.4 The symmetry species of wavefunctions

Any symmetry operation **R** of the group to which a molecule belongs interchanges identical nuclei. This leaves the potential energy unaltered, and since the Laplacian ∇^2 is unchanged by any symmetry operation, it follows that the Hamiltonian operator **H** is invariant to any of the group operations. That is:

$$\mathbf{H} = \mathbf{R}^{-1}\mathbf{H}\mathbf{R} \quad \text{or} \quad \mathbf{RH} = \mathbf{HR} \tag{2.85}$$

so all the symmetry operations commute with **H**.

If we apply a symmetry operation to both sides of the Schrödinger equation:

$$\mathbf{H}\psi_i = E_i\psi_i$$

we obtain

$$\mathbf{RH}\psi_i = \mathbf{R}E_i\psi_i$$

which we may rearrange as

$$\mathbf{H}(\mathbf{R}\psi_i) = E_i(\mathbf{R}\psi_i)$$

from which it follows that $\mathbf{R}\psi_i$ is also a wavefunction corresponding to the eigenvalue E_i.

If the set of wavefunctions corresponding to E_i constitutes an n-dimensional function space, a linearly independent set of base functions $\psi_1, \psi_2, \ldots, \psi_n$ can always be chosen, in terms of which $\mathbf{R}\psi_i$ may be given:

$$\mathbf{R}\psi_i = c_1\psi_1 + c_2\psi_2 + \ldots + c_n\psi_n$$

Since any solution with eigenvalues E_i can be so expressed, the n base functions define a subspace invariant under all the transformations induced by the symmetry operations of the group.

The effect of the operations $\mathbf{R}$ upon these base functions enables us, in the usual way, to obtain a set of $n \times n$ matrices $\mathbf{R}$, in accordance with Equation (2.37), which we may write as:

$$\mathbf{R}\boldsymbol{\psi} = \boldsymbol{\psi}\mathbf{R}$$

where $\boldsymbol{\psi}$ is the row matrix $[\psi_1 \ \psi_2 \ \psi_3 \ldots \psi_n]$.

This set of matrices will be an irreducible representation of the point group because reducibility, as we have seen, requires the discovery of invariant subspaces. In the case of solutions of the Schrödinger equation, these are distinguished according to eigenvalue. Apart from "accidental" coincidences of energy therefore, *the degeneracy n of an eigenvalue E_i will be equal to the dimension of the irreducible representation of its wavefunctions.*

This is a very useful result because it enables us to classify and label molecular states by their symmetry species. Thus the ammonia molecule can only have non-degenerate A_1 or A_2 states, or doubly degenerate E states.

The above result is valid not only for an exact Hamiltonian but also for a partial or approximate one; the solutions obtained will have the symmetry species of the group of the Hamiltonian operator actually used. If the functions are sought which describe the motion of a *single* electron in an electrostatic field with the full molecular symmetry of NH_3, then the resultant *molecular orbitals* will have the same possible symmetry species and will be denoted as a_1, a_2, or e types.

The relationship between the symmetries of the occupied molecular orbitals (MOs) and the symmetry of the overall electronic state is of some importance. If the Hamiltonian operators for which the above MOs are the eigenfunctions are denoted as $\mathbf{H}_i$ then:

$$\mathbf{H} = \sum_i \mathbf{H}_i + \mathbf{H}' \tag{2.86}$$

where the summation is taken over all electrons present and $\mathbf{H}'$ corresponds to that part of the electrostatic interaction between the electrons and the field which has been ignored in the $\mathbf{H}_i$. This must also have the full symmetry of the molecule, and its inclusion will not therefore change the *symmetry species* of those approximate functions which are eigenfunctions of $\sum_i \mathbf{H}_i$ alone, but only their detailed form and their eigenvalues. Now the eigenfunctions of $\sum_i \mathbf{H}_i$ will be (if for the present we ignore electron spin) *products* of the MO functions ψ_i, and our problem reduces to one of finding the symmetry of such products, the symmetries of whose factors are known. (If, to satisfy the Pauli principle, the MOs are combined as Slater determinants, *sums* of such products are involved instead, and the ways in which the spin states may be associated with particular symmetry species will be limited.)

2.5.5 Direct product representations

Suppose we have one function space of dimension α with base functions ψ_i yielding representative matrices **A** under the symmetry operations **R**, and another of dimension β with base functions ϕ_j, yielding matrices **B**.

We now form products, taking one factor from each function space, without distinguishing the order of the factors. By taking all possible products in this way, we form a product space of dimension $\alpha\beta$ which yields a set of *direct product matrices* **C**, whose constitution may be understood from Equations (2.87) and (2.88). Writing the products in sequence as a row matrix:

$$[\psi_1\phi_1 \quad \psi_1\phi_2 \; \ldots \; \psi_2\phi_1 \quad \psi_2\phi_2 \; \ldots \; \psi_i\phi_j \; \ldots \; \psi_\alpha\phi_1 \quad \psi_\alpha\phi_2 \; \ldots \; \psi_\alpha\phi_\beta]$$

which may be abbreviated as $[\psi_i\phi_j]$, we have:

$$\mathsf{R}[\psi_i\phi_j] = [\psi_i\phi_j]\mathbf{C} = [\psi_k'\phi_l'] \tag{2.87}$$

where

$$\mathbf{C} = \begin{array}{r|ccccc|}
 & 11 & 12 & 21 & 22 & \text{: column } kl \\
11 & a_{11}b_{11} & a_{11}b_{12} \;\ldots & a_{12}b_{11} & a_{12}b_{12} & \ldots \\
12 & a_{11}b_{21} & a_{11}b_{22} \;\ldots & a_{12}b_{21} & a_{12}b_{22} & \ldots \\
 & \vdots & \vdots & \vdots & \vdots & \\
21 & a_{21}b_{11} & a_{21}b_{12} \;\ldots & a_{22}b_{11} & a_{22}b_{12} & \ldots \\
22 & a_{21}b_{21} & a_{21}b_{22} \;\ldots & a_{22}b_{21} & a_{22}b_{22} & \ldots \\
\text{row: } ij & \vdots & \vdots & \vdots & \vdots &
\end{array} \tag{2.88}$$

The direct product is written symbolically as: **C** = **A** × **B** and it is evident that:

$$c_{ij,kl} = a_{ik}b_{jl} \tag{2.89}$$

The set of matrices **C** forms generally a *reducible* representation of the group. For our purposes we are interested in the traces of these matrices. Using (2.89) we have:

$$\mathrm{Tr}\,\mathbf{C} = \sum_i\sum_j c_{ij,ij} = \sum_i\sum_j a_{ii}b_{jj} = \mathrm{Tr}\,\mathbf{A}.\mathrm{Tr}\,\mathbf{B} \tag{2.90}$$

The products of the corresponding characters of the individual representations form the characters of the direct product representation. Although we have described the direct product **A** × **B**, this result can be generalized to any number of factors.

To illustrate the direct product we set out in Table 2.10 the character system for $E \times E$ of the point group $\mathcal{C}_{3v}$. Hence

$$E \times E = A_1 + A_2 + E \tag{2.91}$$

The complete direct product multiplication table for this group is shown in Table 2.11. Note that A_1, the totally symmetric representation, occurs only on

Table 2.10

$\mathcal{C}_{3v}$	E	$2C_3$	$3\sigma_v$
$\chi_E(R)$	2	-1	0
$[\chi_E(R)]^2$	4	$\hat{1}$	0

Table 2.11

$\mathcal{C}_{3v}$	A_1	A_2	E
A_1	A_1	A_2	E
A_2	A_2	A_1	E
E	E	E	$A_1 + A_2 + E$

the diagonal and is present there in every entry. This is a feature common to all such tables.

Thus if, in a $\mathcal{C}_{3v}$ molecule, all the electrons were in a_1- or a_2-type MOs, except for two in a doubly-degenerate pair of e-type MOs, this electronic *configuration* would correspond to four electronic *states*, with the symmetry species given by (2.91). The Pauli principle requires however that the A_1 and E states are singlet states (spins paired) and the A_2 is a triplet state (spins parallel). (See Schonland [2.6], p.233.) If there were 3 or 4 electrons in the same pair of e-type MOs the Pauli principle allows only one doubly-degenerate pair of doublet states (^2E) in the former case and only one singlet state ($^1A_1$) in the latter.

Somewhat similar limitations apply in other cases when we no longer wish to distinguish between $\psi_i\phi_j$ and $\psi_j\phi_i$ in forming products between two sets of base functions (or vectors), as above. This will only arise when both sets span the same space, corresponding to the same irreducible representation. We may therefore put $\mathbf{A} = \mathbf{B}$. The dimension of the new *symmetrical product* space will be reduced from α^2 to $\frac{1}{2}\alpha(\alpha+1)$. We now write a row matrix of the *distinct* products, omitting those for $i > j$. The resultant representative matrix, given by the effect of a symmetry operation on this row matrix, is the *symmetrical direct product*, $\mathbf{A}_S^{[2]}$:

$$\mathbf{A}_S^{[2]} = \begin{bmatrix} a_{11}a_{11} & a_{11}a_{12} & \cdots & a_{12}a_{12} & \cdots \\ (a_{11}a_{21}+a_{21}a_{11}) & (a_{11}a_{22}+a_{21}a_{12}) & \cdots & (a_{12}a_{22}+a_{22}a_{12}) & \cdots \\ \vdots & \vdots & & \vdots & \\ a_{21}a_{21} & a_{21}a_{22} & \cdots & a_{22}a_{22} & \cdots \\ \vdots & \vdots & & \vdots & \end{bmatrix} \tag{2.92}$$

The general elements are given by:

$$\left.\begin{aligned} (\mathbf{A}_S^{[2]})_{ii,kl} &= a_{ik}a_{il} \\ (\mathbf{A}_S^{[2]})_{ij,kl} &= (a_{ik}a_{jl} + a_{jk}a_{il}) \quad (i<j) \end{aligned}\right\} (k \leqslant l) \tag{2.93}$$

Clearly

$$\begin{aligned}\text{Tr}\,\mathbf{A}_S^{[2]} &= \sum_{i<j} (a_{ii}a_{jj} + a_{ji}a_{ij}) + \sum_i a_{ii}a_{ii} \\ &= \tfrac{1}{2}\sum_i\sum_j (a_{ii}a_{jj} + a_{ji}a_{ij}) \\ &= \tfrac{1}{2}[(\text{Tr}\,\mathbf{A})^2 + \text{Tr}\,\mathbf{A}^2] \end{aligned} \tag{2.94}$$

Thus the symmetrical direct product $(E \times E)_{\text{sym}}$ for the point group $\mathcal{C}_{3v}$ may be found as follows in Table 2.12, since the trace of $\mathbf{A}^2$ corresponds to the trace of $\mathbf{\Gamma}(\mathbf{R}^2)$. In this case $\mathbf{E}^2 = \mathbf{E}$, $\mathbf{C}_3^2 = \mathbf{C}_3^{-1}$, and $\sigma_v^2 = \mathbf{E}$. Hence

$$(E \times E)_{\text{sym}} = A_1 + E \tag{2.95}$$

Table 2.12

$\mathcal{C}_{3v}$	E	$2\mathbf{C}_3$	$3\sigma_v$
$\chi_E(\mathbf{R})$	2	-1	0
$[\chi_E(\mathbf{R})]^2$	4	1	0
$\chi_E(\mathbf{R}^2)$	2	-1	2
$\chi_{(E\times E)}^{\text{sym}}$	3	0	1

For a symmetrical direct product involving several factors, $(E \times E \times E \ldots)_{\text{sym}}$, Equation (2.94) becomes:

$$\text{Tr}\,\mathbf{A}_S^{[n]} = \tfrac{1}{2}[\text{Tr}\,\mathbf{A}.\text{Tr}\,\mathbf{A}_S^{[n-1]} + \text{Tr}\,\mathbf{A}^n] \tag{2.96}$$

but if these factors involve triply-degenerate functions, $(F \times F \times F \ldots)_{\text{sym}}$, a more complicated expression is required, for which see Heine [2.8], p.263.

2.5.6 Vibrational and vibronic states

In Section 2.4.3 we examined the symmetry of molecular vibrations, accepting a classical mechanical description in which any arbitrary vibration is seen as a superposition of $(3N-6)$ independent normal modes, with coordinates q_i. In fact, this corresponds to the ideal situation in which $2V = 4\pi^2c^2 \sum_i \omega_i^2 q_i^2$ and $2T = \sum_i \dot{q}_i^2$ and all the motions are simple harmonic. Real molecules are anharmonic, although the departure from ideality is usually very small for small displacements from equilibrium.

The quantum mechanical calculation of vibrational states and their energies is in many ways analogous to that of molecular electronic states, discussed in Section 2.5.4. Equation (2.86) may be applied to this case with $\mathbf{H}_i$ being the Hamiltonian for a simple harmonic vibration in the single coordinate q_i, and $\mathbf{H}'$ being the anharmonic perturbation. It is important to note that the latter, in this case also, must correspond to the full symmetry of the molecule because, as already noted, V must be invariant to all the symmetry operations of the group.

As a result *we may determine the symmetry of the vibrational states on the basis of the simple harmonic wavefunctions alone.* $\mathbf{H}'$ may mix these and change the eigenvalues of the energy but it cannot change the symmetry species. In the harmonic approximation some states (those with several quanta in a degenerate mode) coincide in energy although belonging to different symmetry species. When $\mathbf{H}'$ is introduced we expect such degeneracies to be raised, but only to that extent, since states belonging to degenerate symmetry species of the molecular point group will remain degenerate. The analogy with electronic states derived from the same electronic configuration, or occupation of MOs, should be apparent.

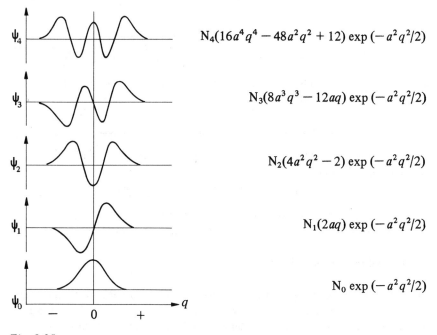

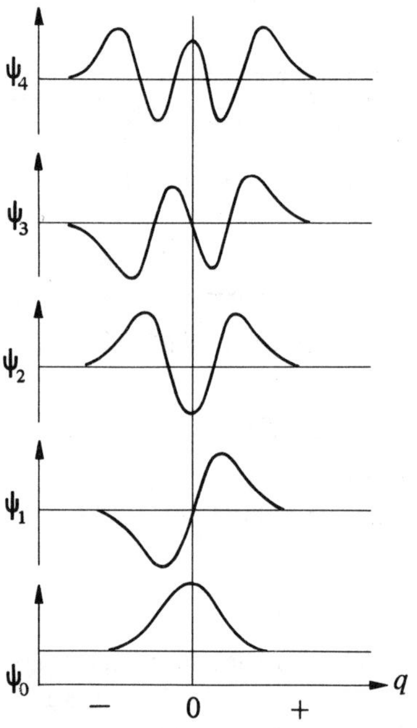

$$N_4(16a^4q^4 - 48a^2q^2 + 12)\exp(-a^2q^2/2)$$

$$N_3(8a^3q^3 - 12aq)\exp(-a^2q^2/2)$$

$$N_2(4a^2q^2 - 2)\exp(-a^2q^2/2)$$

$$N_1(2aq)\exp(-a^2q^2/2)$$

$$N_0\exp(-a^2q^2/2)$$

Fig. 2.35

The simple harmonic wavefunctions $\psi_v(q_i)$, where $v = 0, 1, 2, \ldots$ is the vibrational quantum number, are displayed graphically and analytically in Fig. 2.35, where $a^2 = 2\pi c\omega_i/h$ and the N's are normalization constants. A clear distinction exists between those functions for which v is even and those for which v is odd:

$$\psi_{\text{even}}(q_i) = \psi_{\text{even}}(-q_i) \tag{2.97}$$

$$\psi_{\text{odd}}(q_i) = -\psi_{\text{odd}}(-q_i) \tag{2.98}$$

For non-degenerate modes:

$$\mathbf{R}q_i = \pm q_i \tag{2.99}$$

from which it follows that:

$$\mathbf{R}\,\psi_{\text{even}}(q_i) = \psi_{\text{even}}(q_i)$$

so these functions are totally symmetric. On the other hand:

$$\mathbf{R}\psi_{\text{odd}}(q_i) = \pm\,\psi_{\text{odd}}(q_i),$$

the choice of sign following that in Equation (2.99), so these functions have the same symmetry as q_i.

The corresponding vibrational *states* are soon determined since one or more quanta may only be assigned unambiguously to these non-degenerate modes. If two or more modes are involved the symmetry of their product wavefunctions corresponds to the direct product of their irreducible representations.

For degenerate modes:

$$\mathbf{Rq} = \mathbf{Rq} \tag{2.100}$$

where $\mathbf{q}$ is the column matrix $\{q_i\ \ q_j \ldots\}$ of the degenerate set of normal coordinates, and the matrices $\mathbf{R}$ form an irreducible representation of the group. We have already seen that $(q_i^2 + q_j^2 + \ldots)$ is invariant to all the symmetry operations. If we write a state wavefunction as:

$$\Psi_{v_i, v_j, \ldots} = \prod_i \psi(q_i),$$

then we may say that *the ground state* $\Psi_{00\ldots}$ *of such modes, and therefore of all modes, is totally symmetric and non-degenerate*, as we should expect from its relationship to the classical vibrationless state.

When *one* quantum is assigned to a degenerate mode we may form as many distinct wavefunctions $\Psi_{10\ldots}$, $\Psi_{01\ldots}$, . . . as there are members in the degenerate set. Since each of these has the form:

$$\Psi_{10\ldots} = \psi_1 . \psi_0 \ldots = \mathrm{N}(2aq_i)\exp[-a^2(q_i^2 + q_j^2 + \ldots)/2]$$

the set transforms in the same way as $\mathbf{q}$.

For non-degenerate and degenerate modes alike, therefore, *the vibrational state with one quantum has the same symmetry and degeneracy as the excited vibrational mode.*

When *more than one* quantum is assigned to a degenerate mode the number of distinct ways in which this may be done is less than when two or more different modes are involved. Thus two quanta can only be distributed between doubly-degenerate coordinates q_i, q_j in 3 distinguishable ways, not 4. It follows that the symmetrical direct product must be used to determine the symmetries of these vibrational states.

Thus two quanta in $\tilde{\nu}_4$ of NH_3, an *e*-mode, results, according to Equation (2.95), in an A_1 and an E state. In the harmonic approximation these would coincide, but they need not do so if the anharmonic perturbation $\mathbf{H}'$ raises this degeneracy.

If several quanta are given to more than one mode the possible excited states may be found from the direct product of the symmetry species involved, except that the symmetrical direct product must be used for degenerate modes with more than one quantum. Thus the vibrational states of NH_3 corresponding to the following excitations are as given:

$$\tilde{\nu}_1(a_1) + \tilde{\nu}_4(e) = a_1 \times e = E$$

$$\tilde{\nu}_1(a_1) + 2\tilde{\nu}_4(e) = a_1 \times (e \times e)_{\text{sym}} = A_1 + E$$

$$\tilde{\nu}_3(e) + \tilde{\nu}_4(e) = e \times e = A_1 + A_2 + E$$

$$2\tilde{\nu}_3(e) + 2\tilde{\nu}_4(e) = (e \times e)_{\text{sym}} \times (e \times e)_{\text{sym}} = 2A_1 + A_2 + 3E$$

A further example of the determination of the symmetry species of molecular states on the basis of products of inexact functions of known symmetry arises when there is a significant degree of interaction between electronic and vibrational motions. Such states are then called *vibronic states* and their symmetries are important in connection with selection rules. Since there is no question of equivalence between the electronic and vibrational functions involved, the ordinary direct product is used to find the symmetry species of the vibronic states.

2.6 APPLICATIONS TO SPECTROSCOPY

2.6.1 Selection rules

The probability of a transition taking place between two molecular states is governed by the *transition moment*:

$$\mathfrak{R}^{nm} = \int \psi_n^* \Omega \psi_m \, d\tau \qquad (2.101)$$

where ψ_n and ψ_m are the wavefunctions of the states, Ω is an operator appropriate to the physical perturbation inducing the transition, and $\int \ldots d\tau$ implies that the integration extends over all values of the variables.

If the integral is non-zero the transition is allowed with a probability (and spectroscopic intensity) directly proportional to $|\mathfrak{R}^{nm}|^2$, but if it is zero the transition is forbidden. It is this distinction which constitutes a selection rule; group theory can tell us nothing about the magnitude of $\mathfrak{R}^{nm}$, but it can deduce when it must vanish, if the symmetries of ψ_n, ψ_m, and Ω are each known.

We saw in Section 2.4.1 that base vectors, corresponding to *different* irreducible representations, must necessarily be orthogonal. It follows therefore that the base functions ψ_i and ϕ_j of Section 2.5.5 are mutually orthogonal, and:

$$\int \psi_i^* \phi_j \, d\tau = 0 \qquad (2.102)$$

unless ψ_i^* and ϕ_j have at least the same symmetry species.

Comparing Equations (2.101) and (2.102) it follows that if ψ_n^*, Ω, and ψ_m have the symmetry species Γ_n^*, Γ_Ω, and Γ_m respectively, then Γ_n^* *must be* *contained in* $\Gamma_{\Omega} \times \Gamma_m$ *if the integral is not to vanish* [where $\Gamma_n^* = \Gamma_n$ when ψ_n or the elements of $\boldsymbol{\Gamma}_n(\mathbf{R})$ are all real]. Clearly the subscripts in the above statement may be interchanged, and the statement itself is equivalent, as we have seen in Table 2.11, to one which states that *the integral is zero unless* $\Gamma_n^* \times \Gamma_{\Omega} \times \Gamma_m$ *contains the totally symmetric representation.* This result could also have been deduced from the invariance under the symmetry operations of a scalar product (which is how the integral may be regarded) or of an observable quantity.

The operator Ω takes the form, for electric dipole transitions, of multiplication by $\boldsymbol{\mu} = \sum_i e_i \mathbf{r}_i$ which, from its nature as a vector in physical space, transforms in the same way as the Cartesian unit vectors $\mathbf{e}_1, \mathbf{e}_2, \mathbf{e}_3$, that is as Γ_{xyz}, the decomposition of which is usually recorded with the character tables. We therefore have the rule for electric dipole transitions, in all regions of the spectrum, that *the direct product representation of the two states involved must contain the symmetry species of at least one component* μ_x, μ_y, *or* μ_z *of the electric dipole moment.*

In the case of $\mathcal{C}_{3v}$, $\Gamma_{xyz} = A_1 + E$ (2.49), so $\Gamma_n \times \Gamma_m$ must contain A_1 or E for an allowed transition. Making use of Table 2.11 we see that all states may combine *except* A_1 with A_2. Since μ_z transforms as A_1, while (μ_x, μ_y) transform as E, should $\Gamma_n \times \Gamma_m = A_1$ alone, then only the z-component of $\mathfrak{R}^{nm}$, that is $\mathfrak{R}_z^{nm}$, is non-zero. Likewise, if $\Gamma_n \times \Gamma_m = E$ alone, $\mathfrak{R}_x^{nm}$ and $\mathfrak{R}_y^{nm}$ only are non-zero. If the molecules are all aligned alike (in a crystal) these situations may be distinguished, in absorption, by the use of appropriately orientated plane polarized radiation.

The operator for Raman scattering is multiplication by the time-independent part of the electric dipole induced in the molecule by the electric field of the incident radiation, $\boldsymbol{\mu}^0 = \boldsymbol{\alpha}\boldsymbol{\varepsilon}^0$, where $\boldsymbol{\varepsilon}^0$ is the amplitude of the electric field and $\boldsymbol{\alpha}$ is the polarizability of the molecule. This is a tensor quantity, since we may write:

$$\begin{bmatrix} \mu_x^0 \\ \mu_y^0 \\ \mu_z^0 \end{bmatrix} = \begin{bmatrix} \alpha_{xx} & \alpha_{xy} & \alpha_{xz} \\ \alpha_{yx} & \alpha_{yy} & \alpha_{yz} \\ \alpha_{zx} & \alpha_{zy} & \alpha_{zz} \end{bmatrix} \begin{bmatrix} \epsilon_x^0 \\ \epsilon_y^0 \\ \epsilon_z^0 \end{bmatrix} \tag{2.103}$$

The matrix $[\alpha_{ij}]$ which we will now denote by the letter $\mathbf{A}$, is symmetrical, so there are six distinct elements and $\mathfrak{R}^{nm}$ is non-vanishing if any of the six integrals:

$$\int \psi_n \alpha_{ij} \psi_m \,\mathrm{d}\tau \tag{2.104}$$

is non-zero. *A transition between two vibrational states n and m is Raman active if their direct product representation contains the symmetry species of at least one component* α_{ij} *of the polarizability.*

A symmetry operation $\mathbf{R}$ transforms the matrix equation:

$$\boldsymbol{\mu} = \mathbf{A}\boldsymbol{\varepsilon} \tag{2.105}$$

to $\boldsymbol{\mu}' = \mathbf{A}'\boldsymbol{\varepsilon}'$, since the form of the equation will remain unchanged. Now $\boldsymbol{\mu}' = \mathbf{R}\boldsymbol{\mu}$ and $\boldsymbol{\varepsilon}' = \mathbf{R}\boldsymbol{\varepsilon}$, where $\mathbf{R} = \boldsymbol{\Gamma}_{xyz}(\mathsf{R})$. Therefore $\mathbf{R}\boldsymbol{\mu} = \mathbf{A}'\mathbf{R}\boldsymbol{\varepsilon}$ or $\boldsymbol{\mu} = \mathbf{R}^{-1}\mathbf{A}'\mathbf{R}\boldsymbol{\varepsilon}$, which by comparison with (2.105) gives:

$$\mathbf{A} = \mathbf{R}^{-1}\mathbf{A}'\mathbf{R} \quad \text{or} \quad \mathbf{A}' = \mathbf{R}\mathbf{A}\mathbf{R}^{-1} \tag{2.106}$$

as the transformed polarizability tensor, of which any individual element may be written as:

$$\alpha'_{ij} = \sum_k\sum_l r_{ik}\alpha_{kl}r^{-1}_{lj} = \sum_k\sum_l r_{ik}(r^{-1})^{\mathrm{T}}_{jl}\alpha_{kl} \tag{2.107}$$

in which we have rearranged the order of the terms in the summation to bring the quantity being transformed to the right-hand side. Since $\mathbf{R}$ is orthogonal, $(\mathbf{R}^{-1})^{\mathrm{T}} = \mathbf{R}$, and this becomes:

$$\alpha'_{ij} = \sum_k\sum_l r_{ik}r_{jl}\alpha_{kl} \tag{2.108}$$

The indices of the r elements may be recognized as those of an element of a direct product matrix, and we therefore write out the tensor components as *column-matrices* $\boldsymbol{\alpha}$, to give:

$$\boldsymbol{\alpha}' = (\mathbf{R} \times \mathbf{R})_{\mathrm{sym}}\boldsymbol{\alpha} \tag{2.109}$$

where the symmetrical direct product is required as there are only six distinct components. Therefore:

$$\chi_\alpha(\mathsf{R}) = \tfrac{1}{2}\{[\chi_{xyz}(\mathsf{R})]^2 + \chi_{xyz}(\mathsf{R}^2)\}$$

Since $\chi_{xyz}(\mathsf{R}) = 2\cos\theta \pm 1$ and $\chi_{xyz}(\mathsf{R}^2) = 2\cos 2\theta + 1 = 4\cos^2\theta - 1$, it follows that:

$$\chi_\alpha(\mathsf{R}) = 4\cos^2\theta \pm 2\cos\theta \tag{2.110}$$

From this derivation it may be seen that *the individual elements of* $\boldsymbol{\alpha}$ *transform as the binary products of x, y, and z,* the symmetry species of which are also usually recorded with the point group characters, as in Table 2.7. Thus, in the case of $\mathcal{C}_{3v}$, $\Gamma_\alpha = 2A_1 + 2E$. z^2 has the symmetry A_1; (xz, yz) together have the symmetry E. Since (x, y) transform as E, their binary products transform as $(E \times E)_{\mathrm{sym}} = A_1 + E$. The symmetry-adapted combinations of these are $(x^2 + y^2)$ for A_1 and $(x^2 - y^2, xy)$ for E.

Since the normal modes of NH_3 belong to A_1 and E symmetry species, it follows that the fundamental transitions ($v_i = 0$ to $v_i = 1$) are both infrared and Raman active.

We may note that for those point groups which include the inversion operation i, all the components of $\boldsymbol{\mu}$ must correspond to u-type (odd parity) symmetry species, since x, y, z are transformed to $-x, -y, -z$ by this operation. On the other hand, the components of $\boldsymbol{\alpha}$ must correspond to g-type (even parity) symmetry species, since the binary products of x, y, z, will be unchanged by i. This is the origin of the rule of mutual exclusion: *no fundamental vibrational transition of a molecule with a centre of symmetry is active in both the infrared and the Raman effect.*

2.6.2 Fermi resonance

In Section 2.5.6 we saw that under the influence of the anharmonic part of the potential energy, $\mathbf{H}'$, the harmonic functions are mixed together, although without resultant change in the symmetry species. The corollary of this is that *only wavefunctions of the same symmetry are mixed*. Perturbation theory shows that the corrections to the eigenvalues of the energy and to the form of the eigenfunctions are determined by the values of the integrals:

$$H_{ij} = \int \psi_i^0 \mathbf{H}' \psi_j^0 \, d\tau$$

where ψ_i^0 and ψ_j^0 are the harmonic functions. These integrals will vanish unless $\Gamma_i \times \Gamma_j$ contains the symmetry species of $\mathbf{H}'$, that is, the totally symmetric representation, from which it follows that ψ_i^0 and ψ_j^0 must at least belong to the same symmetry species if the integral is to be non-zero.

The perturbed eigenfunctions ψ_j may be expressed in terms of the harmonic functions:

$$\psi_j = \psi_j^0 + \sum_{i \neq j} c_{ij} \psi_i^0$$

where the coefficients c_{ij} are given, according to a first-order perturbation treatment, by $c_{ij} = H_{ij}/(E_j^0 - E_i^0)$, E_i^0 and E_j^0 being the harmonic eigenvalues $(E_j^0 > E_i^0)$. It follows that the mixing is greatest when two harmonic levels of the same symmetry nearly coincide. The perturbed eigenvalues of the energy, E_i and E_j, will then be given by:

$$2E = (E_i^0 + E_j^0) \pm \sqrt{[4H_{ij}^2 + (E_j^0 - E_i^0)^2]}$$

so that E_i will lie below E_i^0 and E_j above E_j^0. This 'repulsion' of harmonic vibrational levels is known as *Fermi resonance*.

The effect upon the observed intensities is often more marked than the change in the frequencies. In the harmonic approximation $\mathfrak{R}^{nm}$ vanishes, not only in accordance with the symmetry requirements enunciated in the previous section, but also if $\Delta v \neq \pm 1$. Anharmonicity relaxes this restriction, but transitions from the ground state to excited states involving more than one vibrational quantum ('overtones' and 'combination tones') usually remain much weaker than the fundamental transitions. However, if Fermi resonance takes place, e.g. between a level containing one quantum in one mode and another level containing two quanta in a different mode, the resultant mixing of the harmonic wavefunctions allows the overtone to gain intensity at the expense of the fundamental transition.

2.6.3 The correlation of symmetry species

It is frequently necessary to consider the correlation between the irreducible representations of two different point groups, one of which is a subgroup of the other.

Thus we have already seen that the symmetries of molecular states can be deduced on the basis of products of inexact functions which are eigenfunctions of Hamiltonian operators $\mathbf{H}_i$, if the complete Hamiltonian $\mathbf{H}$ is given by Equation (2.86), and $\mathbf{H}'$ has the full symmetry of the molecule. If however $\mathbf{H}'$ has a *lower symmetry*, then $\mathbf{H}$ has this lower symmetry too, and degeneracies existing under the symmetry of $\sum_i \mathbf{H}_i$ may then be removed. The molecular states should then be classified according to the group of $\mathbf{H}'$, and not that of $\sum_i \mathbf{H}_i$, and we need to know the relationship between the irreducible representations of these two groups.

Such a situation arises if a molecule is placed in a physical environment with a symmetry less than its own. This may result from a uniform external field, from the near presence of another molecule in a weakly bound complex, or from its incorporation into a crystalline matrix of some host material.

More drastic changes follow if the molecule is in a crystal of its own kind, because of the increased possibilities of vibrational coupling. Here the overall symmetry will be *higher* than that of the individual molecules. The principles involved are the same but crystal spectra will not be discussed further here; the interested reader is referred to Gilson and Hendra [2.11] and to Turrell [2.12].

However, in addition to those situations in which we alter the environment of the molecule, we may also consider the effect of replacing one or more of a number of symmetrically equivalent atoms by an isotope or even by a chemically different atom. In all these cases the correlation principles involved are the same, and the usefulness of the correlation lies in the help it affords to the interpretation of spectra, especially vibrational spectra.

The subgroup is related to the larger group in having all its symmetry operations in common with the larger group. *An irreducible representation of the larger group, after removal of those matrices representing lost operations, is therefore a representation of the subgroup, although now, in general, a reducible representation.*

We illustrate this by considering a regular tetrahedral molecule, CH_4. See Fig. 2.15. We ask how its vibrational spectrum will be affected (other than just by the change in mass) if one of the hydrogen atoms is replaced by a deuterium atom. The problem is equivalent to one in which a regular tetrahedral molecule suffers a trigonal distortion by being placed at a site with $\mathcal{C}_{3v}$ symmetry. A perchlorate ion, ClO_4^-, close to a metal atom, may be so affected.

The upper part of Table 2.13 contains the result of an analysis of the normal modes of vibration of CH_4, carried out in accordance with Section 2.4.3, and of

Table 2.13

CH_4: $\mathcal{T}_d$	**E**	$8\mathbf{C}_3$	$3\mathbf{C}_2$	$6\mathbf{S}_4$	$6\sigma_d$	n_{3N}	n_{xyz}	n_{rot}	n_{vib}	n_α	Fundl. activity
A_1	1	1	1	1	1	1	0	0	1	1	Ram.
A_2	1	1	1	−1	−1	0	0	0	0	0	
E	2	−1	2	0	0	1	0	0	1	1	Ram.
F_1	3	0	−1	1	−1	1	0	1	0	0	
F_2	3	0	−1	−1	1	3	1	0	2	1	i.r. + Ram.
$\mathsf{U}_\mathbf{R}$	5	2	1	1	3						
θ	0	$2\pi/3$	π	$\pi/2$	0						
$\cos\theta$	1	−1/2	−1	0	1						
$\cos^2\theta$	1	1/4	1	0	1						
$\chi_{\text{rot}} = 1 \pm 2\cos\theta$	3	0	−1	1	−1						
$\chi_{xyz} = 2\cos\theta \pm 1$	3	0	−1	−1	1						
$\chi_{3N} = \mathsf{U}_\mathbf{R} \cdot \chi_{xyz}$	15	0	−1	−1	3						
$\chi_\alpha = 4\cos^2\theta \pm 2\cos\theta$	6	0	2	0	2						
$\mathcal{C}_{3v}$	**E**	$2\mathbf{C}_3$			$3\sigma_v$	n_{A_1}	n_{A_2}	n_E	n_{F_1}	n_{F_2}	
A_1	1	1			1	1	0	0	0	1	i.r. + Ram.
A_2	1	1			−1	0	1	0	1	0	
E	2	−1			0	0	0	1	1	1	i.r. + Ram.

the spectroscopic activity of their fundamental transitions, according to Section 2.6.1. The reader should check these for practice.

In the lower part the character table of $\mathcal{C}_{3v}$ is given, arranged under the appropriate operations of $\mathcal{T}_d$. The non-degenerate representations A_1 and A_2 and the doubly-degenerate E representations correlate directly with each other in the two groups, but the triply-degenerate F_1 and F_2 irreducible representations of $\mathcal{T}_d$ must become reducible representations of $\mathcal{C}_{3v}$ as no higher degeneracy than 2 can exist under this point group. We see that:

$$F_1 \rightarrow A_2 + E \quad \text{and} \quad F_2 \rightarrow A_1 + E.$$

It follows therefore that each of the two f_2-modes of CH_4 will be found split into an a_1- and an e-mode in CH_3D. The experimental data are given in Table 2.14. In these cases there will be no change in spectroscopic activity, but the a_1- and e-modes of CH_4, whose fundamental transitions are active only in the Raman effect, now become, in CH_3D, active in the infrared also. Additional correlations can be deduced concerning overtones and combination tones.

Table 2.14

$\mathcal{T}_d$	CH_4*	cm^{-1}	cm^{-1}	CH_3D†	$\mathcal{C}_{3v}$
a_1	$\tilde{\nu}_1$	2914	2200	$\tilde{\nu}_2$	a_1
			2945‡	$\tilde{\nu}_1$	a_1
f_2	$\tilde{\nu}_3$	3019			
			1300	$\tilde{\nu}_3$	a_1
			3021	$\tilde{\nu}_4$	e
f_2	$\tilde{\nu}_4$	1306			
			1155	$\tilde{\nu}_6$	e
e	$\tilde{\nu}_2$	1534	1471	$\tilde{\nu}_5$	e

* Ref. [2.13].
† Ref. [2.14].
‡ After allowance for Fermi resonance with $2\nu_5$. The observed bands occur at 2973 and 2914 cm^{-1}.

It should be recognized that the 'perturbation', as a result of the isotopic substitution, is large in this example and the forms of all the modes undergo considerable change between the two molecules. We should not seek to identify too closely the 'split' components of the former f_2-modes. However, we may observe that Fermi resonance between $2\tilde{\nu}_5(A_1)$ and $\tilde{\nu}_1(a_1)$ of CH_3D now occurs, whereas this would not have been possible before between $2\tilde{\nu}_2(A_1 + E)$ and $\tilde{\nu}_3(f_2)$ of CH_4, even had the frequencies been closer.

BIBLIOGRAPHY AND REFERENCES

Mathematical texts

2.1 Anderson, J.M., *Mathematics for Quantum Chemistry*, Benjamin, New York (1966).
2.2 Hollingsworth, C.A., *Vectors, Matrices, and Group Theory for Scientists and Engineers*, McGraw-Hill, New York (1967).
2.3 Pettofrezzo, A.J., *Matrices and Transformations*, Prentice-Hall, New Jersey (1966).
2.4 Stephenson, G., *An Introduction to Matrices, Sets and Groups for Science Students*, Longmans, London (1965).

Group theory applied to spectroscopy and other topics; selected texts listed in order of increasing depth of treatment:

2.5 Cotton, F.A., *Chemical Applications of Group Theory*, 2nd edition Wiley-Interscience, New York (1971).
2.6 Schonland, D.S., *Molecular Symmetry*, Van Nostrand, London (1965).
2.7 McWeeny, R., *Symmetry*, Pergamon Press, Oxford (1963).
2.8 Heine, V., *Group Theory in Quantum Mechanics*, Pergamon Press, Oxford (1960).

References

2.9 Longuet-Higgins, H.C., *Mol. Phys.*, **6**, 445 (1963).
2.10 Buerger, M.J., *Elementary Crystallography*, Chapman and Hall, London (1956).
2.11 Gilson, T.R. and Hendra, P.J., *Laser Raman Spectroscopy*, Interscience, London (1970).
2.12 Turrell, G., *Infrared and Raman Spectra of Crystals*, Academic Press, London (1972).
2.13 Feldman, T., Romanko, J. and Welsh, H.L., *Can. J. Phys.* **33**, 138 (1955).
2.14 Wilmhurst, J.K. and Bernstein, H.J., *Can. J. Chem.* **35**, 226 (1957).

3 Microwave spectroscopy

3.1 INTRODUCTION

The term *microwave region* is generally taken to refer to electromagnetic radiation in the wavelength region from about 1 mm to 30 cm. Apart from one experiment in 1934 by Cleeton and Williams [3.1] who investigated the absorption by ammonia vapour at a wavelength of 1.5 cm, this region was not investigated until about 1945, when the development of microwave frequencies for radar detecting systems had made monochromatic klystron oscillators generally available as sources of the radiation.

This spectroscopic region is principally concerned with absorption of electromagnetic radiation, and generally in the case of gases individual molecules are excited to a higher rotational level. These pure rotational changes take place with negligible alteration of the internuclear distance (c.f. the infrared and ultraviolet techniques). In the past ten years microwave emission spectra have played an increasing role in identifying interstellar molecules.

The microwave method is the most accurate way of determining internuclear distances and bond angles, and its accuracy reveals small changes in structural features hardly observable by other methods. For example, in the series of nitrosyl halides, ONF, ONCl, ONBr, the O—N internuclear distance changes are 1.138, 1.139, and 1.149 Å, while the bond angles are 110° 12′, 113°20′, and 114°30′. Studies are made on linear, symmetric-top, and asymmetric-top molecules, and of particular interest are the studies on the planarity and non-planarity of molecules. The method is invaluable in giving accurate molecular electric dipole moments in the gaseous and liquid phases, and yields also nuclear quadrupole coupling constants leading to detailed information on molecular structure. The method has been frequently employed in the study of internal rotation involving low energy barriers up to ~ 15 kJ mol^{-1}, and again its high accuracy is a most attractive feature; an additional use for this type of study is in the deduction of conformations present and of the energy differences of the rotational isomers. Information may also be gained on the rotational energy transfer in collisions. Further, the approach is valuable in the study of rotation–vibration interaction and of the harmonic force fields.

The most useful recent development has been that commercial instruments have become available. Undoubtedly this will broaden the application of the method to chemical analysis and lead to more overlap in the study of a particular structural problem by the microwave and other methods (e.g. in the study of rotation–vibration interaction).

One recent trend has been to examine larger and more flexible molecules. Generally, in the solid and liquid states, owing to the strong intermolecular interaction, the simple quantized form of the rotational levels disappears, and some of the microwave studies in the past few years have been aimed at appreciating the molecular motion possible in the liquid phase. In some cases these results have been correlated with those in the adjacent far-infrared region (10–200 cm^{-1}), and attempts have been made to explain the broad band absorption (see Chapter 5) in this region which appears to be exhibited by all polar liquids and also by non-polar liquids such as the inert cyclohexane. At first sight absorption by non-polar liquids in the microwave (and far-infrared) region would not be expected since the criterion for the absorption of microwaves is that the molecule should possess either a permanent electric or magnetic dipole moment, where in nearly all cases the interaction takes place between the electric field component of the microwave radiation and the rotating electric dipole moment of the molecule. The mechanism of absorption of microwave radiation by these non-polar molecules now appears to be well understood and is related to the ability of the higher moments to induce dipole moments in the surrounding molecules in the liquid phase. In the case of cyclohexane the higher moment is a quadrupole one.

One molecule which does not interact with the electric field component is gaseous oxygen (O_2) which owing to its symmetry has no electric dipole moment. Unlike other symmetrical diatomic molecules, this molecule, as a result of its two unpaired electrons, possesses a permanent magnetic dipole moment which enables it to interact with the magnetic field component of the electromagnetic radiation.

A few books are now available on microwave spectroscopy [3.2–3.6], and there are some recent accounts [3.7–3.9] of certain aspects of microwave spectroscopy.

3.2 EXPERIMENTAL METHOD

Many of the microwave investigations have been carried out in the region of 5 mm to 5 cm wavelength, mainly about 1 cm. The absorption frequencies for a pure rotational change are determined by the moment(s) of inertia of the molecule. For a rigid diatomic molecule the selection rule is $\Delta J = \pm 1$, and the absorbed frequencies are given by the formula:

$$\nu = \frac{h}{4\pi^2 I}(J'' + 1) \tag{3.1}$$

where J'' is the rotational quantum number in the lower rotational state and takes the values 0, 1, 2, 3, ..., and I is the moment of inertia about an axis perpendicular to the molecular axis. The frequency region where the change in rotational energy can be studied is fixed by the quantity $h/4\pi^2 I$ which is the frequency difference between the successive rotational lines. The number of rotational energies a molecule can have is quite large. For example, J might range from 0 to 60. If the $J_{1 \leftarrow 0}$ rotational line occurred at 30 cm wavelength, then lines with equal spacing would stretch right across to 0.5 cm where the $J_{60 \leftarrow 59}$ would occur.

In the microwave region polar gases usually give sharp absorption lines whose widths are in the kHz to MHz range. The spectrometers used for such studies have the following basic features in common: (i) a microwave source which can be made to sweep through a narrow frequency range; (ii) a frequency measuring system; (iii) an absorption cell; (iv) a microwave detector; (v) an amplifier for the detected signal; (vi) a signal indicator; (vii) a frequency modulator for the source.

One of the major aims in a number of studies is to achieve maximum sensitivity in the microwave spectrometer. This would be very important, for example, in the detection of free radicals where normally their lifetimes are short and their concentration inevitably low. The maximum sensitivity of a microwave spectrometer is defined as the weakest detectable absorption line for which the signal-to-noise ratio is 1. The total noise may be regarded as being introduced through the processes of detection and amplification; the sensitivity of some spectrometers is of the order of 10^{-10} cm^{-1}. Thus, if the absorption coefficient of the gas is 10^{-10} cm^{-1}, then if the cell containing the gas were 1 metre long the alteration in the detector power level would be 10^{-6} per cent.

Two types of spectrometer are employed: one is termed the video microwave spectrometer and the other the Stark modulation spectrometer. The latter type which is now widely used for rotational spectra studies employs an electric field in the cell which causes splitting of the rotational energy levels. This leads to the appearance of a fine structure in the rotational spectrum and produces what is termed the Stark effect in a molecular spectrum. This method will be considered later in this chapter, and we shall now examine the video microwave spectrometer for which a block diagram is given in Fig. 3.1; the aim will be to give the basic features of the apparatus since the component parts and theoretical points relating to them are so often outside the range of a chemist's training.

The *reflex klystron* is employed in the centimetre region as the microwave source. The klystron is really a very high frequency radio valve which provides a very stable monochromatic source of radiation. Fortunately, however, the

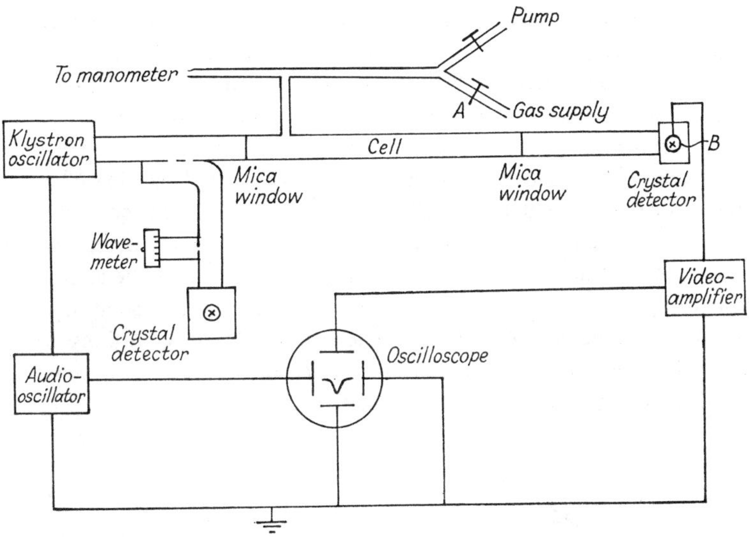

Fig. 3.1 A block diagram of a video spectrometer. (A Pirani gauge is often used for relative pressure measurements).

frequency itself can be varied since, as in most absorption spectroscopy, it is desirable to cover a range of frequencies. This frequency variation may be achieved by either mechanical or electronic tuning; the former leads to a variation in frequency of about 15 per cent, and the latter to a variation of from only 0.1 to 1 per cent. Thus, for a klystron operating at 3.2 cm a mechanical tuning range from about 8500 to 10 000 MHz is obtainable, whilst the electronic tuning range is about 30 MHz. In general, the tuning ranges obtainable are smaller for higher frequency klystrons. By the use of a series of different klystrons it is possible to cover the range 1000 MHz (30 cm) to 37 500 MHz (8 mm). At higher frequencies than this it is possible to use frequency multiplication techniques. In addition, more klystrons are available at 70 and 140 GHz.

More recently microwave power sources tunable over a full half-octave band have become available up to $\sim 10^{10}$ Hz while backward wave oscillator tubes are used at higher frequencies. In fact, cavity tuning is no longer necessary, and the band may be covered in one long scan.

The radiation is fed from the microwave source by wave-guides which are in effect hollow metallic conductors through which the electromagnetic radiation is propagated. The dimensions of the wave-guide are fixed by the frequency employed, neglecting, of course, the 15 per cent variation over which the klystron functions. The patterns of the electric and magnetic fields of the microwave radiation inside a rectangular wave-guide are shown in Fig. 3.2. It will be noted that the direction of flow of the surface currents is at right-angles to the line of magnetic field at the surface of the wave-guide.

A small fraction of the radiation is fed out of the main wave-guide system and into the frequency measurement system by means of a directional coupler (see Fig. 3.3). This power is coupled from the main wave-guide by means of the two holes (about a quarter of a wavelength apart) in the two lengths of wave-guide which are joined together along their narrow sides. For such a type of coupler the power flow in the auxiliary guide is in the same direction as that in the main guide. The frequency can be estimated to about 1 part in 10^4 by a calibrated wave-meter which is usually a circular cylindrical cavity, one end of which is closed by a piston that can be moved axially by a micrometer screw. The resonance condition for the cavity is that its length should be an integral multiple of $\lambda_g/2$ where λ_g is the wavelength of the microwave radiation in the cavity. The micrometer screw adjusts the length of the cavity until resonance is achieved. Resonance can be detected since, when it occurs, microwave power is transmitted through the auxiliary crystal detector.

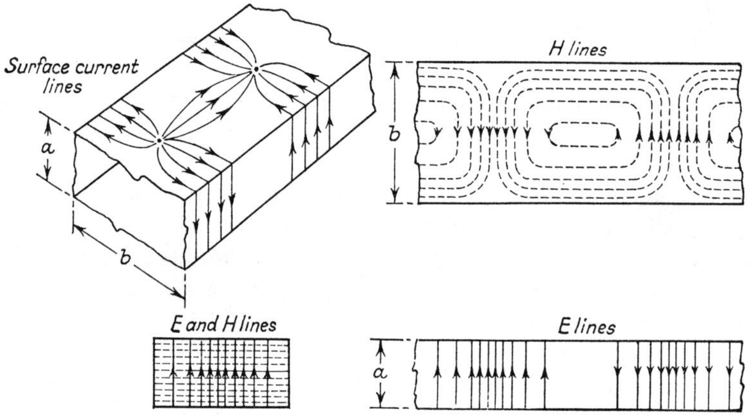

Fig. 3.2 Direction of surface currents and patterns of electric and magnetic fields inside a rectangular wave-guide (Reproduced by permission from Ref. [3.2]).

When a precise value of the frequency is required, the following devices are employed. (a) A quartz crystal oscillator is brought, for example, to within 5 MHz of a standard frequency such as a standard signal broadcast frequency which is known accurately to at least 1 part in 10^8. (b) Electronic multipliers are then employed to yield a series of higher frequencies at, for example, 60, 180, 340, ... MHz which are fed simultaneously into a silicon crystal in the side arm of the wave-guide. (c) Higher harmonics are generated by this crystal and produce beat-frequency signals with the microwave frequency.

Such a set-up provides a series of accurate marker frequencies at multiples of 60 MHz which are accurate to about 1 part in 10^7. Normal techniques employed for radiofrequencies can then be used to determine the frequency difference

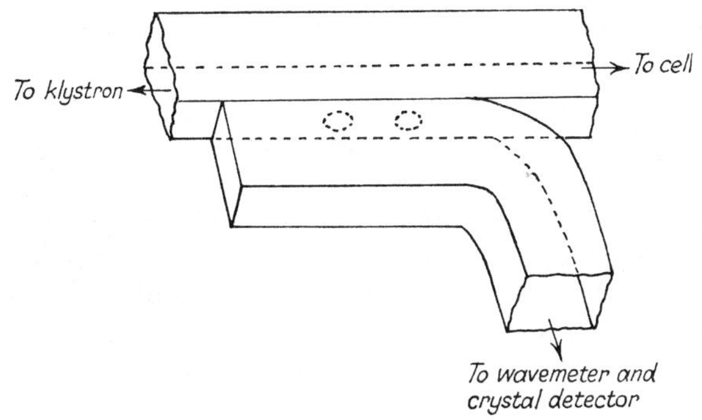

Fig. 3.3 Two-hole directional coupler.

between an absorption line and an adjacent known harmonic of 60 MHz. The frequencies of the absorption lines are quoted to seven figures, and such accuracy is considerably greater than that in, say, the infrared regions.

The absorption cell consists of a length of wave-guide which is separated from the remainder of the system by mica windows. The cell can be evacuated and the gas introduced at the required pressure. The lengths of the cell which have been employed have ranged from one to a hundred feet. In fact, the cell length must be sufficiently long to enable a detectable absorption of the microwave power to occur at a low pressure of gas (to avoid pressure broadening see later).

The radiation transmitted through the cell is detected by means of a silicon crystal, the silicon being a semiconducting material. Directly in contact with the silicon is a fine wire, and these together form the crystal detector. When an alternating voltage is received rectification occurs. For video detection operating at low power the rectified output is dependent on the input power.

In order that a.c. methods of amplification can be used, the frequency of the klystron is swept over a limited range of approximately 30 MHz by applying a saw-tooth voltage from a suitable audio oscillator to the reflector electrode (electronic tuning). The same voltage is used to sweep the spot of the cathode-ray tube horizontally. The repetition frequency of the sweep is usually low and sometimes as little as one cycle in several seconds. The voltage detected by the crystal is modulated (owing to absorption in the cell), and the absorption line is identified with a decrease in the power when the klystron sweeps through the absorption frequency. The amplified output of the crystal is displayed on the cathode-ray oscilloscope. Thus, the actual line shape is rapidly displayed on the screen, where the horizontal scale of the cathode-ray tube is a frequency scale, and the vertical scale gives a measure of the intensity of absorption. On the frequency scale is a set of frequency markers obtained by mixing a sample of the microwave radiation with a known reference frequency. If no absorption occurs, a horizontal line is obtained on the screen, but when absorption takes place a peak occurs for each absorbed frequency provided that the resolution is sufficiently great.

One of the snags with the video microwave spectrometer is that the detected signal is amplified by a wide band system which has the limiting effects of detector noise. This can be overcome by applying modulation to the cell, and this is done in the Stark modulation spectrometer (see later) which has operating characteristics that permit detection of the unperturbed absorption line. In this case modulation of the energy results from modulation of the energy levels, while amplification and detection are achieved by a narrow band system of which the detection circuit is synchronized with the modulation.

3.3 INFORMATION DERIVED FROM WORK ON GASES

Most simple polar molecules undergo pure rotational energy changes somewhere in the microwave region by the absorption of electromagnetic radiation. Exceptions are light molecules such as H—F and H—Cl which have very low moments of inertia and have all their absorption frequencies in the infrared region. However, slightly heavier molecules such as DBr and HI absorb both in the microwave and infrared regions. Some of the most important information derived from the microwave studies of small molecules is the determination of their moments of inertia and consequently their internuclear distances.

For molecules at sufficiently low pressures to be regarded as isolated molecules the factors which determine their pure rotational transitions are: (a) the spacing of the energy levels which for linear molecules is governed by an inverse dependence on the moment of inertia; (b) selection rules; (c) the magnitude of the electric or magnetic dipole moment which interacts with the electromagnetic radiation where the former would interact with the electric field component and the latter with the magnetic field component of the electromagnetic radiation; transitions involving magnetic dipole moments are normally appreciably weaker in intensity than those involving electric dipole moments (μ), and in the latter case the absorption intensity is proportional to μ^2; (d) the molecular population of the level from which the transition takes place. From the Boltzmann distribution law for a system consisting of N molecules the fraction N_J/N will populate a particular energy level E_J where:

$$\frac{N_J}{N} = (2J+1)\frac{\exp\left(-E_J/kT\right)}{f} \tag{3.2}$$

where k is the Boltzmann constant, T is the absolute temperature, and the $(2J+1)$ term takes into account the fact that each rotational level is $(2J+1)$-fold degenerate. f is a partition function (see Vol. 3):

$$f = \Sigma \exp(-E_r/kT) \tag{3.3}$$

and may be looked upon as a proportionality constant; since N is by definition constant, then:

$$N_J \propto (2J+1)\exp(-E_r/kT) \tag{3.4}$$

In Chapter 1 we saw that for a rigid linear molecule:

$$E_r = J(J+1)Bhc \tag{3.5}$$

and therefore on substitution for E_r in Equation (3.4) we obtain:

$$N_J \propto (2J+1)\exp[-BhcJ(J+1)/kT] \tag{3.6}$$

(see Vol. 3). Hence, the number of molecules in a particular level E_J is dependent on the absolute temperature, the value of the moment of inertia (since $B = h/8\pi^2 cI$), and the actual value of J. From Equation (3.6) it follows that the value of N_J increases with increasing temperature but decreases rapidly with increasing J. Differentiation of Equation (3.6) gives the maximum populated level (J_{max}) to be:

$$J_{max} = \sqrt{(kT/2hcB)} - 1/2 \tag{3.7}$$

where J_{max} is taken to the nearest integral value.

3.3.1 Classification of rotors

Rotating molecules (rotors) may be classified in terms of their moments of inertia. The rotor may be regarded as being composed of mass points (atoms) $m_1, m_2, m_3, \ldots$ rigidly connected and at perpendicular distances $x_1, x_2, x_3, \ldots,$ respectively, from an axis. The moment of inertia about this axis (say A) is given by $I_A = \Sigma_i m_i x_i^2$. A molecule has three such axes, termed the principal axes, where each of these axes is perpendicular to the others, all of which pass through the centre of gravity of the molecule. I_A, I_B, and I_C are so chosen that: I_A is the smallest moment of inertia; I_C is the largest; I_B lies between I_A and I_C unless $I_A = I_C$.

Molecules are then classified into the following types.

(a) Linear molecules, such as CO_2, CS_2, and OCS, where $I_A = 0$ and $I_B = I_C$.

(b) Spherical-top molecules, such as CCl_4 and SF_6, where $I_A = I_B = I_C$.

(c) Symmetric-top molecules, where two of the moments of inertia are equal, and the third is different and not zero. Such molecules are divided into two types which have different equations characterizing their rotational energies: (i) prolate (cigar-shaped) symmetric-top molecules, such as the tetrahedral structure in CH_3Cl, where $I_A < I_B = I_C$. (ii) oblate (pancake-shaped) symmetric-top molecules, such as the planar symmetrical structure of BCl_3, where $I_A = I_B < I_C$. Symmetric-top molecules have a 3-fold or higher symmetry axis.

(d) Asymmetric-top molecules, such as H_2O, F_2O, SO_2, CH_2F_2, and CH_3CHO, where $I_A < I_B < I_C$.

The application of microwave spectroscopy to the rotational spectra of such types will now be considered. It will be noted that linear molecules with a centre of symmetry such as CO_2 and the spherical-top type, where both the magnetic and electric dipole moments are zero, do not exhibit a pure rotational spectrum.

3.3.2 Linear molecules

For a rigid linear molecule in its ground vibrational state with no resultant electronic angular momentum, and in the absence of an applied field, the quantized energy states for its rotation are given by:

$$E_r/hc = BJ(J+1) \tag{3.8}$$

where J is the rotational quantum number and has integral values 0, 1, 2, 3, ..., and:

$$B = \frac{h}{8\pi^2 cI}\ \mathrm{cm}^{-1} \tag{3.9}$$

where I is the moment of inertia. Formula (3.8) applies at very low values of J, but when centrifugal stretching occurs at higher values of J the formula has to be modified to:

$$E_r/hc = BJ(J+1) - DJ^2(J+1)^2 \tag{3.10}$$

where D is the centrifugal distortion constant. For a diatomic molecule D may be related to B and the fundamental vibrational frequency ω of the molecule by the equation $D = 4B^3/\omega^2$.

In excited vibrational states the constants B and D would have different values, and for linear polyatomic molecules their values in the vth vibrational level are given by equations (3.11) and (3.12), respectively:

$$B_v = B_e - \Sigma_i \alpha_i \left(v_i + \frac{d_i}{2}\right) \tag{3.11}$$

$$D_v = D_e + \Sigma_i \beta_i \left(v_i + \frac{d_i}{2}\right) \tag{3.12}$$

where α and β are interaction constants which take into account the interaction between the rotational and vibrational energies. Σ_i indicates that the summation is carried out for all the fundamental modes of vibration where v_i is the quantum number for the ith vibrational mode and d_i is the degeneracy of that mode. B_e and D_e refer to a hypothetical vibrationless state and are known as equilibrium values. In the case of a diatomic molecule in the vth vibrational level, Equations (3.11) and (3.12) would reduce to:

$$B_v = B_e - \alpha(v + \tfrac{1}{2}) \tag{3.13}$$

$$D_v = D_e + \beta(v + \tfrac{1}{2}) \tag{3.14}$$

A linear triatomic molecule (XYZ) has three fundamental vibrational modes, given in Fig. 3.4, which are governed by the three vibrational quantum numbers

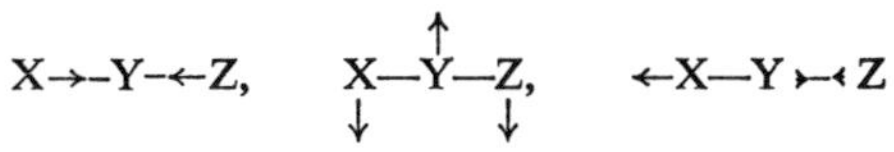

Fig. 3.4 Fundamental vibrational modes of a linear triatomic molecule.

v_1, v_2, and v_3 respectively. The non-linear mode is doubly degenerate, that is its d_i value is 2. It follows from Equation (3.11) that:

$$B_v = B_e - \alpha_1(v_1 + \tfrac{1}{2}) - \alpha_2(v_2 + 1) - \alpha_3(v_3 + \tfrac{1}{2}) \qquad (3.15)$$

In principle, by measurement of the frequencies of the pure rotational transitions in the fundamental modes of vibration and at least one excited state for each mode of vibration, it is possible to determine the interaction constants and the B_e value.

The selection rule for the interaction of μ with electromagnetic radiation requires that the quantum number J should change by only one unit in the absorption of rotational energy, i.e. $\Delta J = J' - J'' = +1$. For a rigid linear molecule when J changes to $(J+1)$ the energy change ΔE_r and the frequency of the spectral line ν are given by:

$$\Delta E_r = E_r' - E_r'' = h\nu = hcB\,[(J+1)(J+2) - J(J+1)] \qquad (3.16)$$

$$\nu = 2cB(J+1) \qquad (3.17)$$

where $J = J'' = 0, 1, 2, 3, \ldots$. For the non-rigid rotator, where centrifugal forces occur, the rotational frequency is obtained by substituting for E_r' and E_r'' from Equation (3.10) into the equation:

$$E_r' - E_r'' = h\nu \qquad (3.18)$$

On insertion of $(J''+1)$ for J':

$$\nu = c\,[2B(J+1) - 4D(J+1)^3] \qquad (3.19)$$

where $J'' = J = 0, 1, 2, 3, \ldots$. In order to achieve the accuracy provided by the microwave method it is necessary to employ formula (3.19) in preference to (3.17) for all except the lowest J values.

The spectrum of a rigid linear molecule should consist of evenly spaced lines with frequencies in the ratios 1:2:3:4:..., corresponding to the J values 0, 1, 2, 3, ... Such a series of lines with J values from 4 to 10 was first confirmed for hydrogen chloride the pure rotational spectrum of which occurred very approximately between 40 and 100 μ, that is, outside the microwave region. Microwave work on $^{12}C^{16}O^{32}S$ has confirmed the integral frequency ratio; the observed frequencies for four rotational transitions $J' \leftarrow J''$ are shown in Table 3.1. It may be observed from Table 3.1 that the frequencies for these transitions are almost an integral multiple of 12 162.9 which is the value in MHz for the $1 \leftarrow 0$ transition. The very small variation of 0.1 MHz in the value for the $2 \leftarrow 1$ transition is all that is required to make the data the correct multiple of 12 162.9 MHz. This 0.1 MHz discrepancy may be attributed to centrifugal stretching of the molecule whose moment of inertia increases with increasing J.

Table 3.1 Rotational frequencies of COS (MHz)

	$2 \leftarrow 1$	$3 \leftarrow 2$	$4 \leftarrow 3$	$5 \leftarrow 4$
	24 325.9	36 488.8	48 651.6	60 814.1
i.e.	2 x 12 162.9	3 x 12 162.9	4 x 12 162.9	5 x 12 162.8

For a linear triatomic molecule such as OCS two internuclear distances, r_{CO} and r_{CS}, are involved while the masses of the atoms m_O, m_C, and m_S are known. The procedures involved in the determination of the internuclear distances are: (a) Determination of the moment of inertia for two isotopic species by carrying out two separate experiments on, say, $^{16}O^{12}C^{32}S$ and $^{16}O^{12}C^{34}S$. From the frequency difference between consecutive rotational lines moments of inertia I_1 and I_2 are determined for the two cases. (b) These moments of inertia may then be employed in an equation which relates I to the masses of the atoms and the internuclear distances, to yield two equations from which r_{CO} and r_{CS} can be calculated. This equation can be deduced as follows.

The arrangement of the atoms in the carbonoxy sulphide molecule is shown in Fig. 3.5. Point A is the centre of gravity of the molecule, about which the rotation of the molecule takes place. On taking moments about point A we get:

$$m_O r_O + m_C r_C = m_S r_S \tag{3.20}$$

while the moment of inertia is given by:

$$I = m_O r_O^2 + m_C r_C^2 + m_S r_S^2 \tag{3.21}$$

From Fig. 3.5 it follows that:

$$r_O = r_{CO} + r_C \tag{3.22}$$

and

$$r_S = r_{CS} - r_C \tag{3.23}$$

On substitution of (3.22) and (3.23) into (3.20) we obtain:

$$M r_C = m_S r_{CS} - m_O r_{CO} \tag{3.24}$$

where $M = m_C + m_O + m_S$ = total mass of molecule. On substitution of (3.22) and (3.23) into (3.21) it follows that:

$$I = M r_C^2 + 2 r_C (m_O r_{CO} - m_S r_{CS}) + m_O r_{CO}^2 + m_S r_{CS}^2 \tag{3.25}$$

and on insertion of the value of r_C from Equation (3.24) into (3.25) we get:

$$I = m_O r_{CO}^2 + m_S r_{CS}^2 - (m_O r_{CO} - m_S r_{CS})^2/M \tag{3.26}$$

For the case $^{16}O^{12}C^{32}S$ substitution of $m_C = 12.01$, $m_S = 32.00$ and $m_O = 16.00$ in equation 3.26 gives a value of I_1 while for another isotopic species $^{16}O^{12}C^{34}S$ substitution of $m_C = 12.01$, $m_S = 34.00$ and $m_O = 16.00$ leads to a value I_2. From both these equations the two internuclear distances, r_{CO} and r_{CS} are obtained.

In COS studies the following isotopic combinations have been used:

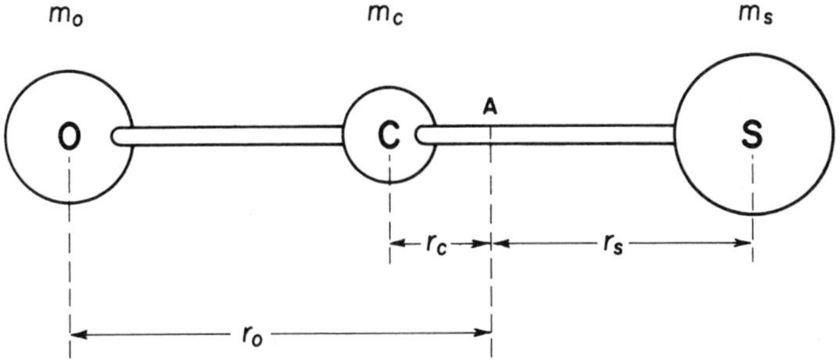

Fig. 3.5 Diagram which indicates for carbonoxy sulphide the distances of the atoms from the centre of gravity A.

$^{16}O^{12}C^{34}S$; $^{16}O^{13}C^{32}S$; $^{16}O^{13}C^{34}S$; $^{18}O^{12}C^{32}S$, and the values of r_{CO} and r_{CS} in the ground vibrational state for the $J_{2\leftarrow 1}$ transition ranged from 1.165 and 1.558 Å to 1.155 and 1.565 Å respectively from the study of the following pairs of molecules:

	r_{CO}	r_{CS}
$^{16}O^{12}C^{32}S$ and $^{16}O^{12}C^{34}S$	1.165	1.558
$^{16}O^{12}C^{32}S$ and $^{16}O^{13}C^{32}S$	1.163	1.559
$^{16}O^{12}C^{34}S$ and $^{16}O^{13}C^{34}S$	1.163	1.559
$^{16}O^{12}C^{32}S$ and $^{18}O^{12}C^{32}S$	1.155	1.565

This variation in r may be attributed to different zero-point energy in these various molecules. In the most accurate work corrections have to be made for the variation of internuclear distances resulting from different zero-point energies. The zero-point amplitude of $^{16}O^{12}C^{34}S$ (the heavier molecule) is slightly less than that of $^{16}O^{12}C^{32}S$, and since the internuclear distances depend on the amplitude of vibration, then, on the amplitude of the zero-point vibration being altered, the internuclear distances will be slightly changed. This may perhaps be better appreciated by considering two isotopes of a diatomic molecule in terms of its potential-energy curve. Thus, in the case of H—^{35}Cl and D—^{35}Cl the potential-energy curve is given in Fig. 3.6, and because of the asymmetry of the potential-energy curve it follows that the mean internuclear distance must differ for the two types of molecules in their lowest vibrational state.

For diatomic molecules of the type X—Y there is no need to use isotopes, since the internuclear distances can be determined directly from the moment of inertia and the atomic masses:

$$I = \frac{m_X m_Y}{m_X + m_Y} r_{XY}^2 \qquad (3.27)$$

Many other linear molecules have had their internuclear distances determined by the microwave method; these include FCl, HCN, HCCCN, and NNO. In

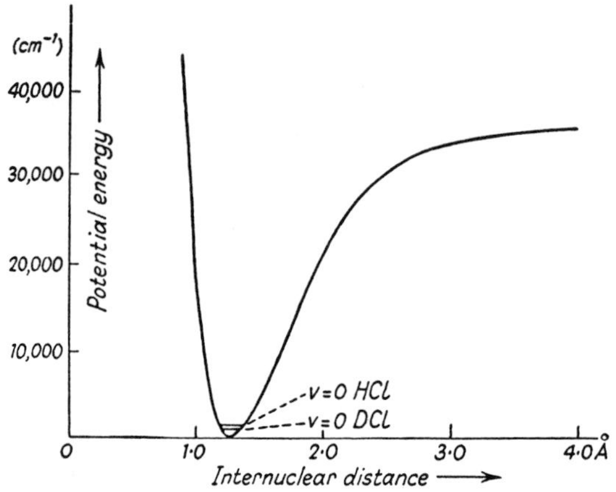

Fig. 3.6 Potential-energy curve illustrating the zero-point energy of $H^{35}Cl$ and $D^{35}Cl$.

general, for an unsymmetrical linear molecule containing n atoms, $(n - 1)$ internuclear distances have to be evaluated, and $(n - 2)$ isotopic substitutions would be required for a complete structural determination; only one isotopic substitution is made per atom since by this means an independent equation is obtained.

One diatomic molecule which differs from all other symmetrical diatomic molecules in having a microwave spectrum under normal conditions is O_2. The oxygen molecule differs from the other symmetrical diatomic molecules, since in the ground state it has two unpaired electrons with parallel spins. Consequently, it has a permanent magnetic dipole moment which can couple with the magnetic field component of the microwave radiation and give rise to absorption. Furthermore, the magnetic dipole can couple with the rotational energy of the molecule through the small magnetic moment which results from the molecular rotation. Thus, a series of energy levels which are dependent on the rotational and spin momenta will be produced. The type of coupling is that of Hund's case (b) (see Vol. 3). If the total angular momentum quantum number (exclusive of nuclear spin) is J, and S is the total electron spin quantum number, and if N is the quantum number corresponding to end-over-end rotation of the molecule, then J may take only the values:

$$J = (N+S), (N+S-1), \ldots, |(N-S)|$$

where $N = 0, 1, 2, \ldots$. In the case of a triplet state, $S = 1$, that is for each value of N there are three values of J. The observed microwave transitions are governed by the selection rules $\Delta J = \pm 1$ and $\Delta N = 0$. This results in a group of closely spaced absorption lines which may be attributed to transitions between the fine structure components of the rotational levels.

The absorption of energy occurs when the electron spin magnetic moment alters its orientation by interaction with the magnetic vector of the microwave radiation; the rotational momentum vector remains unchanged. Work on oxygen at low pressure has resolved part of the spectrum into twenty-six lines, which lie in the region of 5 mm wavelength. The N values of these lines range from 1 to 25. From the frequencies of these lines the magnetic dipole moment of the oxygen molecule has been evaluated.

The recent trend in the study of diatomic molecules in the microwave region has been to examine (a) less volatile molecules such as AgF and CuF [3-10], and (b) diatomic free radicals such as ClO [3.11] and NS [3.12] using fast flow spectrometers.

3.3.3 Spherical-top molecules

These are molecules where $I_A = I_B = I_C$, such as SF_6 and CCl_4. Such molecules have no permanent electric dipole moment, and as a result of this it is not possible to observe changes in their rotational energy in the microwave region. This is to be contrasted with the infrared region where I can be determined from certain rotation–vibration transitions in spherical-top molecules.

3.3.4 Symmetric-top molecules

The molecules examined have been mainly of the pyramidal type, as in NH_3 and PCl_3, or of the tetrahedral type, as in CH_3Cl and $CHCl_3$ where $I_B = I_C \neq I_A$, and none of these values is zero. For the tetrahedral and pyramidal type of molecule I_A is the moment of inertia along the main axis of symmetry (the z-axis). I_B and I_C are the moments of inertia about the other two perpendicular principal axes. Figure 3.7 shows the axes chosen for the analysis of the rotation of the symmetric-top molecule CH_3Cl, where the main dipole (C—Cl) lies along the molecular axis of symmetry. The z-axis was chosen along the internuclear line of the carbon and chlorine atoms because of the resulting CH_3 group symmetry about this axis. The other two axes were then placed perpendicular to this axis through the centre of mass. The axis of highest symmetry of the molecule is placed along the z-axis. I_A is always chosen so that it is less than I_C, and if the axis of I_A lies along the highest symmetry axis, this type of molecule is termed a prolate symmetric-top. Thus, the CH_3Cl molecule is a *prolate (cigar-shaped) symmetric-top*. However, if the axis of the greater moment of inertia I_C (where $I_A = I_B < I_C$) lies along the symmetry axis as in the plane symmetrical structure of BCl_3, then the molecule is termed an *oblate (pancake-shaped) symmetric-top*. Thus, an oblate symmetric-top has the two smaller moments of inertia equal (i.e. $I_A = I_B$) whereas a prolate symmetric-top has the two larger ones equal (i.e. $I_B = I_C$). In general, only the case of a prolate symmetric-top will be considered, and the equations given will apply to this type.

The rotational energies for the prolate symmetric-top molecule are given by:

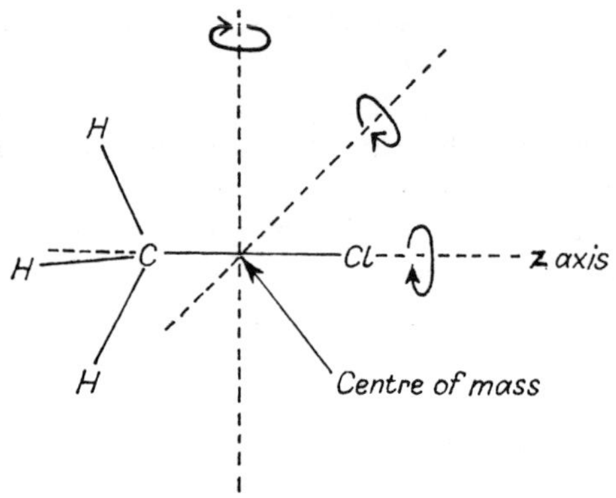

Fig. 3.7 Rotation of a prolate symmetric-top molecule about the three principal axes.

$$E_r = \frac{h^2}{8\pi^2 I_B} J(J+1) + \frac{h^2}{8\pi^2}\left(\frac{1}{I_A} - \frac{1}{I_B}\right)K^2 \tag{3.28}$$

where J is the quantum number governing the total angular momentum $\sqrt{[J(J+1)]}\, h/2\pi$ of the molecule. K is a second rotational quantum number whose value in $h/2\pi$ units is the component of the total angular momentum directed along the symmetry axis where K takes the values:

$$-J, -(J-1), \ldots, 0, \ldots, (J-1), J$$

Equation (3.28) for E_r applies exactly only to a rigid molecule, but the centrifugal effect can be corrected by the use of a more complex expression (see later).

The selection rules are that J changes by unity while K remains unchanged; if ΔE_r is the energy absorbed for the transition $E_r' \leftarrow E_r''$ where $J' = J'' + 1$, then:

$$\Delta E_r = \frac{h^2}{8\pi^2}\left[\frac{J'(J'+1)}{I_B} - \frac{J''(J''+1)}{I_B} + \left(\frac{1}{I_A} - \frac{1}{I_B}\right)K'^2 - \left(\frac{1}{I_A} - \frac{1}{I_B}\right)K''^2\right] \tag{3.29}$$

but $\Delta K = 0 = K' - K''$, i.e. $K'^2 = K''^2$, and by substituting for J' the value $(J''+1)$:

$$h\nu = \Delta E_r = \frac{h^2}{4\pi^2 I_B}(J+1) \quad \text{where} \quad J = J'' = 0, 1, 2, \ldots \tag{3.30}$$

Equation (3.30) is usually accurate only when low values of J are involved. From the spacing of the rotational lines I_B can be determined, whereas I_A cannot be found from the microwave spectrum. However, in a complete analysis of the molecule, internuclear distances and valency angles are involved; that is, in the case of CH_3Cl there would be three unknowns, an angle, r_{CH}, and r_{CCl}, but only one equation. Therefore, either some assumptions would have to be made about two of these values (possibly data from other sources) or three isotopic molecules would have to be employed.

If the necessary isotopic substitutions have been made, the spectra analysed, and the three values of I_B obtained, the I_B values, from moment of inertia considerations, may be related to the masses of the atoms and certain distances within the molecule; then from simple geometrical considerations the bond angles and internuclear distances can be calculated.

In practice the analysis is not as simple as Equation (3.30) may suggest, since interaction occurs between the rotational and vibrational energies. Owing to centrifugal stretching effects during the molecular vibration, the rotational energies for a symmetric-top molecule are affected, and for a non-rigid symmetic rotator:

$$E_r = hc\,[BJ(J+1) + (A-B)K^2 - D_J J^2(J+1)^2 - D_{JK}J(J+1)K^2 - D_K K^4] \tag{3.31}$$

where D_J, D_K, and D_{JK} are centrifugal stretching constants, and:

$$A = h/8\pi^2 cI_A \quad \text{and} \quad B = h/8\pi^2 cI_B$$

If Equation (3.31) is substituted into the equation $E' - E'' = h\nu$ for the transitions $J' \leftarrow J''$ and $K' \leftarrow K''$, and the selection rules $J' - J'' = 1$ and $K' - K'' = 0$ are applied, the following equation for the absorption frequencies is obtained:

$$\nu = c\,[2B(J+1) - 4D_J(J+1)^3 - 2D_{JK}(J+1)K^2] \tag{3.32}$$

where $J = J''$. Equation (3.32) should be contrasted with Equation (3.30) as the former contains a K^2 term. Hence, on the basis of Equation (3.32) for a particular transition between two rotational levels, $(J+1) \leftarrow J$, the number of rotational lines observed, except for the $(J-1) \leftarrow 0$ transition, would be greater than 1. This follows since K takes the values:

$$0, \pm 1, \pm 2, \ldots, \pm J$$

Thus, when J is equal to 8, K would have the values:

$$0, \pm 1, \ldots, \pm 8$$

For the $J = 9 \leftarrow 8$ transition when these K values are substituted into the formula:

$$\nu = c[2B(8+1) - 4D_J(8+1)^3 - 2D_{JK}(8+1)K^2] \tag{3.33}$$

nine lines would be expected. From formula (3.30), however, only one line would be predicted, at the frequency:

$$\nu = c[2B(8+1)] \tag{3.34}$$

In practice Equation (3.32) is obeyed and the formula (3.30) is inadequate. Thus, in general for a $(J+1) \leftarrow J$ transition in a symmetric-top molecule, the rotational line is split into $(J+1)$ components. These lines lie very close together but can usually be resolved. As an example of this type of resolution the work by Anderson, Trambarulo, Sheridan, and Gordy [3.13] on the

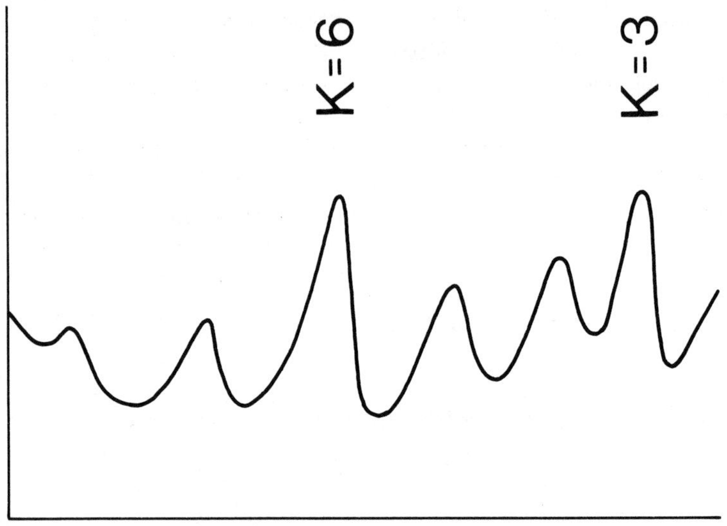

Fig. 3.8 Part of a $J = 9 \leftarrow 8$ transition in a symmetric-top molecule CF_3CCH (After Ref. [3.13]).

symmetric-top molecule CF_3CCH may be considered. They studied the $J = 9 \leftarrow 8$ transition in this molecule. Part of the $J = 9 \leftarrow 8$ transition is given in Fig. 3.8 where the lines from $K = 3$ to 8 may be observed.

Another example of the study of a symmetric-top molecule was one on methyl mercuric chloride (CH_3HgCl). From work on the $J = 17 \leftarrow 16$ transition the centrifugal stretching constants and I_B were evaluated. In addition, the fact that this spectrum could be analysed by the symmetric-top formulae showed unequivocally that the C—Hg—Cl grouping was strictly linear.

In the same way as for linear molecules, the formula relating the value of B (or A) to the vibrational mode is

$$B_v = B_e - \Sigma_i \alpha_i (v_i + d_i/2) \tag{3.35}$$

although it is rarely possible to get sufficient data on higher vibrational states to evaluate B_e. The small variation of the centrifugal stretching constants with the vibrational state is normally neglected.

Many symmetric-top molecules have now been studied including quite a number of halides, for example halides of CH_3, $(CH_3)_3C$, Cs, CH_3Hg, As, and P. The halides have been studied in such detail because they are frequently gaseous or have a sufficiently high vapour pressure at room temperature. This is one of the requirements of the method, although it can be overcome to some extent by the construction of a high-temperature absorption cell.

Three more examples of molecules whose structure has been determined are listed in Table 3.2 together with the values of their internuclear distances and bond angles and their quoted accuracy.

Table 3.2 Values of internuclear distances and bond angles of three symmetric-top molecules

Molecule	*Bond*	*Internuclear distance/(Å)*	*Bond angle value*
CH_3I	C—H	1.100 ± 0.010	
	C—I	2.140 ± 0.005	HCH 110°58′ ± 1°
PCl_3	P—Cl	2.043 ± 0.003	ClPCl 100°6′ ± 20′
ReO_3Cl	Re—Cl	2.230 ± 0.004	OReO 108°20′ ± 1°

3.3.5 Asymmetric-top molecules

This is the type of molecule where $I_A \neq I_B \neq I_C$, and none of these moments is zero. The rotation of a planar asymmetric-top molecule, such as H_2O, about the three principal axes is illustrated in Fig. 3.9. These molecules usually present a complex pure rotational spectrum where the number of observed lines is much greater than for the symmetric-top molecules, and the frequencies for the rotational transitions cannot be generally expressed in terms of simple equations. Even worse is the fact that the equations usually apply only for certain low J values, and centrifugal distortion effects have to be neglected. It is therefore not surprising that this method cannot be applied to complex molecules and that even for small molecules the method of analysis seems most involved.

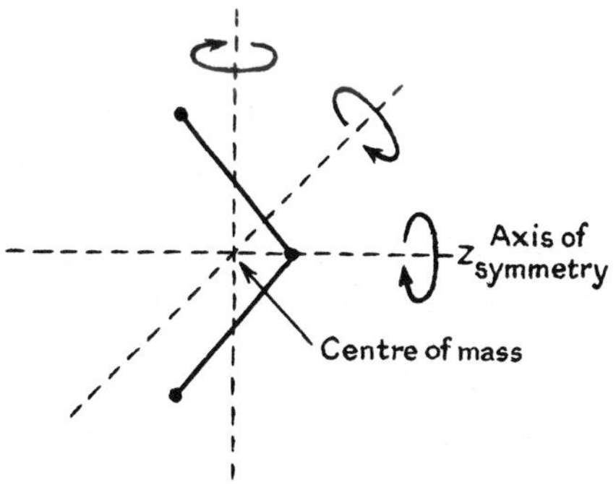

Fig. 3.9 Rotation of an asymmetric-top molecule about the three principal axes.

The asymmetric-top type of molecule presents a much more complex problem than the symmetric-top type for several reasons. Apart from the obvious point that there are now three unknown moments of inertia, where previously there were only two, the following factors have to be considered: (a) the centrifugal distortion is larger; (b) for each $J' \leftarrow J''$ transition each energy level is split into $(2J + 1)$ sub-levels, and the selection rule corresponding to $\Delta K = 0$ does not apply; in fact, the selection rule is $\Delta J = 0, \pm 1$ where all three types of transition may be obtained.

Thus, it may be appreciated that the allocation of energy levels can be very involved and that the identification of the J value for an observed transition, although generally possible, is more difficult than for linear and symmetric-top molecules.

To gain some appreciation of the rotational energy levels of an asymmetric-top molecule we may regard the molecule as lying in between the shapes of a prolate (cigar-shaped) and an oblate (pancake-shaped) symmetric-top. The energy levels (see Fig. 3.10) of the symmetric-top may then be regarded as lying between those of a prolate and an oblate symmetric-top. If $B = h/8\pi^2 cI_B$, $A = h/8\pi^2 cI_A$, and $C = h/8\pi^2 cI_C$, then, since for a prolate symmetric-top $B = C$ and for an oblate one $A = B$, the modification in the energy level pattern on its proceeding from the prolate to the oblate type may be followed in the figure by increasing the value of B. In the diagram it may be noted that: (a) all symmetric-top energy levels in which k is ± 0 are doubly degenerate; this may be interpreted as rotation about the symmetry axis in opposite direction; (b) on changing from either the prolate or oblate symmetric-top to the asymmetric-top type the degeneracy is removed, and a particular level is characterized by $J_{K_{-1}K_1}$ where K_{-1} and K_1 are the K values of the level for the corresponding case in the limiting oblate and prolate symmetric tops.

A variety of treatments of the asymmetric-top molecule has been given, and for a summary of the theoretical methods see a review by Nielsen [3.14]. It is proposed here merely to quote one of the more readily memorizable rotational energy equations due to Wang [3.15] who obtained the following expression for the energy levels:

$$E_r = \frac{h^2}{16\pi^2}\left(\frac{1}{I_B} + \frac{1}{I_C}\right)J(J+1) + \left[\frac{h^2}{8\pi^2 I_A} - \frac{h^2}{16\pi^2}\left(\frac{1}{I_B} + \frac{1}{I_C}\right)\right] W_\tau \quad (3.36)$$

where W_τ is a very involved function of the moments of inertia.

The equations obtained by the various treatments generally apply to low values of J, and usually for a rigid rotator only to $J = 12$, but when centrifugal effects are included only to $J = 6$.

The various J transitions for an asymmetric-top molecule are generally spread over a fairly wide microwave region. However, in order to determine the moments of inertia it is desirable to study the region where the low J transitions occur, so that the equations may be applied within their limits.

The most satisfactory analyses depend on the magnitude of the $J''(J')$ value and how nearly the molecule approaches a symmetric-top type.

For some of the relatively simple molecules, the three moments of inertia have been determined, and the bond angles and internuclear distances then evaluated by means of the isotope substitution technique; for example, ethylene

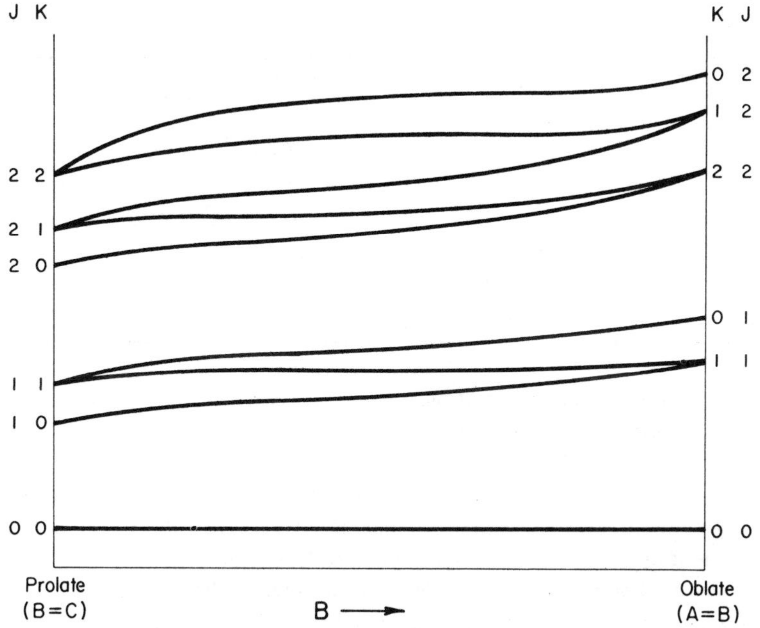

Fig. 3.10 Diagrammatic representation of the energy levels of an asymmetric-top molecule in relation to the limiting forms of prolate and oblate symmetric-top molecules.

oxide had the C—O, C—C, and C—H distances and the angles evaluated with the help of the deuterium isotope. In principle, however, fewer isotopic substitutions are needed for the structural determination of an asymmetric-top than for any other type of molecule, because all its three principal moments of inertia can be determined from its pure rotational spectrum. For a planar asymmetric-top molecule containing n atoms, a minimum of $(n-2)$ isotopic substitutions is required for a complete structural determination and $(n-3)$ for a non-planar one. However, if some of the internuclear distances or angles are obviously equivalent, then the minimum number of required substitutions is reduced by that number. Perhaps rather surprisingly, many simple molecules have been analysed by the microwave approach, and a few examples are HNCO, CH_2Cl_2, H_2O, and CH_3OH. Some recent studies include those on F_2O [3.16], SO_2 [3.17], and SeO_2 [3.18]. Some idea of the quoted accuracy in recent work and its dependence on the actual molecule itself may be gained from Table 3.3.

Table 3.3 Values of internuclear distances and angles of three asymmetric-top molecules.

Molecule	*Bond*	*Internuclear distance/(Å)*	*Bond angle value*
CH_2Cl_2	C—H	1.068 ± 0.005	HCH 112°0′ ± 20′
	C—Cl	1.7724 ± 0.0005	ClCCl 111°47′ ± 1′
	H—N′	1.021 ± 0.01	
HN_3(HN′N″N‴)	N′—N″	1.240 ± 0.003	HN′N″ 112°39′ ± 30′
	N″—N‴	1.134 ± 0.003	
O_3	O—O	1.278 ± 0.003	OOO 116.49′ ± 30′

3.3.6 Accuracy of moments of inertia and internuclear distances

The absolute accuracy of measuring the microwave frequency employed has been estimated to be of the order of 1 part in 10^8, and in principle, at any rate, the moment of inertia could be determined to this degree of accuracy if Planck's constant were known to this same degree of accuracy. The moment of inertia experimentally determined would generally be for the lowest vibrational level ($v = 0$). Formulae of the type (3.11) are sometimes applied to convert this value into the moment of inertia corresponding to the value of the equilibrium internuclear distance (r_e). This is done by the determination of the (B_v) values in the higher vibrational states; this is often not possible because of the very weak absorption lines. For a diatomic molecule:

$$B_v = B_e - \alpha(v + \tfrac{1}{2}) \tag{3.37}$$

B_v is the rotational constant for the vth vibrational level, and α the rotation vibration interaction constant. If B_v is plotted against v then from an extrapolation to $v = -\frac{1}{2}$, B_e is obtained. From the B_e value the value of r_e is readily calculated. Some equilibrium values of the internuclear distances have been obtained for linear and also for symmetric-top molecules.

In general, the determination of r_e values is difficult. In equations of the type of Equation (3.11) it is necessary to obtain first the value of the rotation–vibration constants (α) for all the normal vibrations. Thus, as a minimum it is necessary to obtain the B values in the first excited vibrational state. However, if one of these normal vibrations is a high-frequency type, as is often the case, then the rotational spectrum in the excited vibrational states is usually too weak to be detected. So far r_e values have been obtained only for a small number of asymmetric-top molecules including F_2O, SO_2, and SeO_2.

The different isotopes of the molecule produce different internuclear distances for a particular vibrational level. It is sometimes assumed that the internuclear distance does not alter by more than about 0.005 Å by isotopic substitution. In addition, errors of 0.005 Å may be introduced by uncertainty of the exact isotopic masses. Many internuclear distances are quoted as accurate to better than ± 0.001 Å and some bond angles to better than ± 1′. The accuracy, of course, depends on the type of molecule being investigated and the particular isotopic substitution.

To gain some appreciation of the increase in accuracy by the microwave approach it is interesting to consider a molecule studied by a particular worker (plus co-workers) on two different occasions. In both 1964 and 1969 Saito [3.19] examined SO_2. The values obtained on the two occasions and their quoted accuracies are:

	1964 work	1969 work
r_e	1.4308 ± 0.002 Å	1.4307_6 ± 0.00013 Å
θ	119°19′ ± 2′	119°19′ ± 0.7′

The structural parameters from the 1964 and 1969 work are almost identical. The accuracy quoted in the most recent work is better than that obtained by any other available method for the structural determination of SO_2.

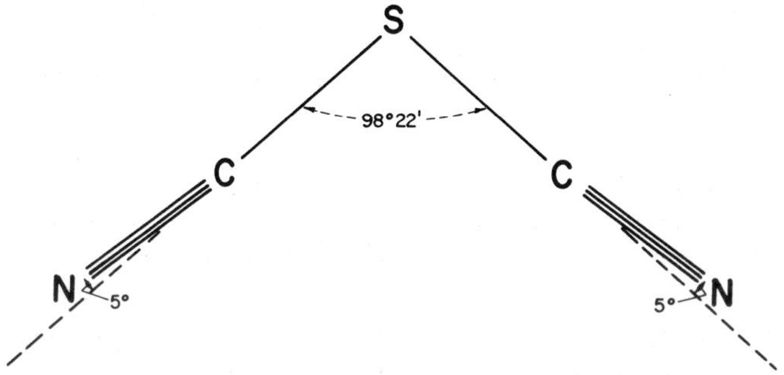

Fig. 3.11 Geometry of $S(CN)_2$.

3.3.7 Some trends in the determination of molecular geometry

Microwave spectroscopy has made invaluable contributions in the study of the molecular geometry of small molecules, and in some cases has revealed subtleties which could hardly have been anticipated. Some of the more recent contributions have been:

(a) A study of how the geometry varies in the five-membered ring systems cyclopentadiene, furan, thiophen, and pyrrole [3.20].

(b) Examination of small distortions in molecular structure which have not been revealed by other methods, as in, for example, $S(CN)_2$ where the CN groups are tilted from the line of the S—C bond by 5° (see Fig. 3.11).

(c) On replacement of an H atom by another atom in the benzene ring the CCC angles become no longer equal; this is illustrated in Fig. 3.12 for fluorobenzene.

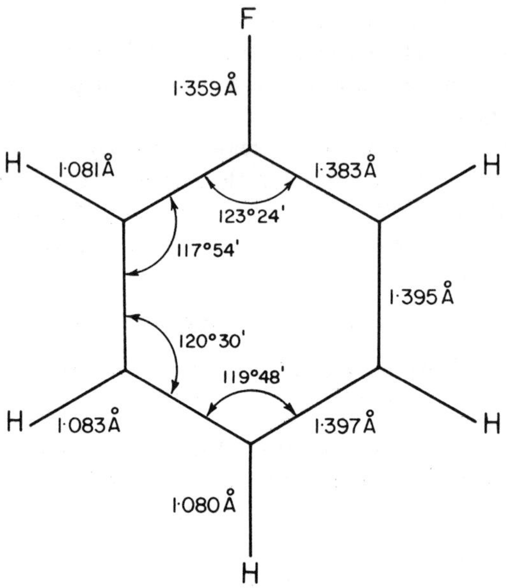

Fig. 3.12 The geometry of fluorobenzene as determined from its microwave spectrum [3.22].

(d) Examination of the non-planarity of small-ring compounds. One type of study has been carried out on rings (e.g. saturated four-and five-membered ring systems) which have a puckered equilibrium configuration. The case of trimethylene oxide is considered in Chapter 5, and in Fig. 5.8 it may be noted that the potential function is symmetric about the planar configuration and has a barrier between the two minima. In this case the barrier is very low. However, in other cases, if the barrier is sufficiently high for such a molecule, then it behaves as a permanently puckered molecule. From its spectrum the rotational constants can be determined and the structure may follow.

3.4 HYPERFINE STRUCTURE AND QUADRUPOLE MOMENT

For the majority of molecules in the ground state the magnetic field due to the motion of the electrons almost completely cancels out because the electrons are paired; this results in hardly any magnetic field effect due to them at the nucleus. However, the internal electric field in the molecule, due mainly to the bonding electrons, is often quite appreciable, and if the electric charge distribution of the nucleus is non-spherical in shape the nucleus orients itself in certain possible quantized directions with respect to the direction of this field. Its particular orientation governs the hyperfine structure of the microwave spectrum. Fundamental information is obtained from a study of such hyperfine structure. Before considering this further, though, it is necessary to attach some meaning to the term *quadrupole moment.*

The distribution of positive charge in an atomic nucleus is only spherical in shape when the nuclear spin quantum number is 0 or $\frac{1}{2}$ as in Fig. 3.13(a) and (b); generally, the shape is assumed to be represented by a prolate or an oblate spheroid [Fig. 3.13(c) and (d), respectively].

An excess of positive charge at the north and south poles of the spinning nucleus is indicated in these figures by means of plus signs, whereas a deficiency of positive charge is indicated by the minus sign.

The quantity which measures the deviation in the charge distribution from spherical symmetry is known as the *electric quadrupole moment of the nucleus*, denoted by eQ, and is defined by:

$$eQ = \int \rho r^2 (3\cos^2\alpha - 1)\,\mathrm{d}\tau \qquad (3.38)$$

where e is the charge on the proton, r is the distance from the centre of gravity of the positive charge to the element of volume $\mathrm{d}\tau$, α is the angle between r and the spin axis, and ρ is the nuclear charge density. When Q has a positive value the positive charge in the nucleus is elongated along the spin axis as in Fig. 3.13(c) while the negative value for Q indicates that it is flattened along the spin axis as in Fig. 3.13(d). Q is in units of cm^2. The electric field associated with the spinning nuclear charge extends outside the nucleus, and the interaction of the quadrupole with the electric fields inside molecules and crystals leads to information on the surrounding charge distribution. A typical example of interaction between the nuclear field and the surrounding electric field occurs when the electrons around a particular quadrupole have a non-spherical charge distribution. This is generally the case for bonding electrons, for example, the valence shell formed from an sp-hybrid. However, if the electric potential (V) at the nucleus due to the electrons and the other nuclei in the molecule were symmetrical, a quadrupole effect would not be observed in the spectrum. It is only asymmetrical electric fields which produce these effects. The magnitude of the asymmetrical field is indicated by the electric field gradient $\partial^2V/\partial z^2$ along the axis of symmetry of the molecule (z-axis). The z-axis is generally defined as the axis along which the total angular momentum has its maximum projection (i.e. when $M_J = J$). In equations involving the quadrupole interaction energy an average value of the second derivative of the potential at the nucleus is employed and is written $\langle(\partial^2V/\partial Z^2)\rangle_{\mathrm{av}}$. This quantity may be related to the distances of the quadrupole from the surrounding electric charges by means of the equation:

$$\left\langle\left(\frac{\partial^2 V}{\partial Z^2}\right)\right\rangle_{\mathrm{av}} = \left\langle \sum_k e_k \left(\frac{3\cos^2\theta_k - 1}{r_k{}^3}\right)\right\rangle_{\mathrm{av}} \qquad (3.39)$$

e_k is the unit of electronic charge, r_k the radius vector connecting the nucleus and the kth extranuclear charge; and θ_k the angle between r_k and the chosen

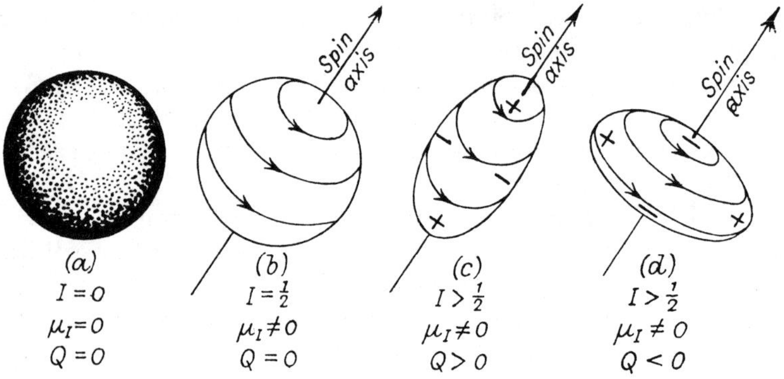

Fig. 3.13 Charge distribution for a nucleus (a) which does not spin (i.e. $I = 0$), (b) which has $I = \frac{1}{2}$, and (c) and (d) which has $I > \frac{1}{2}$.

Z-axis,† while Σ indicates the summation, taking into account all contributing extranuclear charges. Because there is an inverse cube dependence on r, only the electronic and nuclear charge within the order of an Ångström unit need be considered in the summation. In Fig. 3.14 one of the quantized orientations of the quadrupole in the inhomogeneous electric field of the molecule is illustrated, and the spin axis is the broken line AB.

The difference orientations of the nucleus result in different interaction quadrupole energy levels, the values of which are dependent on the nuclear spin and the molecular rotational quantum numbers. These different energy states are responsible for the hyperfine structure of the rotational line.††

Take now a particular case where the electric quadrupole moment has to be considered to account for the spectrum. The case of a linear triatomic molecule will be examined, where two of the nuclei have quadrupole moments. In the microwave region from 23 860 to 23 910 MHz for the $J_{2\leftarrow 1}$ transition of $^{35}ClC^{14}N$ several lines are observed which obviously cannot be explained without considering the nucleus. These lines may be accounted for by the fact that the orientations of the nuclear spin axes are quantized relative to the axis of rotation of the whole molecule. This results in rotational energy sub-levels, and thus the hyperfine structure is due to transitions from the sub-levels for $J = 1$ to the sub-levels where $J = 2$.

In the particular case of $^{35}ClC^{14}N$ the major splitting is attributed to the ^{35}Cl

† This expression was derived for a Z-axis fixed in space, but in microwave work it is customary to refer to the z-axis in the molecule, which for a linear molecule is the bond axis.

†† In this chapter we shall consider this type of hyperfine structure of rotational lines. This is to be contrasted with the chapter on nuclear quadrupole resonance (Vol. 1) which is termed pure quadrupole resonance and involves transitions between the M_I levels (i.e. reorientation of the nuclear quadrupole) themselves without a change in the rotational energy as well.

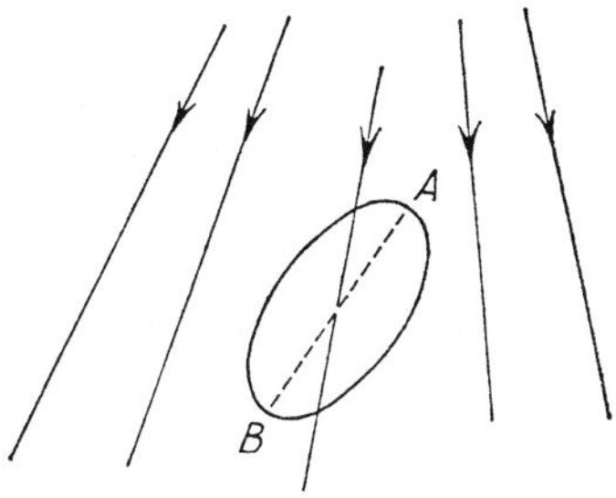

Fig. 3.14 Orientation of the electric quadrupole in an inhomogeneous electric field.

nucleus and the finer structure to the weaker ^{14}N interaction, the nuclear spin quantum numbers of ^{14}N, ^{35}Cl, and ^{12}C being 1, 3/2, and 0, respectively. Since that of ^{12}C is zero its nucleus is spherical, and it has no quadrupole moment. Thus, the carbon nucleus does not contribute to any hyperfine structure. In a simpler case than $^{35}ClC^{14}N$, where the triatomic molecule is still linear, but where only one nucleus has spin, the type of splitting may be accounted for in terms of the coupling of the nuclear spin with the rotation of the molecule as a whole. Thus, if J is the rotational quantum number governing the total angular momentum, then this forms a resultant (F) with the nuclear spin quantum number (I) where F is the quantum number for the total angular momentum of the whole molecule and includes the nuclear spin. F may take the values:

$$F = (J+I), (J+I-1), \ldots, |(J-I)| \tag{3.40}$$

while in the presence of a magnetic field a corresponding magnetic quantum number M_F is given by:

$$M_F = F, (F-1), \ldots, -F \tag{3.41}$$

In addition, since J and I are quantum numbers, then the angle between J and I is quantized, and hence so must be their energy of interaction. To appreciate the nuclear hyperfine splitting of the rotational lines an equation giving the quantized values of the interaction energy will be considered. For a molecule in which only one nucleus has a quadrupole moment, when quadrupole coupling occurs; the interaction energies are given by the equation:

$$E_Q = +eQ\left\langle\frac{\partial^2 V}{\partial Z^2}\right\rangle_{av} \frac{(3/8)G(G+1)-(1/2)I(I+1)J(J+1)}{JI(2I-1)(2J-1)} \tag{3.42}$$

where $G = F(F+1) - I(I+1) - J(J+1)$ and E_Q is the interaction energy of the quadrupole with the surrounding internal electric field. For a linear molecule with one nucleus taking part in quadrupole coupling the $(\partial^2 V/\partial Z^2)_{av}$ term has been evaluated as:

$$\left\langle\frac{\partial^2 V}{\partial Z^2}\right\rangle_{av} = -\frac{2J}{2J+3}\left(\frac{\partial^2 V}{\partial z^2}\right) \tag{3.43}$$

where the bond axis is the z-axis. It may be observed from formulae (3.42) and (3.43) that the quadrupole coupling energy of a single nucleus is directly proportional to eQq where $q = \partial^2 V/\partial z^2$, and q measures the divergence of the electric field from spherical symmetry, where the contributing factors are the extranuclear electrons and adjacent nuclei in the particular molecule. The product eQq is called the *nuclear quadrupole coupling constant*, and from microwave spectra alone only this coupling constant can be evaluated. Unless the value of q is evaluated from other sources, the quadrupole moment cannot be determined by this method.

To calculate the frequencies of the quadrupole hyperfine structure for a given rotational transition, the appropriate interaction energies given by formula (3.42) have to be added to the energies of the unperturbed rotator, and we obtain:

$$E = E_r + E_Q \tag{3.44}$$

where in the case of a rigid linear rotator the value of E_r would be given by Equation (3.8). If now the energy difference is taken between the two energy levels between which a transition may take place, we obtain an equation connecting the wavenumbers of the lines with the change in rotational and quadrupole interaction energies, that is:

$$\Delta E = (E_r' - E_r'') + (E_Q' - E_Q'') = h\nu \tag{3.45}$$

For a rigid linear molecule it follows from the formula (3.17) that Equation (3.30) would become:

$$h\nu = 2hB(J'' + 1) + (E_Q' - E_Q'') \tag{3.46}$$

The $(E_Q' - E_Q'')$ term can be evaluated from Equation (3.42) if it is taken into account that I does not change and that for an absorption transition:

$$J' - J'' = +1 \quad \text{and} \quad F' - F'' = 0, \pm 1 \tag{3.47}$$

To explain the number of lines obtained in the nuclear hyperfine structure of a linear molecule, where only one nuclear spin is involved, it is necessary (a) to know the $J' \leftarrow J''$ rotational transition involved; (b) to know the value of the nuclear spin quantum number of the nucleus which couples with the rotational quantum numbers; (c) to evaluate the F values from:

$$F = (J+I), (J+I-1), \ldots, |(J-I)| \tag{3.48}$$

and (d) to apply the selection rule $\Delta F = 0, \pm 1$.

The quoted example of $^{35}ClC^{14}N$ for the $J = 2 \leftarrow 1$ transition is a case where both the Cl and N nuclei possess electric quadrupole moments and can interact with the molecular rotational energy to produce small splitting of the energy levels. The spin quantum number of ^{35}Cl is 3/2; therefore, even if the nitrogen did not have a nuclear spin, and the coupling took place through the chlorine quadrupole alone, the possible levels would be:

(1) In the state where $J = 1$ and $I = 3/2$: $F = 5/2, 3/2, 1/2$

(2) In the state where $J = 2$ and $I = 3/2$: $F = 7/2, 5/2, 3/2, 1/2$
and the permitted theoretical transitions would be when

$$\Delta F = 0: 1/2 \leftarrow 1/2,\ 3/2 \leftarrow 3/2,\ 5/2 \leftarrow 5/2$$
$$\Delta F = +1: 3/2 \leftarrow 1/2,\ 5/2 \leftarrow 3/2,\ 7/2 \leftarrow 5/2$$
$$\Delta F = -1: 1/2 \leftarrow 3/2,\ 3/2 \leftarrow 5/2$$

Thus, solely on account of the ^{35}Cl nucleus, eight lines might be expected.

The microwave spectrum of the $^{35}ClC^{14}N$ molecule at low resolution is given in Fig. 3.15(a). If initially the nitrogen quadrupole contribution were ignored, then the theoretical pattern due to the chlorine quadrupole alone might be interpreted in terms of the above given theoretical transitions. The positions of eight transitions are indicated in Fig. 3.15(a). Figure 3.15(b) is the same spectrum at higher resolution where the three peaks in Fig. 3.15(a) have split into several smaller peaks, and these may all be explained by taking into account the nitrogen quadrupole as well.

In the case of a linear molecule with two coupling nuclei, if the nuclear spin quantum numbers are I_1 and I_2, these may be regarded as coupling with J according to the formulae:

$$F_1 = (J + I_1), (J + I_1 - 1), \ldots, |(J - I_1)| \qquad (3.49)$$

$$F_2 = (J + I_2), (J + I_2 - 1), \ldots, |(J - I_2)| \qquad (3.50)$$

If F is the total angular momentum quantum number for the whole molecule, then:

$$F = (J + I_1 + I_2), (J + I_1 + I_2 - 1), \ldots, |(J - I_1 - I_2)| \qquad (3.51)$$

It follows from Equations (3.49) and (3.50) that:

$$F = (F_1 + I_2), (F_1 + I_2 - 1), \ldots, |(F_1 - I_2)| \qquad (3.52)$$

The nuclear hyperfine structure for the $(J + 1) \leftarrow J$ transition results from changes in F. These changes are governed by the selection rule:

$$\Delta F = 0, \pm 1 \qquad (3.53)$$

For the molecule $^{35}ClC^{14}N$ where $I_1 = I_{Cl} = 3/2$, in the $J = 2$ level the F_1 values are given by substitution in Equation (3.49). This gives:

$$F_1 = 7/2, 5/2, 3/2, 1/2 \qquad (3.54)$$

while for the $J = 1$ level:

$$F_1 = 5/2, 3/2, 1/2 \qquad (3.55)$$

These $_{J=2}F_1$ values and $_{J=1}F_1$ values are the ones which have to be substituted into formula (3.52) to give a corresponding set of F values for each $_{J=2}F_1$ and $_{J=1}F_1$ value. For example, when $_{J=2}F_1 = 7/2$, $_{J=1}F_1 = 5/2$, and $I_2 = I_{\text{nitrogen}} = 1$, then the corresponding F values are:

(i) for $J = 2$, $F' = 9/2, 7/2, 5/2$

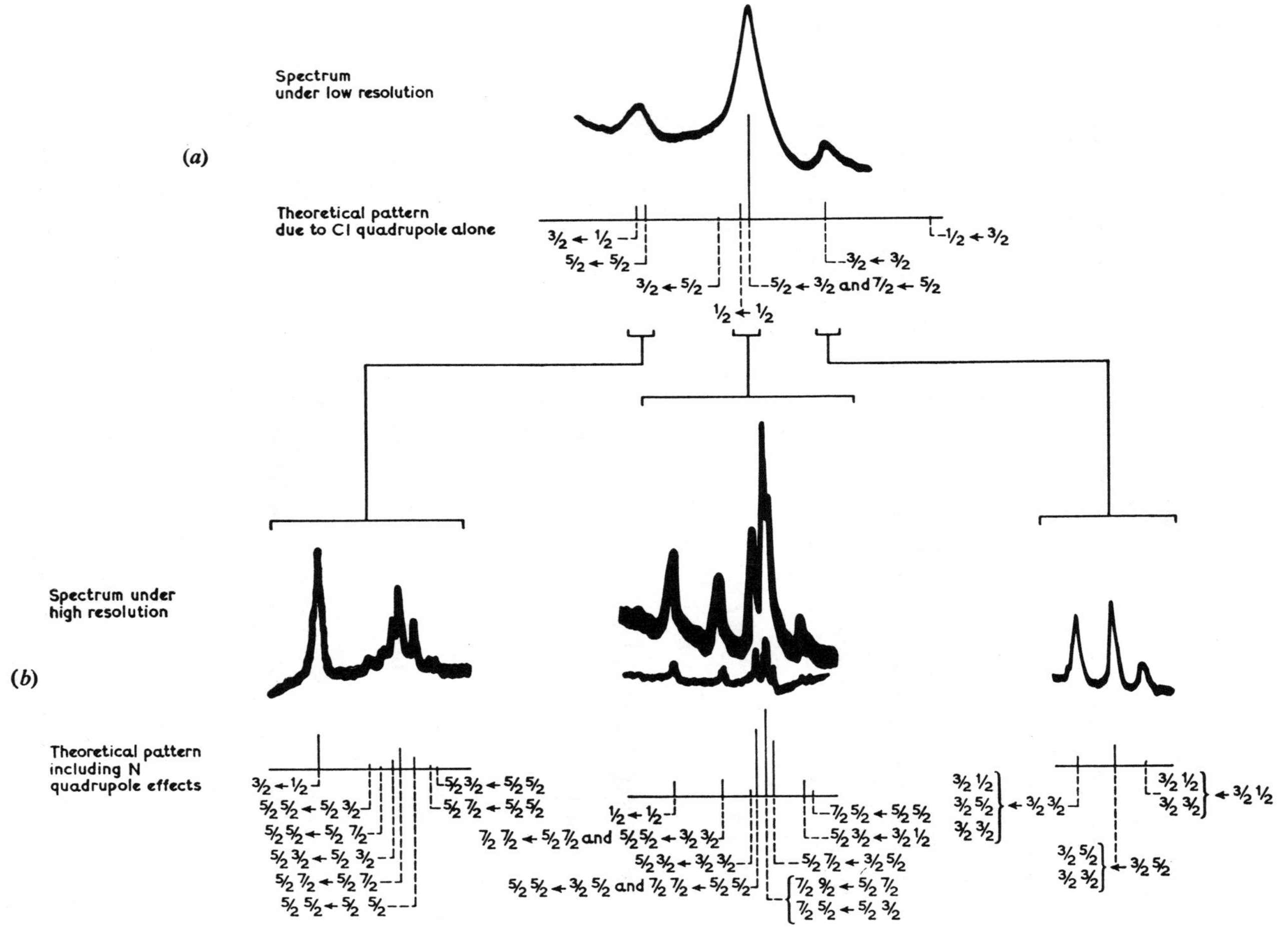

Fig. 3.15 Spectrum of $^{35}ClC^{14}N$ $J = 2 \leftarrow 1$ transition at (a) low and (b) high dispersion. A comparison is made with the theoretical pattern. (After Townes, Holden, and Merritt [3.23].

(ii) for $J = 1, F'' = 7/2, 5/2, 3/2$
and $F' \leftarrow F''$ transitions are permitted as follows:

$$9/2 \leftarrow 7/2,\ 7/2 \leftarrow 5/2,\ 5/2 \leftarrow 3/2 \quad \text{where} \quad \Delta F = +1$$
$$7/2 \leftarrow 7/2,\ 5/2 \leftarrow 5/2 \quad \text{where} \quad \Delta F = 0$$
$$5/2 \leftarrow 7/2 \quad \text{where} \quad \Delta F = -1$$

All these transitions are for the case where ${}_{J=2}F_1 = 7/2$ and ${}_{J=1}F_1 = 5/2$. Thus, each transition is characterized by the values:

$$(7/2, F') \leftarrow (5/2, F'')$$

In general, the transitions are characterized by:

$$({}_{J=2}F_1, F') \leftarrow ({}_{J=1}F_1, F'')$$

where each F change is governed by the selection rule $\Delta F = 0, \pm 1$.

The positions of these lines and their corresponding

$$({}_{J=2}F_1, F') \leftarrow ({}_{J=1}F_1, F'')$$

values are indicated in Fig. 3.15(b) which is the higher resolution spectrum for $^{35}ClC^{14}N$ for the $J = 2 \leftarrow 1$ transition. It is interesting to note that the agreement between the measured and the theoretical spectrum is so extensive and detailed that this is considered to be excellent confirmation that the nuclear spin quantum number of ^{35}Cl is 3/2.

3.4.1 Quadrupole coupling constants

For a linear molecule with only one nucleus with a quadrupole moment, if the nuclear spin quantum number and the moment of inertia are known, and if the F and J values for the lines in the hyperfine structure of a rotational transition can be determined, then it follows from Equations (3.46), (3.42), and (3.43) that the quadrupole coupling constant can be evaluated. Quite a number of linear molecules and symmetric-top molecules have had their quadrupole coupling constants evaluated by such means, and the values are used to obtain information on the electronic structure of the molecules. Examples of the application of the quadrupole coupling constants are listed below.

3.4.2 To obtain information on the valence state of atoms in molecules

For example, in a triply bonded nitrogen atom, such as that in NH_3, the ^{14}N quadrupole coupling constant is much greater than that for a quadrivalent nitrogen atom of the type in CH_3NC. The low coupling constant in the quadrivalent nitrogen may be accounted for by the fact that the valence shell electrons surrounding the nitrogen atom are almost spherically distributed, and hence the electric field gradient is much smaller in this case.

A wide variety of cases has now been considered where quadrupole coupling constants obtained by microwave spectroscopy have been applied to elucidate problems of molecular structure. For example, when the coupling constants for $CF_2{=}CHCl$ are considered, then the relative weights of the various possible resonance structures turn out to be:

$$\begin{array}{ccc} CF_2{=}CHCl & {}^{-}CF_2{-}CH{=}Cl^{+} & CF_2{=}CH^{+}Cl^{-} \\ 85\% & 9\% & 6\% \end{array}$$

It is interesting to note that the proportions of ionic structures in vinyl derivatives are much less than in the corresponding methyl derivatives.

Another interesting case is the quadrupole splitting of N_2O which is consistent with almost equal proportions of the following hybrids:

$$N{\equiv}N^{+}{-}O^{-} \quad \text{and} \quad N^{-}{=}N^{+}{=}O$$

3.4.3 To relate certain ranges of quadrupole coupling values to definite structural units.

In the case of HCN, CH_3CN, and HC≡CCN the coupling constants for ^{14}N range from − 4.58 to − 4.20 MHz and are considered to be of the same order. Thus, this order of magnitude characterizes a particular type of carbon–nitrogen linkage. In the cases of CH_3NC and HNCS the quadrupole coupling constants are + 0.5 and + 1.20 MHz respectively; these values are taken to indicate different types of carbon–nitrogen linkage.

3.4.4 To yield information on the hydridization of a certain atom in a molecule.

Quadrupole coupling data have shown that there is no doubt that certain sulphur bonds involve hybridized sulphur atoms, and also the presence of s-character in the bonding orbitals of compounds of the Group V elements.

For the treatment of nuclear quadrupole coupling and chemical bonding a paper by Orville-Thomas [3.24] and a book by Sugden and Kenney [3.5] should be consulted.

3.4.5 To test electronic wave-functions in the neighbourhood of the nucleus

Since quadrupole coupling constants are sensitive to the electronic distribution near the nucleus, it is considered feasible now to be able to test wave-functions in the neighbourhood of the nucleus. In this region there was previously no experimental means of testing the wave-functions.

3.4.6 Some more recent studies

Stark [3.25] (in 1967) tabulated the molecular constants from microwave spectroscopy, and these include the electric quadrupole coupling constants for molecules with one or more nuclei having quadrupole moments from microwave spectroscopy.

Flygare and Gwinn [3.26] showed for CH_2Cl_2, where the two Cl atoms are identical quadrupolar nuclei, that the principal axis of the quadrupole tensor at the chlorine nucleus coincides with the C—Cl internuclear axis.

Wolf, Williams, and Weatherly [3.27] have developed a theory which accounts for the hyperfine structure of the rotational spectrum of symmetric-top molecules, such as $CH^{35}Cl_3$ and $CF^{35}Cl_3$ which contain three identical quadrupolar nuclei.

Flygare and Weiss [3.28] have obtained several deuterium nuclear quadrupole coupling constants by employing a high-resolution technique, and have considered a linear relationship between the force field of the molecule and the deuterium field gradient.

3.4.7 Nuclear quadrupole moments

A number of quadrupole moments has been determined from microwave spectra by using a $\partial^2 V/\partial z^2$ value obtained from other sources. This estimation of $\partial^2 V/\partial z^2$ usually involves wave-mechanical considerations of the degree and kind of bond orbital hybridization and the type of resonance between the different molecular structures.

However, it is possible to determine the value of Q by the atomic beam technique; hence, if the quadrupole coupling constant is determined by the microwave approach, then $\partial^2 V/\partial z^2$ is obtained. The value of $\partial^2 V/\partial z^2$ is of fundamental importance since it depends on the environment of the quadrupole (nucleus) and therefore reflects the nature of the surrounding bonding electrons. Since s-electrons and completed inner shells have spherical symmetry, and d- and f-electrons do not in general move near the nucleus, then it follows that $\partial^2 V/\partial z^2$ is mainly influenced by the bonding p-electrons.

3.5 DETERMINATION OF DIPOLE MOMENTS BY THE STARK EFFECT

In atomic spectra the Zeeman effect is used much more than the Stark effect to gain fundamental information, yet in molecular spectra in the microwave region the reverse is true. The reason for this is that molecules in their ground state have no unpaired electron(s) (notable exceptions are O_2, ClO_2, and NO) and no resultant electron magnetic moment. Even very strong magnetic fields cause very small splitting of the spectral lines, whereas even weak electric fields may completely resolve the individual line components. This is to be contrasted with atomic spectra where very high electric fields are required.

We shall now see that a much weaker electric field applied to polar gas molecules causes splitting of the rotational energy levels, and in the subsequent absorption fine structure in the rotational spectrum is observed and yields what is termed the Stark effect in a molecular spectrum.

Two types of Stark effect are employed.

3.5.1 Second-order Stark effect

This is the case where the splitting of the rotational levels by an electric field of intensity E is proportional to E^2. Such behaviour is exhibited by linear molecules where the dipole moment is perpendicular to the total angular momentum of molecular rotation (see Fig. 3.16).

3.5.2 First-order Stark effect

This is the case when the splitting of the rotational levels is directly proportional to E, and applies to molecules which have a component of the molecular dipole moment along the direction of the total angular momentum of rotation. An example would be a symmetric-top molecule where μ has a component along $\sqrt{[J(J+1)]}\, h/2\pi$ except for the case where $K = 0$.

Application of these effects permits the evaluation of the most accurate electric dipole moments for molecules in the gaseous state. This is achieved by means of equations which relate the Stark shift of a rotational line ($\Delta\nu$) to μ and E provided that the rotational states and their appropriate quantum numbers and moments of inertia are also (or have previously been) determined. Experimentally $\Delta\nu$ and E have to be measured as accurately as possible. $\Delta\nu$ can be measured to within 0.01 Hz while the shift could be as large as 10 MHz. In order to determine $\Delta\nu$ accurately it is desirable that the lines be sharp, and therefore the electric field has to be uniform. The magnitude of an electric field which is uniform can be deduced by calibration with carbonoxy sulphide the dipole moment of which in the ground vibrational state is known most accurately to be 0.71499 ± 0.00005 D. The $J = 1$ to $J = 2$ transition is most intense,

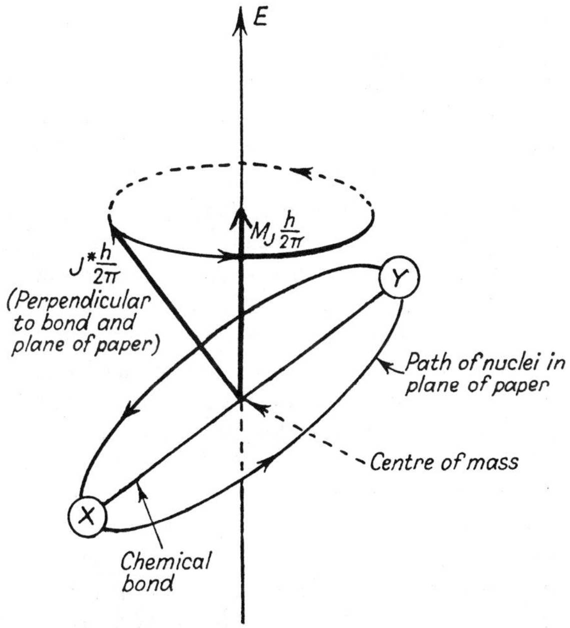

Fig. 3.16 Vector diagram for rotation of the diatomic molecule X—Y about an applied field.

and being a linear molecule it exhibits a second-order Stark effect. Thus, $\Delta\nu$ is determined experimentally for carbonoxy sulphide, while its μ is known, and E is calculated from the appropriate equation.

For a gaseous, linear molecule whose rotational quantum number is J, in the presence of a suitable magnetic field the corresponding magnetic quantum number (M_J) would have the following $(2J + 1)$ values:

$$M_J = J, (J - 1), \ldots, -J$$

In the absence of an external electric or magnetic field this J level is $(2J + 1)$-fold degenerate; that is, there are $(2J + 1)$ equivalent states identical in energy. When an electric field is applied part of this degeneracy is removed, and for linear molecules $(J + 1)$ of these levels now have slightly different energies corresponding to the $(2J + 1)$ orientations that the total angular momentum takes up with respect to the direction of the electric field. The transition $(J + 1) \leftarrow J$, which in the absence of an electric field was one line, now involves transitions from a set of energy levels related to J to a corresponding set of levels in $(J + 1)$ and splitting of the line into a number of components occurs. For a given value of J the corresponding set of energy levels in an electric field is characterized by a quantum number M_J which takes the values:

$$0, 1, 2, \ldots, J$$

Thus, there are $(J + 1)$ values for each value of J, and except for $M_J = 0$, the M_J levels will be doubly degenerate.

In order to gain a clear picture of the factors involved, consider now the vector diagram for the rotation of an asymmetric diatomic molecule XY about the axis of a given electric field.§ The angular momentum of the rotating molecule is $\sqrt{[J(J+1)]}\, h/2\pi$, while the value of its component in the direction of the applied field is $M_J H/2\pi$, as illustrated in Fig. 3.16. The vector representing angular momentum is perpendicular to the bond at the centre of mass, i.e. the point about which the molecule rotates, and the vector processes about the direction of the applied field. The dipole moment acts along the direction of the bond.

Which transitions are permitted depends on the relative orientations of the applied electric field and the high-frequency electric field of the microwave radiation. If these electric fields are parallel the selection rule is $\Delta M_J = 0$, but when they are perpendicular it becomes $\Delta M_J = \pm 1$. The electric field is applied by fixing a thin strip of metal down the centre of the rectangular wave-guide forming the cell, the strip of metal being insulated from the wave-guide at the edges; connection is made to it by a wire through a hermetic seal in the side wall.

If a steady direct potential is applied between the strip and the wave-guide, the electric field produced is parallel to the direction of the polarized electric vector of the microwave radiation. In this case the selection rule for M_J would therefore be $\Delta M_J = 0$.

We shall now consider the Stark modulation spectrometer itself which makes use of these principles, and then follow this by application of the $\Delta \nu$ and E data to the determination of μ of a linear molecule.

3.5.3 The Stark modulation spectrometer

As is the case in some other types of spectrophotometers, some form of 'chopping' and synchronous or 'lock-in' detection is required. This is achieved in a microwave spectrometer by Stark modulation of the molecular spectrum which results in modulation of the absorption as the monochromatic radiation frequency is varied across the spectral range. Square-wave type Stark modulation is employed with electrostatic field strengths up to 2–300 V mm^{-1}. This type of spectrometer, known as the Stark modulation spectrometer, is widely used for the study of rotational absorption lines. These instruments are now available commercially, one such being the Camspek microwave spectrometer produced by Cambridge Scientific Instruments Ltd. This instrument works in

§ A few hundred to a few thousand volts per centimetre field strength are employed.

the 26 500–40 000 MHz range and the power is supplied by means of a backward-wave oscillator.

The Stark modulation spectrometer may be used for both structural work and chemical analysis. A diagrammatic representation of this type of spectrometer is given in Fig. 3.17. The cell consists of a 3–5 metre length of wave-guide with Teflon windows at the end with a Stark electrode fitted along its length and parallel to and equidistant from the wide faces of the wave-guide. A zero-based square-wave voltage is applied to the Stark electrode with a frequency of ~ 100 kHz. In the first half-cycle the applied field may be zero, whereas in the second half the field may be several hundred volts/cm. In this latter case the Stark effect takes place, and the line(s) is displaced and possibly split. Thus, in practice, when a range of frequency in an absorption frequency area is swept through by the klystron (to which a saw-tooth sweep of a few Hz is applied) then both the undisplaced line and its Stark components may be presented on the oscilloscope.

The output from the crystal detector goes to a tuned preamplifier and then to a tuned amplifier which may be a radio receiver. The sensitivity can be increased by using phase-sensitive detection. The phase-sensitive detector can be incorporated into the amplifier and take its lock-in signal from the Stark modulator.

A directional coupler taps off a fraction of the microwave power which is then directed to a mixer crystal which also receives power from a crystal oscillator. The frequency of the radiation from the klystron is monitored by a method of successive comparisons with harmonics of crystal-controlled local oscillation.

The power from the klystron may be tuned to the frequency of an absorption line. If a periodic field is then applied to the Stark electrode, the power detected at the crystal varies, since the electric field produces shifts and splittings of the rotational levels and hence changes in the values of the absorbed frequencies. This can be achieved electronically at a high frequency, and the net result is that the microwave power absorbed by the gas is modulated. This modulation is effected by changing the absorption frequency of the molecule (via the Stark effect) and not by varying the frequency of the klystron. This procedure provides a means of reducing the crystal noise and of producing a more sensitive spectrometer. This Stark modulation spectrometer is thus preferable to the video instrument even in the study of the pure rotational spectra of gases. One snag with the Stark modulation spectrometer, however, is that if only low powers are available, then the source modulation technique is preferable.

The highest sensitivity so far achieved with a spectrometer employing the Stark effect is $\alpha = 4 \times 10^{-11}$ cm.

Various types of spectrometers have been developed for work up to 1300° K which considerably widens the application of the method to only slightly volatile substances.

As Stark modulated microwave spectrometers have now become available

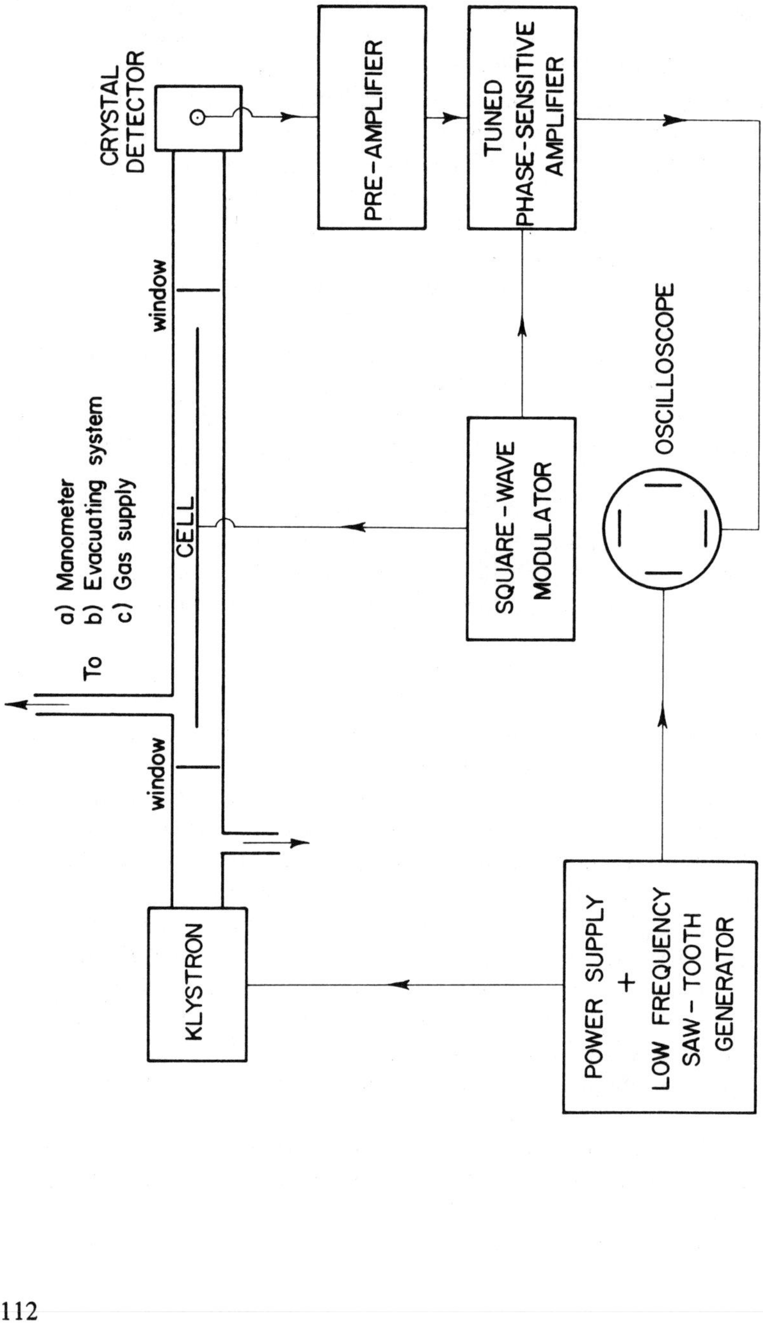

Fig. 3.17 Diagrammatic representation of the Stark modulation spectrometer.

commercially, it seems appropriate to consider the main features of one such apparatus with a view to appreciating its potentiality.

The Camspek microwave spectrometer (see Fig. 3.18) is basically similar to a single-beam spectrophotometer with extremely high resolution. Monochromatic radiation, the frequency of which may be set or scanned within the quoted range, is directed through an absorption cell containing the sample to be examined. Part of the radiation is absorbed by the sample, and it is from this absorption signal that measurements are made.

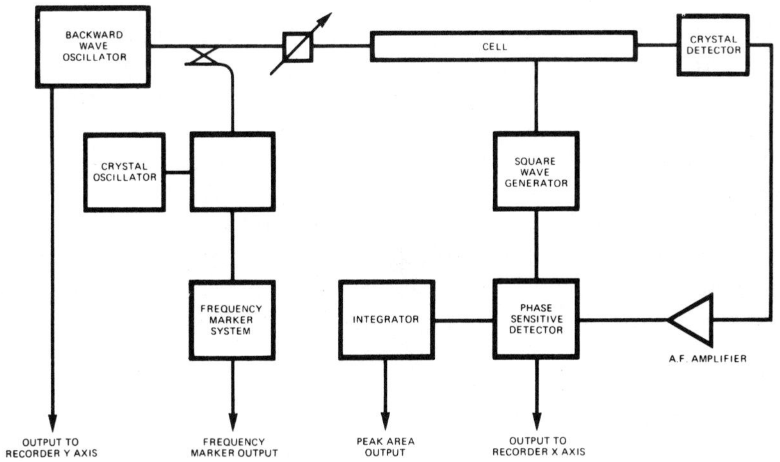

Fig. 3.18 The Camspek microwave spectrometer which is a Stark-modulated spectrometer operating in the *Q*-band (26 500–40 000 MHz). (Courtesy of Cambridge Scientific Instruments Ltd.)

Since the amount of radiation absorbed by the sample is very small (a few parts per million) the absorption signal is modulated by applying a switched electrostatic field to the sample. The modulation frequency is 40 kHz. The modulated signal is now detected by a crystal diode and the detected signal amplified in a narrow-band amplifier. The 40 kHz signal corresponding to the absorption is then separated from noise etc. in a phase-sensitive detector, taking its reference from the modulator. The result is a d.c. output proportional to the absorption of the microwave radiation by the sample. This is displayed against a frequency baseline on an X-Y recorder. The X-deflection corresponds to the scan set on the microwave source and is proportional to frequency. The scan fills the whole width of the recorder so that preprinted charts may be used to facilitate comparisons between spectra. The chart is automatically calibrated in frequency during a scan.

A sampling system introduces the substance into the absorption cell in the form of a gas at a pressure ranging between 1 and 500 millitorr, and an automatic gas sample pressure regulator ensures that reproducible doses of gas samples can be put into the cell. Solid samples can be introduced by means of a small furnace attached to one gas inlet. The simplest mode of operation is a vacuum distillation of the solid into the spectrometer. Samples are pumped out by a vacuum system which can achieve an ultimate vacuum better than 0.1 millitorr. In order to facilitate sample removal from the sampling system and the absorption cell, warm air up to 150° can be passed through. In addition, the sample cell may be cooled to solid carbon dioxide temperatures. As a result of cooling, any vibrational satellite lines which complicate the spectrum will probably disappear at low temperatures, as may lines due to higher energy conformations of the molecule.

An integrator is incorporated which measures the area under the absorption line. Over a wide range of pressures and microwave power density, covering conditions in which appreciable pressure broadening and power saturation occur, line areas are simply related to partial pressure, even in mixtures. Thus, one of the strong features of this commercially available instrument is its use in quantitative analysis.

Since a microwave spectrum may contain several thousand absorption lines even in the frequency range of the spectrometer, the spectrometer has been designed for simple interfacing to a computer for logging spectra automatically. To do this manually would in some cases be formidable.

3.5.4 Evaluation of the electric dipole moments of gaseous linear molecules

Dakin, Good, and Coles [3.29] made the first Stark effect measurements in microwave spectroscopy when they observed the splitting of $J_2 \leftarrow J_1$ rotational line for COS. On application of an electric field in a direction parallel to the direction of the polarized electric vector of the microwave radiation, the transitions $J_{2\leftarrow1}, M_{J0\leftarrow0}, J_{2\leftarrow1}$, and $M_{J1\leftarrow1}$ were observed (see Fig. 3.19).

The rotational energy of a linear molecule in an electric field is:

$$E_{M_J} = J(J+1)hBc - \frac{1}{2}\frac{\mu^2}{2hBc}\left[\frac{3M_J - J(J+1)}{J(J+1)(2J-1)(2J+3)}\right]E^2 \qquad (3.56)$$

where $B = h/8\pi^2 cI$, E is the static field strength, and μ is the electric dipole moment. It should be noted that: (i) when $E = 0$ the normal rotational energy equation for a linear rigid rotator is obtained; (ii) when $J = 0$ and $M_J = 0$ the second term on the right-hand side of Equation (3.56) must be replaced by $-\mu^2E^2/6hcB$.

The second term on the right-hand side of Equation (3.56) is dependent on the square of the electric field strength, i.e. this is a second-order Stark effect. Equation (3.56) applies also to symmetric-top molecules for the case $K = 0$. When $K \neq 0$, then the second-order effect can usually be neglected in comparison with the first-order effect for symmetric-top molecules.

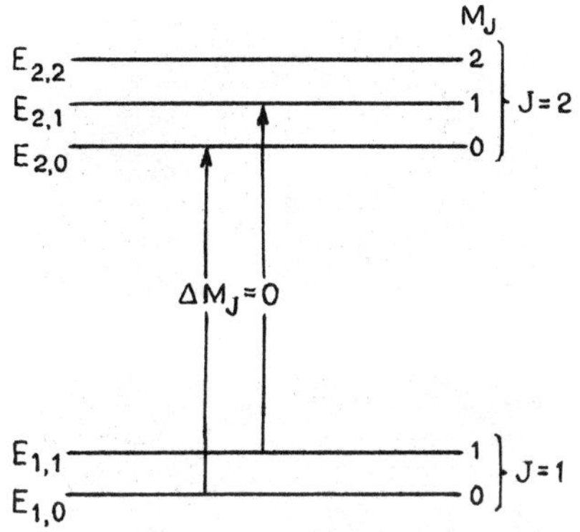

Fig. 3.19 Transitions illustrating the Stark effect splitting of the $J_2 \leftarrow J_1$ of COS when the d.c. electric field is applied in a direction parallel to the polarized electric vector of the microwave radiation.

In the studied transitions for COS, by reference to Fig. 3.20(a) it may be readily appreciated that the difference in energy between the two transitions is given by:

$$(E_{2,1}-E_{1,1})-(E_{2,0}-E_{1,0}) = h\Delta\nu \tag{3.57}$$

where $\Delta\nu$ is the frequency separation between the two observed peaks in Fig. 3.20(b) and (c) which result from applying the electric field. Figure 3.20(a) illustrates the absorption in the absence of an electric field. By inserting the appropriate J and M_J values into Equation (3.56), then substituting for the E values in Equation (3.57), it may be shown that:

$$\Delta\nu = \left(\frac{3}{20}-\frac{1}{84}\right)\frac{8\pi^2 I\mu^2 E^2}{h^3} \tag{3.58}$$

The value of $\Delta\nu$ for a field of 1070 V cm^{-1} is ~ 5 MHz.

From inserting this value of $\Delta\nu$ into Equation (3.58) where $E = 1070$ V cm^{-1}, and by using the moment of inertia determined from its microwave spectrum (or by any other means), the dipole moment of COS can be calculated. Recently, several accurate measurements have been made on COS, and it is frequently regarded as a standard dipole moment for microwave measurement, a value of $\mu = 0.71499 \pm 0.00005$ D (D = Debye unit) being adopted. It is interesting to note that this figure is in agreement with the value obtained by the heterodyne method, although the accuracy of the former is much greater.

The general procedure in determining the dipole moment is to plot $\Delta\nu$ against the square of the electric field strength. This produces a straight line from whose slope and Equation (3.58) the dipole moment can be calculated.

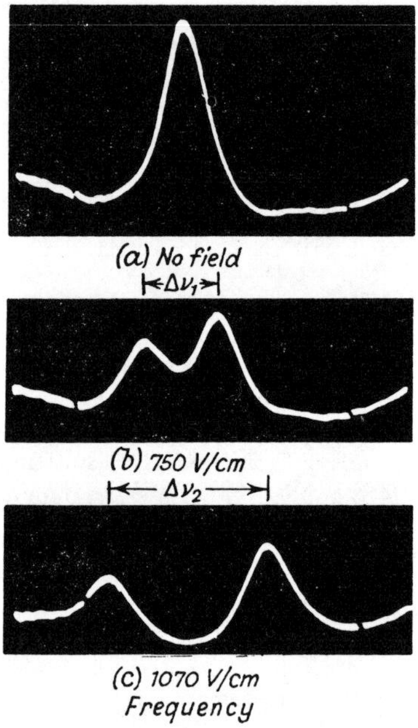

Fig. 3.20 Microwave absorption of the $J_{2\leftarrow 1}$ transition in COS: (a) with no electric field; (b) with a parallel electric field of 750 V cm^{-1}; (c) with a parallel electric field of 1070 V cm^{-1}. In the absence of an electric field the rotational line lies at a frequency of 24 320 MHz. The frequency markers in the oscilloscope pictures are 6 MHz apart in each case and appear as points of different intensity at each side of the oscilloscope trace. (After Dakin, Good, and Coles [3.29]. Courtesy of the 'Physical Review'.)

3.5.5 Conclusions on application of Stark effect to molecules

A comparison of a few of the values of the electric dipole moments obtained from microwave spectra and the dielectric constant method, both involving measurements on gases, is given in Table 3.4.

These two very different methods (which are regarded as being the most accurate methods of determining electric dipole moments) lead to figures in very close agreement. In range of applicability to simple molecules the microwave method has the advantage over the dielectric constant method that it can be used at the very low pressure of $\sim 10^{-3}$ torr and hence to investigate substances that have not sufficient vapour pressure to be examined by the dielectric constant method in the vapour phase. The microwave method also yields the dipole moment of a free radical if its microwave spectrum can be obtained.

The dipole moments of a few free radicals and a considerable number of linear or symmetric-top molecules have now been evaluated by the microwave procedure; a number of small asymmetric-top molecules has been determined.

Table 3.4 Comparison of dipole moments by the microwave and the dielectric constant methods [3.30]

Compound	$\mu(D)$	
	Stark method	*dielectric constant method*
HCN	2.96	2.93
N_2O	0.166	0.17
H_2O	1.85	1.84
CH_3Br	1.80	1.79
SO_2	1.59	1.61
NH_3	1.47	1.47
$(CH_2)_2O$	1.88	1.89

However, if the dipole moment does not lie along a principal axis, the procedure is difficult and complex. In such cases, from the analysis of a sufficient number of transitions, it is possible to obtain the components of the dipole moments along these principal axes, and from vector addition the molecular dipole moment can be deduced.

The method is also available in that it is possible to determine the direction of the dipole moments with respect to the principal axes within the molecule. The microwave method is unique in that it yields the dipole moment not only in the ground vibrational states but in the excited states as well, and, in addition, may determine the dipole moments of different isotopic species where the change in μ is caused by the change in zero-point amplitude. The method is also suited to studying small differences in dipole moments as, for example, in a homologous series, and also very small amounts can be evaluated; for example, the μ of n-propane is shown to be 0.083 D and not zero as was once thought to be the case.

One most valuable feature of the Stark spectrometer is its application in identifying transitions; in fact, this procedure is employed in commercially available spectrometers used for the study of rotational transitions. The procedure is to search a particular frequency region for rotational transitions, then to measure them, and finally to identify them by studying characteristic Stark effects.

For a more detailed and wider treatment of most of these advanced topics the reader should consult books by Gordy, Smith, and Trambarulo [3.2], Ingram [3.3], Townes and Schawlow [3.4], and Sugden and Kenney [3.5].

3.6 INTRAMOLECULAR VIBRATIONS

3.6.1 Study of hindered internal rotation

Microwave spectroscopy has proved most successful in the study of low energy barriers up to the order of 17 kJ mol^{-1}. Numerous studies have been made on the barrier opposing internal rotation in systems such as:

$$H_3C{-}C\lessdot \quad H_3C{-}Si\lessdot \quad H_3C{-}NO_2$$

and added interest in such work is in the study of the types and proportions of rotational isomers.

A considerable amount of data is now available on the energy difference between rotational isomers and also on the energy barriers opposing group rotation. We shall now consider the latter in more detail, since it is of considerable interest to chemists to have knowledge of barriers opposing rotation. Ultimately it must be the hope to formulate adequate theories to predict the magnitude of these energy barriers. So far, ethane, n-propane, and n-butane have been adequately treated theoretically and their energy barriers predicted as well as the energy difference of their rotational isomers.

In a molecule such as $CH_3{-}CHO$, on rotation of the methyl group through 360° about the axis of symmetry, the potential function passes through three potential-energy minima and maxima as is illustrated in Fig. 3.21. The potential-energy barrier to internal rotation about a single bond in a molecule can usually be expressed in terms of a Fourier series of terms:

$$V(\alpha) = \frac{V_1}{2}(1 - \cos\alpha) + \frac{V_2}{2}(1 - \cos 2\alpha) + \frac{V_3}{2}(1 - \cos 3\alpha) + \ldots \quad (3.59)$$

If this equation is applied to rotation about the C—O bond in phenol, and the two stable conformations are taken to be those where the O—H is planar with the benzene ring, the internal rotation could be expressed in terms of a two-fold energy term V_2 and:

$$V(\alpha) = \frac{V_2}{2}(1 - \cos 2\alpha) \quad (3.60)$$

A four-fold energy term $(V_4/2)(1 - \cos 4\alpha)$ could also be added to the right-hand side of Equation (3.60), but this should be very small and would normally be neglected, and Equation (3.60) would be adequate. The potential energy is dependent on the internal rotational angle α (see Fig. 3.21), and for acetaldehyde this is a case of a three-fold symmetry axis which is expressed by means of a three-fold energy term:

$$V(\alpha) = \frac{V_3}{2}(1 - \cos 3\alpha) \quad (3.61)$$

The six-fold energy term $(V_6/2)(1 - \cos 6\alpha)$ would be small and is neglected. The internal rotation is quantized, and for acetaldehyde in Fig. 3.21, when $V > 2$, then the methyl group has sufficient energy to rotate freely about the C—C axis. However, when $V = 0$, 1, and 2, the methyl group is restricted to one of the three potential-energy valleys and can then execute only a torsional vibration about equilibrium positions. When the barrier height is low (e.g. $V_3 = 4.940\,\mathrm{kJ\,mol^{-1}}$) for acetaldehyde, then splitting of the vibrational levels occurs

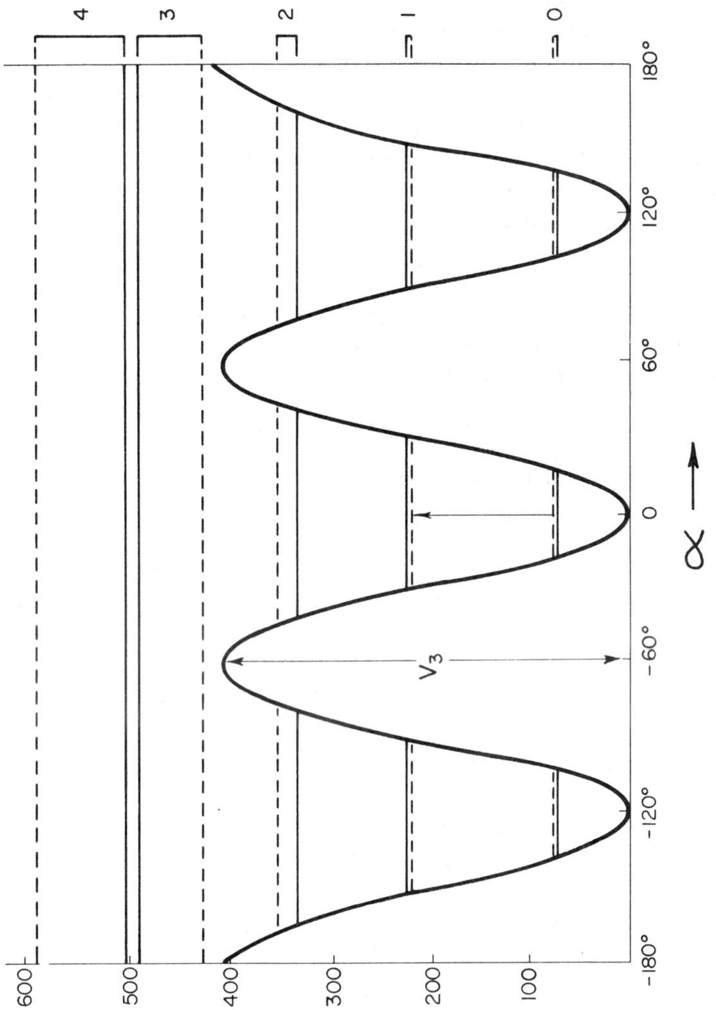

Fig. 3.21 Potential function and energy levels for the hindered internal rotation of acetaldehyde.

(see Fig. 3.21). For acetaldehyde the transition from the 1←0 level for the methyl torsional oscillation occurs at ~ 150 cm^{-1}.

Interaction occurs between the torsional oscillation motion and the overall rotation of the molecule, and coupling takes place between the overall rotational angular momenta and the momentum associated with the restricted internal motion. Consequently, the energy levels are modified, and this can be recognized in: (a) the fine structure superimposed on the normal rotational spectrum; (b) the fine structure in the torsional spectrum.

From the splitting which results from rotation–torsion interaction, the barrier height can be obtained by what is known as the relative intensity method which is as follows. If for a particular rotational transition in the ground torsional state the intensity of the line is I_0 and its population is n_0, while in the first excited torsional state its intensity is I_1 and its population is n_1, then:

$$n_1/n_0 = I_1/I_0 \tag{3.62}$$

If ΔE is the torsional energy separation and ω is the torsional oscillation frequency, then for the 1←0 torsional transition $\Delta E = h\nu = h\omega$.

From the Boltzmann Distribution Law:

$$n_1/n_0 = \exp(-\Delta E/kT) \tag{3.63}$$

and therefore it follows that from these equations:

$$I_1/I_0 = \exp(-h\omega/kT) \tag{3.64}$$

and from the experimental values of I_1 and I_0, then ω is obtained.

If for the torsional oscillation a harmonic oscillation approximation is taken to apply, then:

$$\omega = \frac{1}{2\pi}\left(\frac{V}{2I_r}\right)^{1/2} \tag{3.65}$$

where I_r is a reduced moment of inertia for the torsional motion about the bond and V is the potential barrier.

I_r can be calculated if the geometry of the molecule is known. If, for example, the molecule is of the type H_3C—CCl_3, where the torsional oscillation occurs about the C—C bond, then the CH_3 and CCl_3 groups may be regarded as coaxial symmetric tops with moments of inertia I_{CH_3} and I_{CCl_3} about a principal axis through the C—C bond. In this case:

$$I_r = I_{CCH_3}I_{CCl_3}/(I_{CCH_3} + I_{CCl_3}) \tag{3.66}$$

Hence, from I_r and ω (obtained from I_1 and I_0) V results.

Thus, the evaluation of V by this method is straightforward. The procedure requires accurate determination of intensities. This is a limitation to this approach, as is the necessity of measuring relative intensities for low-lying transitions where the harmonic oscillator approximation is most likely to hold. A more precise and involved microwave method than this exists. Nevertheless, a good number of barrier values has been obtained by the relative intensity method. Three examples are acetaldehyde ($4618 \pm 250\,J\,mol^{-1}$), ethyl cyanide ($3.73 \pm 1.21\,kJ\,mol^{-1}$), and ethyl chloride ($14.2 \pm 2.5\,kJ\,mol^{-1}$), which give some appreciation of the order of accuracy to be expected.

Many studies have been made on molecules which have different rotational isomers of the *cis*, *trans*, and *gauche* forms. The most stable form is the one which has the lowest potential. For example, the potential function for n-propyl fluoride may be diagrammatically represented as in Fig. 3.22, and the potential function has the form:

$$V = \sum_n (\tfrac{1}{2} V_n)(1 - \cos n\theta\alpha) \tag{3.67}$$

where α is the angle of rotationa bout the central C—C bond. Thus, the form with the lowest potential is the *gauche* one, and this is the more populated form. In the case of acetaldehyde the methyl group has three-fold symmetric, and $n = 3$ and also an integer multiple of 3. However, for n-propyl fluoride this is no longer the case, and terms which are not a multiple of 3 may be very important. From relative intensity measurements it is possible to determine which rotational form is the more stable, and their energy differences. Hirota [3.31] has studied the *trans* and *gauche* forms of n-propyl fluoride.

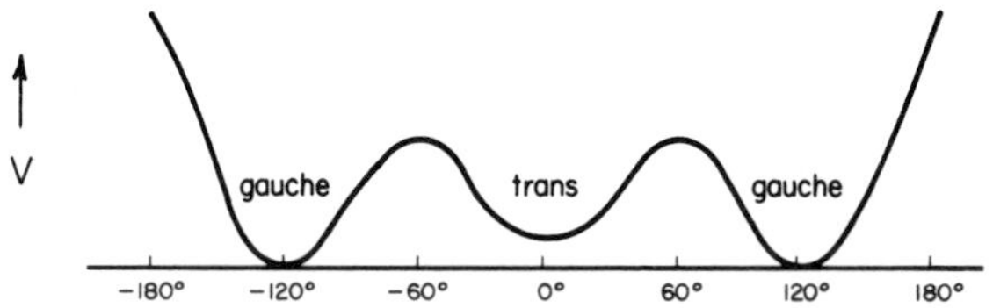

Fig. 3.22 The potential function of n-propyl fluoride.

Many microwave studies on substituted alkanes have shown the coexistence of rotational isomers. Interesting general features which emerge are that when the internal axis of rotation is of the type C—C (i.e. two carbon atoms linked together by sp^3 bonds) then the isomers are always of the *trans* and *gauche* forms, whereas when a double bond is adjacent to the single bond, as in allyl chloride, the forms are of the *cis* and skew types.

3.6.2 Inversions

In addition to the usual forms of rotational and vibrational energy changes, a pyramidal molecule has the interesting feature of being able to turn itself inside out as is illustrated in Fig. 3.23(a). The potential barrier opposing the passage of the nitrogen atom through the plane of the hydrogen atoms is $24.3\,kJ\,mol^{-1}$. The frequency at which the inversion occurs happens to fall in the microwave region.

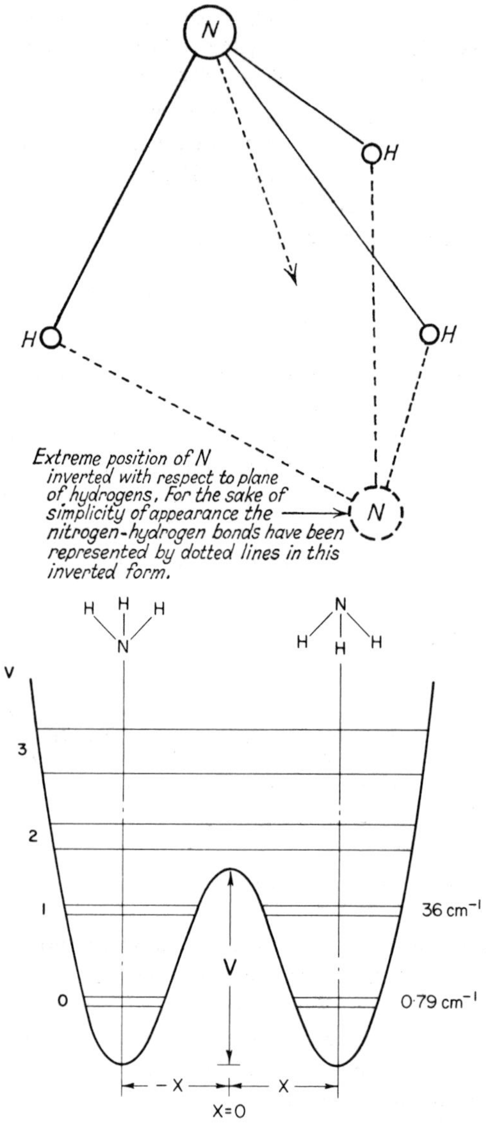

Fig. 3.23 (a) Inversion of the ammonia molecule. (b) Potential function for the inversion coordinate of ammonia.

The potential function diagram for the inversion coordinate of ammonia is given in Fig. 3.23(b). When the N atom is above the plane of the three H atoms, the coordinate is allotted a positive value and negative when below the plane. When $x = 0$, this corresponds to the case when all four atoms are planar, and this leads to the hump in the potential-energy curve; the height of this maximum is 2070 cm^{-1} above the minimum. Classically a molecule at the minimum level requires this amount of energy to reach the top of the barrier for the N atom to pass through the plane of the three H atoms. However, when the barrier is low, as is the case in NH_3, and the mass of the atoms attached to the N is small, then there is an appreciable quantum mechanical tunnel effect. The consequence is that the N atom passes through the plane of the three H atoms, and inversion takes place, although the available thermal energy available is appreciably less than the barrier height. The inversion process thus involves a change in the x-coordinate of the N atom from say $+x$ to $-x$, that is the inversion of only one coordinate.

Each of the vibrational levels is split into two and corresponds to slightly different energy levels. This splitting is very dependent on the barrier height. The transitions which take place in the $v = 0$ levels occur at 0.79 cm^{-1} (i.e. at 24 GHz) and are between the split $v = 0$ levels on opposite sides of the maximum having a difference in energy of 0.79 cm^{-1}. In this transition the N atom has tunnelled from one side of the plane of the three H atoms to the other and gives what is known as the pure inversion spectrum which occurs in the microwave region. This mechanism also applies to the $v = 1$ vibrational level. In this case the separation of $v = 1$ levels is 36 cm^{-1} [see Fig. 3.23(b)]. These separations may be deduced from either the infrared rotation–vibration spectrum or the fine structure in the rotational spectrum in the far-infrared region.

The inversion frequency in such systems is highly sensitive to the potential-energy barrier (V), and the reduced mass (μ) of the inverting systems, and for the lowest pair of levels depends roughly on $\exp(-\mu^{\frac{1}{2}} V^{\frac{1}{2}})$. Thus, for ND_3, since its reduced mass is greater than that of NH_3, the inversion frequency is less and is, in fact, ~ 1600 MHz (c.f. NH_3 value of 23.786 GHz).

Manning [3.32] has developed a formula which relates the barrier height (V in cm^{-1}) to the distance (x) of the N above the plane of the three hydrogen atoms and the reduced mass [$\mu = mM/(3m + M)$ where m and M are the masses of the H and N atoms respectively]. His formula is:

$$V = 66\,551 \,\text{sech}^2\left(\frac{x\mu^{\frac{1}{2}} 10^8}{13.96}\right) - 109\,619 \,\text{sech}^2\left(\frac{x\mu^{\frac{1}{2}} 10^8}{13.96}\right) \tag{3.68}$$

When this V is substituted in the Schrödinger equation for inversion, and the latter equation is solved to obtain the various energy levels (E), then from these the separation of the levels follows. The estimated value of the inversion frequency which follows from this by determining the separation of the $v = 0$ levels for $^{14}NH_3$ is 25 000 MHz, and this is in reasonable agreement with the measured value of 23 786 MHz.

Costain and Sutherland [3.33] have developed a formula which relates V to the changing N–H bond length (Δr) and changing HNH bond angle ($\Delta\theta$) during the inversion:

$$V = (3/2)k_r(\Delta r)^2 + (3/2)k_\theta(\Delta\theta)^2 \tag{3.69}$$

where k_r and k_θ are the force constants corresponding to changing bond length and angle. Thus, the motion may be looked upon as being of the NH bending type, and the bending fundamental would then be 932 cm^{-1}. This is to be contrasted with the inversion frequency of 0.79 cm^{-1} [see Fig. 3.23(b)] which is between the $v = 0$ components of the ground vibrational state; this is accompanied by the movement of the N atom from the $v = 0$ level on one side of the plane of the H atoms to the $v = 0$ level on the other side.

Both the Costain and Sutherland and the Manning formulae yield values of V in good agreement, being 2077 cm^{-1} and 2072 cm^{-1} respectively.

The Costain and Sutherland formula has also been employed to predict the lowest inversion frequencies of PH_3 and AsH_3 to be 0.14 MHz and 0.5 c/year. These very low frequencies are a consequence of the increased barrier heights in these steeper pyramids.

The inversion frequency is dependent on the rotational energy of the molecule, and molecules with the higher rotational quantum number values are subject to centrifugal forces which disturb the molecule; for example, if the rotation is about the symmetry axis, then the centrifugal effects cause the hydrogen N atoms to move outwards; this reduces the barrier. Hence, since different centrifugal effects are produced for differing rotational quantum number values, slightly different inversion frequenices will be observed, resulting in the band-like appearance of its spectrum.

The separation (0.79 cm^{-1}) between the ground state doublet vibrational level of ammonia was employed in evolving the first maser. These vibrational levels have different electric quadrupole moments, and consequently, when the ammonia molecules are passed through a strong inhomogeneous electric field, the two moments are deflected differently. As a result of this the molecules in the higher vibrational state may be separated from those in the lower state and then passed into a microwave cavity the resonance frequency of which is set at 0.79 cm^{-1}. Microwave power at this frequency is supplied to the cavity; this then interacts with these molecules in the upper state, and consequently they give up their energy. The result is that the output microwave radiation from the cavity is greatly enhanced in power, and hence the name – microwave amplification by stimulated emission of radiation (maser). Masers have found extensive use in radio-astronomy. Of much wider use has been the laser (light amplification by stimulated emission of radiation) which may be looked upon as a maser functioning at optical frequencies.

Another interesting application of the frequency of one of these intense ammonia absorption lines (which has been measured most accurately) is to control a clock to an accuracy of one part in 10^9. That is the line itself is made a frequency standard and is used to check the radio oscillator.

A number of substituted ammonias has now been studied by the microwave technique. Methylamine is one such case and has an inversion barrier [3.34] of 20.1 kJ mol^{-1} which is to be compared with the recent value for NH_3 of 24.3 kJ mol^{-1}. In formamide the barrier is reduced to 4.44 kJ mol^{-1}; the low barrier in this case is attributed to a flattening of the pyramidal structure.

For molecules of the type XNH_2 the degree of non-planarity varies considerably with the substituent where ϕ is the angle between the bisector of the HNH group and the extension of the N–X bond. Thus, for NH_3, CH_3NH_2, $C_6H_5NH_2$, *p*-$FC_6H_5NH_2$ and NH_2CHO, the ϕ values are 59°, 52°, 38°, 46°, and 17° respectively. All of these molecules exhibit inversion splitting in their microwave spectra.

3.6.3 Puckering motion of nearly planar rings

A number of microwave and far-infrared studies has now been made on three-, four-, and five-membered ring systems with a view to deciding whether the rings are planar or non-planar. Such work is considered in more detail in Chapter 5, and it will be seen that for rings which have a puckered equilibrium structure the potential function is symmetric about the planar configuration with a barrier lying between the two minima. From the intensities of the satellites in the microwave spectrum, information on the spacing of the vibrational levels of the puckering mode may be deduced.

Four-membered ring systems, for which a number of energy barriers has been measured, have only one out-of-plane vibrational mode. These energy barriers are of the type $(CH_2)_3X$ where X = O, S, and CH_2, and the barriers are 0.180, 3.28, and 5.376 kJ mol^{-1} for the puckered rings [3.35–3.37]. These varying values are interpreted in terms of torsional interaction and ring strain. Torsional interaction works in favour of a puckered conformation while ring strain favours a planar form. Thus, in these four-membered ring systems, when CH_2 is replaced by O, the torsion is reduced and the barrier lowered, while on passing from S to O, since the COC angle is greater than the CSC and the strain less important in the latter, torsion in the sulphide is greater than that for the oxide.

Some five-membered rings such as pentanone have two out-of-place vibrations and undergo bending and twisting vibrations independently. In some five-membered ring systems, such as tetrahydrofuran, the phase of the puckering rotates more or less freely around the ring, and this is termed pseudo-rotation, a form of inversion which does not require the passage through a relatively high barrier as, say, N does in NH_3 in passing through the plane of the three H atoms. Thus, in tetrahydrofuran the barrier to inversion in the pseudo-rotation process is only 57 cm^{-1}, whereas for direct inversion through the plane it would be much higher (1220 cm^{-1}). Both the microwave and far-infrared techniques have been applied to such studies.

3.7 QUALITATIVE AND QUANTITATIVE ANALYSIS

Gaseous microwave spectroscopy has the following advantages in chemical analysis. (a) It has very high resolving power and can, moreover, detect easily different isotopic species of the same molecule. In theory, because of the improbability of overlap of a large proportion of the lines which characterize the individual substances, the method should be capable of detecting a large number of polar substances in a gaseous mixture. As a result of the high resolution (> 100 000) of the microwave spectrometer and accurate spectral line position (1 p.p.m.), it is in principle preferable to the infrared or the highest resolution mass spectrometer for qualitative analysis. (b) The amount of gas required in the cell is very small. (c) The sample is not destroyed in the course of analysis. (d) The absorption may be recorded automatically.

The limitations of microwave spectra in chemical analysis are as follows. (a) The substance must have a permanent electric or magnetic dipole moment. (b) In general, the substance must be examined in the gaseous phase. However, if the substance has a vapour pressure greater than about 10^{-2}–10^{-3} torr at some realizable temperature, then liquids or solids can be examined. The microwave spectra of the vapours in equilibrium with many solids have been examined at room temperature. (c) The microwave spectrum, unlike the infrared spectrum, only identifies the molecule itself and does not pick out structural units or groupings which are sometimes a useful feature of infrared analysis of an unknown substance.

3.7.1 Qualitative analysis

The intensity of a rotational absorption line depends on the square of the molecular dipole moment, the particular rotational transition, and the cube of its frequency. The sensitivity of the present-day spectrometers corresponds to an absorption of about one part in 10^{10} per cm path length. A typical absorption intensity for an asymmetric-top molecule is 10^{-7} cm^{-1}.

As a result of the high resolving power and the accurate spectral line position, the microwave technique has a considerable advantage over other physical methods in some qualitative analysis problems. Thus, each substance in a mixture can be identified by measuring some of its absorption lines. Further, there is no need to remove impurities since the spectra are strictly additive. In addition, it becomes possible to identify the impurities. Further, the whole molecule is identified; that is, the method does not rely on identifying characteristic groupings and supplementary information.

The National Bureau of Standards has compiled a list of microwave absorption line frequencies as standards for analysis. Since most substances have a number of lines stretching completely across the microwave region, a relatively narrow frequency range may be chosen for experimental investigation. The *Q*-band frequency range of 26 500–40 000 MHz has been a popular one.

A refined use of microwave spectroscopy in qualitative analysis is for the ready detection of the presence of different isotopes in similar molecules; for example, this may be done in the following pairs:

(a) $H_3{}^{13}C-C\equiv CH$ and $H_3C-C\equiv CH$
(b) $H_3C-C\equiv CH$ and $H_3C-C\equiv CD$
(c) $H_3{}^{13}C-C\equiv CH$ and $H_3C-C\equiv {}^{13}CH$

A small part of the spectra for (c) is given in Fig. 3.24; it will be noted that the rotation lines for these two species are easily resolved, and, in fact, their frequencies differ by about 92 MHz.

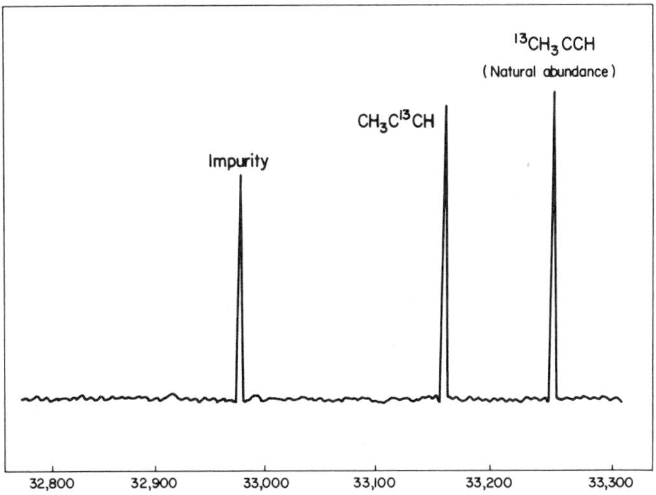

Sample: $CH_3C{\equiv}CH$
Pressure: 50 μ Hg (5×10^{-2} torr) at room temperature
Frequency range: 32·1 – 33·2 GHz (1·07 – 1·18 cm^{-1})
Microwave power level: 2·5 mW
Stark field: 1800 V/cm
Detector time constant: 1 s
Recorder output is approximately logarithmic

Fig. 3.24 Part of the microwave spectrum of a commerical sample of methylacetylene. (Courtesy of Hewlett-Packard).

Not only can microwave spectroscopy be used to distinguish between different isotopic species in molecules but it can also be employed to identify the different conformational isomers. For example, the *trans* and *gauche* forms of n-propyl iodide have different moments of inertia which lead to different rotational frequencies. This is illustrated in Fig. 3.25 for 1-iodopropane.

The frequency of microwave radiation can be measured to 1 part in 10^7 by standard electronic counting techniques, while a few parts in 10^6 accuracy is usually adequate to identify a pure sample by comparison with a known spectrum or spectral atlas. With a totally new spectrum, identification is possible, if the probable structure is known, since computer programmes can now be employed to evaluate moments of inertia, and from these the approximate spectral line positions can be estimated.

The identification of components in a mixture of gaseous molecules can be straightforward by microwave spectroscopy. This is illustrated in Fig. 3.26 for a

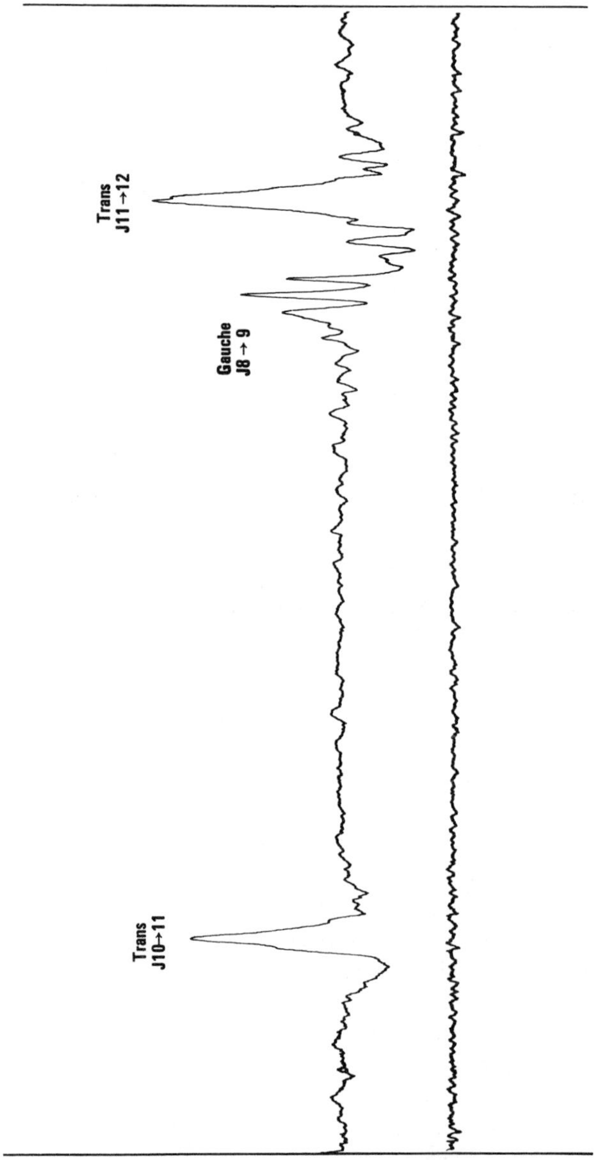

Fig. 3.25 Some of the rotational frequencies for the *trans* and *gauche* forms of 1-iodopropane. (Courtesy of Cambridge Scientific Instruments Ltd.)

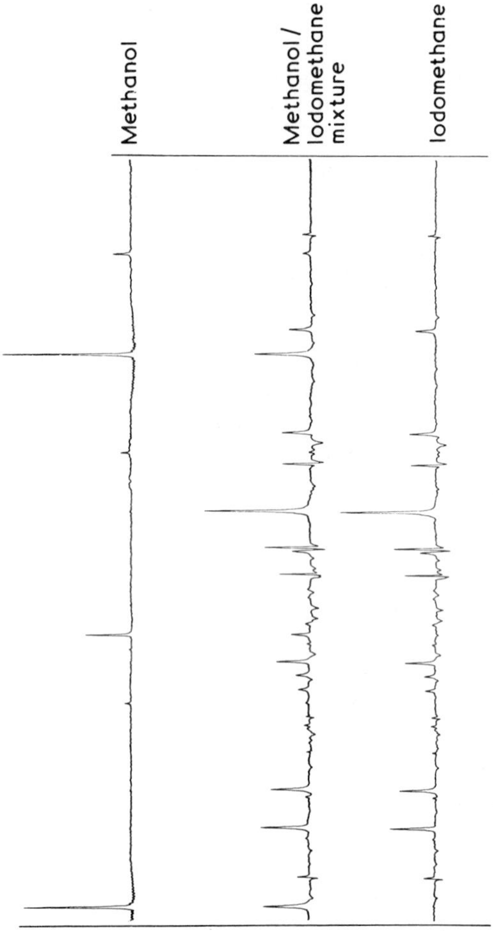

Fig. 3.26 Microwave spectra in the 29 500–30 500 MHz range of (a) methanol, (b) a mixture of methanol and iodomethane, and (c) iodomethane (Courtesy of Cambridge Scientific Instruments Ltd.)

mixture of methanol and iodomethane where comparison of the spectra in the frequency range (39 500–30 500 MHz) of the mixture with that of the pure components readily identifies their presence. In general, the only change in the spectra of the mixtures compared with those of the pure components is the broadening of the lines due to a pressure increase.

The data from the Camspek microwave spectrometer may be used in the following way in qualitative analysis. A computer may be employed to simplify the logged spectrum to facilitate comparisons with spectra previously recorded. A simple listing of the twenty most intense lines would be sufficient to identify unambiguously a pure compound, but this would not allow for identification of low concentration components in a mixture. One procedure is to list the two most intense lines in each 100 MHz interval across the frequency range of the spectrometer. Studies of spectra so far obtained show that this technique can successfully separate mixtures of up to five components present in varying concentrations.

3.7.2 Quantitative analysis

In quantitative estimations the most straightforward way is to use the *method of relative intensities*; this consists of measuring the intensities of the absorption lines and comparing them with the intensities of those in a mixture of known concentrations and similar compositions. An accuracy of a few per cent can be readily attained. This comparison method has been carried out in a number of isotopic mixtures, and for certain mixtures of $^{15}NH_3$ and $^{14}NH_3$ the $^{15}N/^{14}N$ ratio was determined to an accuracy within about 3 per cent. Other mixtures which have been quantitatively analysed are NH_3 in the presence of NH_2D, and deuterations such as $CHF_3 \rightarrow CDF_3$ have been followed.

Another quantitative application has been to follow the concentration of a given component in a flowing gaseous mixture by allowing the gas to travel at very low pressure through the absorption cell. Any change in concentration can easily be detected. The application of microwave spectra as a method for quantitative analysis has been reviewed by Hughes [3.38]. A recent account reviewing both quantitative and qualitative analysis by Cuthbert and Denney [3.9] should be consulted.

In a mixture containing two or more types of polar molecules the use of peak intensities in quantitative analysis is unsatisfactory since each substance has its own specific broadening effect on any one particular absorption line. As a consequence of this the height of the absorption line does not give a direct measure of concentration. This is illustrated in Table 3.5 where the absorption due to methanol gas is compared with the absorption of the same amount of methanol in water. Thus, a comparison of columns 2 and 4 reveals a big discrepancy in the peak heights where clearly the peak heights do not vary linearly with the partial pressure of methanol. However, the integrated areas under the 'absorption line' closely correspond, and a suitable calibration curve can be constructed from the plot of volume concentration against peak area.

Cuthbert and Denney [3.9] have shown how to carry out quantitative analyses using the fact that the integrated absorption intensity of the line is proportional to partial pressure over a very wide range of power saturation, pressure broadening, and mixture composition. Figure 3.27 illustrates the results from one of their quantitative analyses on the impurities present in proprietary laboratory grade 99 per cent ethanolamine based on their procedure. The accuracy of this approach is ~ 2–3 per cent.

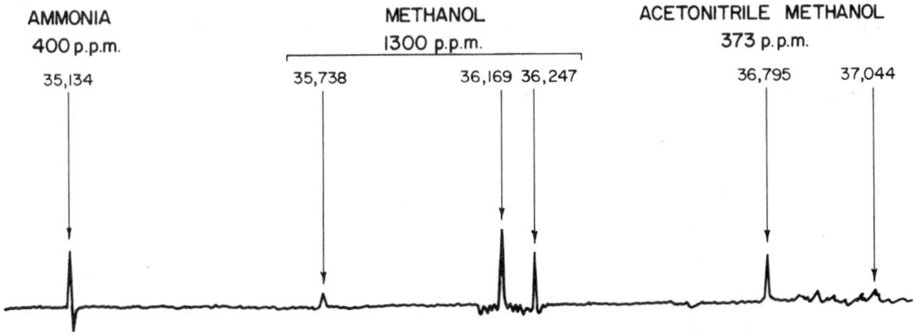

Fig. 3.27 Part of the microwave spectrum of proprietary laboratory grade ethanolamine. (After Cuthbert and Denney [3.9]. Courtesy of the Institute of Petroleum).

Table 3.5 Quantitative analysis by means of the methanol absorption at 36 169 MHz for solutions of methanol in water

Volume of methanol, μl	*Methanol alone*		*Methanol in water*	
	peak height	*peak area*	*peak height*	*peak area*
0.1	52.5	78	27.2	72
0.2	106.1	149	51.3	151
0.3	136.2	225	73.9	228
0.4	156.7	307	91.0	300
0.5	170.5	377	115.0	369
0.6	177.2	452	124.9	450
0.7	182.8	527	133.8	522
0.8	186.0	602	149.0	597
0.9	188.2	688	156.1	682
1.0	191.0	756	167.2	750

3.8 SOME ADDITIONAL AND RECENT STUDIES

So far the information which the microwave procedure yields has been dealt with separately. However, a particular investigation may involve, for example, determination of (a) molecular geometry, (b) electric dipole moment, and (c) quadrupole coupling constants. In addition, the work may be extended to the study of rotation–vibration interaction and also the determination of harmonic force fields. This section aims to give a brief indication of the potentiality of the method this way. Further, it gives a short account of some other applications of microwave spectroscopy, which are rotational energy transfer, identification of free radicals, and the detection of interstellar atoms and molecules.

3.8.1 Microwave and infrared spectra study of sulphur dicyanide

The study by Pierce, Nelson, and Thomas [3.21] on sulphur dicyanide has already been quoted as a case which reveals a small distortion in molecular structure where the CN groups are tilted from the line of the S—C bond by 5°. However, this study is also interesting in the breadth of the information derived which includes the following.

(a) Internuclear distances and bond angles in $S(CN)_2$, $S^{13}CCN_2$, $SC_2N^{15}N$, and $^{34}S(CN)_2$ by microwave studies in the 8–30 GHz region where isotope shifts yield the following structural parameters: CS = 1.701 Å, CSC = 98°22′, CN = 1.156 Å, and $2\phi = 108°22'$, where 2ϕ is the angle between the two C–N internuclear lines; i.e. the SCN angle deviates from linearity by ~ 5°.

(b) From Stark effect measurements the dipole moment was determined to be 3.04 ± 0.03 D.

(c) Quadrupole hyperfine structure partially resolved in several transitions of $S(CN)_2$ and $SC_2{}^{15}NN$ yielded the coupling constants in the two isotopic species.

(d) Centrifugal distortion constants and the inertial defect for a low-lying vibrationally excited state ($v_4 = 1$) for the CSC deformational frequency (ω_4) were of great value in making assignment of normal frequencies.

3.8.2 Rotation–vibration interaction and force field studies

The overlap of microwave with spectral studies in other areas is a profitable one, particularly in the examination of rotation–vibration interaction and the evaluation of rotation–vibration interaction constants such as Coriolus constants and centrifugal distortion constants for the determination of harmonic force fields. The determination of the distortion constants is important; e.g. for non-linear molecules such as SO_2 the four centrifugal distortion constants evaluated from the experimental data can be used to yield the four harmonic potential constants in the general valence force field (see Vol. 3). Thus, it is possible to evaluate the force field from centrifugal data alone, and reasonable agreement is found with force fields calculated by the use of two isotopic species of SO_2. The force field can also be deduced for these triatomic non-linear molecules [3.39] from the three vibrational frequencies and the inertial defect in different vibrational states which for a planar molecule is $\Delta = I_C - (I_A + I_B)$, whereas from theory Δ should be zero. Δ is important in force field considerations.

For linear triatomic molecules the rotational constants (B_v) for a particular vibrational state v may be related to the vibrational quantum number by the equation:

$$B_v = B_e - \alpha_1(v_1 + 1) - \alpha_2(v_2 + \tfrac{1}{2}) - \alpha_3(v_3 + \tfrac{1}{2}) \qquad (3.70)$$

where v_1, v_2, and v_3 characterize the three fundamental modes, and α_1, α_2, and α_3 are the rotation–vibration constants. A study of the α values which can be useful in the evaluation of the potential function from the α values is too involved to be considered in detail here. However, it is an important trend in microwave and infrared studies. Millen and Burden [3.40] considered the potential function of the linear triatomic molecule in terms of the three α values for each of the two isotopic species.

3.8.3 Rotational energy transfer

When one gaseous molecule interacts with another by virtue of the force field of the second, then the rotational levels of the first molecule become perturbed. If strong interaction occurs, rotational energy may be transferred during collision, i.e. the rotational energies of both molecules are altered as a result of collision, and hence the term *rotational energy transfer*. The interaction itself may be of the electric or magnetic type, although the former is more usual and is normally a stronger type of interaction. The interaction may be of the dipole–dipole, dipole–quadrupole, quadrupole-induced dipole (and so on) types. For example, for symmetric-top molecules the line broadening arises almost entirely from dipole–dipole interaction. For linear molecules, though, such as COS, other short-range forces have to be included, and in this molecule the molecular quadrupole term has also to be included when the calculated line widths are being fitted. The interaction energy is formulated in terms of the appropriate interdipole distances (r), and the interaction potential normally includes a $1/r^6$ term.

The width of a rotational line is dependent on the following factors: (a) molecular collisions; (b) natural line width of $\sim 10^{-4}$ Hz which comes about as a result of the uncertainty principle and is caused by the interaction of the zero-point energy of the electromagnetic radiation with the molecule; (c) collisions with the walls; (d) the Doppler effect; (e) the spectrometer itself.

Fortunately, it is possible to separate (a) from (b)–(e), and this is achieved by making a series of line-width measurements at different pressures. From the plot of line half-width against pressure a straight line is obtained, and the slope is known as the line-broadening parameter.

Boggs [3.7] has reviewed the factors involved and indicated the theoretical treatment pertinent to such studies. The first step is to estimate theoretically the width of the rotational line (for comparison with the measured one), and this involves estimating the probability that a molecule will undergo a rotational transition when it closely passes another molecule in a particular rotational state. This probability is dependent on the distances apart of the molecules. One case which Boggs considers is the probability of a COS molecule making the $J = 2 \leftarrow 1$ transition as a result of colliding with another COS molecule in the $J = 10$ state. As the approach distance diminishes, then the probability increases; e.g. when the approach distance is 15 Å the probability of a transition is low, but at 10 Å the probability is quite high. It is interesting to note that the kinetic theory diameter of COS is 4 Å. Rotational energy transfer work is interesting in that it should lead to a better understanding of molecular force fields, and of one form of energy transfer in gas-phase kinetics.

3.8.4 Identification of species

(*a*) *Detection of free radicals.* Potentially, microwave spectroscopy should be capable of detecting free radicals, since the absorption lines of a free radical produced by the removal of one or more atoms from the parent molecule would be quite different from those of the undissociated molecule. The considerable experimental difficulties encountered are reviewed by Mays [3.41]. The main difficulties facing the detection of free radicals by the microwave technique are their low concentrations and their short lifetimes. In 1953 the OH and OD radicals were detected, but it was not until 1964 that the next free radical was identified by this technique. Improvements in microwave techniques – especially sensitivity and fast flow of gases – has improved this approach, and more recent work has established the radicals ClO, BrO, and NS. This method is attractive in that not only does it yield the rotational constant which is sufficient to characterize the radical but it gives information on the separation of charge in the free radical since the electric dipole moment is also evaluated.

(*b*) *Study of chemical reactions.* Since the position of a particular isotope in the molecule may affect the value of the moment of inertia, different isotopic molecules have different pure rotational absorption frequencies. This principle was used to demonstrate that when $^{15}NH_4{}^{14}NO_3$ decomposes, the produce is $^{15}N^{14}NO$; this was done by the evaluation of the moment of inertia of this linear molecule from the observed spectral lines and by comparison with its value known from other sources (e.g. infrared). It follows that the nitrogen atom from the ammonium does not, as a result of the decomposition, become finally bonded to the oxygen.

Chemical reactions at very low pressures have sometimes been unexpectedly detected in microwave investigations. For example, when methyl mercuric bromide vapour was admitted to a silver absorption cell containing a trace of silver chloride, the spectrum of methyl mercuric chloride was observed, which suggested the following reaction:

$$CH_3HgBr + AgCl \rightarrow CH_3HgCl + AgBr$$

3.8.5 Applications of microwaves to atomic spectra

Some of the frequencies required for transitions between fine and hyperfine levels in certain atoms fall in the microwave region. However, apart from inert gases it is difficult to obtain a sufficient concentration of atoms in the cell to study their microwave absorption. In general, the microwave transitions are of low intensity, and so far only a few microwave studies have been made. One of the most outstanding was made by Lamb and Retherford [3.42] who performed a microwave experiment employing an atomic beam method and studied the separation of the $2^2P_{1/2}$ and $2^2S_{1/2}$ levels in the hydrogen atom.

3.8.6 Detection of interstellar atoms and molecules

Another outstanding study was that of the 21 cm emission line resulting from the transition between two hyperfine levels of interstellar hydrogen atoms. For the hydrogen atom in its ground state $J = \frac{1}{2}, I = \frac{1}{2}$, and F has the values of 0 and 1. The separation of these hyperfine levels was also measured by the atomic beam resonance approach and found to be virtually identical with the microwave value. This 21 cm emission line may be used to identify interstellar H atoms and, of course, corresponds to the $F = 1 \rightarrow 0$ transition in the ground state. In fact, one of the principal ways of gaining information on the matter which lies between the stars in our galaxy is by studying the microwave emission spectra. Until 1963 only the 21 cm emission line of the H atom had been detected by this approach. In 1963, though, the OH line was detected first by its absorption line at 18 cm and then later by the corresponding emission line. Both absorption and emission 18 cm OH lines are detected in a large number of radio sources in the galaxy.

In 1968 some emission lines of NH_3 were detected from clouds in the centre of the galaxy. The observed transitions lies around 1.25 cm and correspond to transitions from the metastable $J = K$ rotational states. From the relative intensities of these lines the temperature of the emitter may be deduced. Hence, NH_3 is considered to be a particularly useful probe for the study of conditions inside interstellar dust clouds.

A water vapour emission line was detected at 1.3 cm and a formaldehyde absorption line at 6 cm in 1969, since when about a score of molecules and radicals has been detected inside the dark clouds of our galaxy. Thus, from a pattern of the interstellar molecules in various parts of our galaxy, we shall gain a better understanding of it and its chemistry, especially since molecules give information on interstellar temperatures, mass, and radiation densities.

An up-to-date review on Interstellar Molecules and the Interstellar Medium by Rank [3.7] should be consulted.

The application of spectroscopy to astrophysics is considered in much more detail in Vol. 3, but there the experimental data are obtained in the main from electronic spectra.

REFERENCES

3.1 Cleeton, C.E. and Williams, N.H., *Phys. Rev.* **45**, 234 (1934).

3.2 Gordy, W., Smith, W.V. and Trambarulo, R.F., *Microwave Spectroscopy*, p. 21, Wiley, New York, Chapman and Hall London (1953).

3.3 Ingram, D.J.E., *Spectroscopy at Radio and Microwave Frequencies*, Butterworths, London, (1955), Second Edition, Plenum, New York (1967).

3.4 Townes, C.H. and Schawlow, A.L., *Microwave Spectroscopy*, McGraw-Hill New York (1955).

3.5 Kenney, N.C. and Sugden, T.M., *Microwave Spectroscopy of Gases*, Van Nostrand, London (1965).

3.6 Wollrab, J.E., *Rotational Spectra and Molecular Structure*, Academic Press, New York (1967).

3.7 Rao, K.N. and Mathews, C.W., *Molecular Spectroscopy in Modern Research*, Academic Press, New York and London (1972).

3.8 Gordy, W. and Cook, C.W., *Molecular Spectra*, Ed. W. West, in *Chemical Applications of Spectroscopy*, Part II, Vol. 2. Wiley, New York (1970).

3.9 Molecular Spectroscopy, Ed. P. Hepple, Institute of Petroleum, London (1971).

3.10 Törring, T. and Häft, J., *Z. Naturforsch.*, **25a**, 35 (1970).

3.11 Amano, T., Saito, S., Hirota, E., Morina, Y., Johnson, D.R. and Powell, F.X., *J. Molec. Spectros.*, **30**, 275 (1969).

3.12 Amano, T., Hirota, E. and Morino, Y., *J. Molec. Spectros.*, **32**, 97 (1969).

3.13 Anderson, W.E., Trambarulo, R.F., Sheridan, J. and Gordy, W., *Phys. Rev.*, **82**, 58 (1951).

3.14 Nielsen, H.H., *Rev. Mod. Phys.*, **23**, 90 (1951).

3.15 Wang, S.C., *Phys. Rev.*, **34**, 243 (1929).

3.16 Morino, Y. and Saito, S., *J. Molec. Spectros.*, **19**, 435 (1966).

3.17 Saito, S., *J. Molec. Spectros.*, **30**, 1 (1969).

3.18 Takev, H. and Hirota, E., *J. Molec. Spectros.*, **34**, 370 (1970).

3.19 Saito, S., *J. Molec. Spectros.*, **30**, 1 (1969).

3.20 Nyguard, L., Nielsen, J.T., Kircheiner, J., Mattesen, G. and Rastrup-Anderson, J., *J. Molec. Struct.*, **3**, 491 (1969).

3.21 Pierce, L., Nelson, R. and Thomas, C., *J. Chem. Phys.*, **43**, 3243 (1965).

3.22 Nyguard, L., Bojesen, I., Pedersen, T. and Rastrup-Anderson, J., *J. Molec. Struct.*, 2, 209 (1968).

3.23 Townes, C.H., Holden, A.N. and Merritt, F.R., *Phys. Rev.*, **74**, 1113 (1948).
3.24 Orville-Thomas, W.J., *Quart. Rev.*, **11**, 162 (1957).
3.25 Starck, B., *Landolt-Bernstein*, Vol. 4, Springer-Verlag, Berlin and New York (1967).
3.26 Flygare, W.H. and Gwinn, W.D., *J. Chem. Phys.*, **36**, 787 (1962).
3.27 Wolf, A.A., Williams, Q. and Weatherly, T.L., *J. Chem. Phys.*, **47**, 5101 (1967).
3.28 Weiss, V.W. and Flygare, W.H., *J. Chem. Phys.*, **45**, 8, 3475 (1966).
3.29 Dakin, T.W., Good, W.E. and Coles, D.K., *Phys. Rev.*, **70**, 560 (1946).
3.30 Smith, J.W., *Electric Dipole Moments*, Butterworths, London (1955).
3.31 Hirota, E., *J. Chem. Phys.*, **37**, 283, 2918 (1962).
3.32 Manning, M.R., *J. Chem. Phys.*, **3**, 136 (1935).
3.33 Costain, C.C. and Sutherland, G.B.B.M., *J. Phys. Chem.*, **56**, 321 (1952).
3.34 Tsuboi, M., Hirakawa, A.K. and Tamagoke, K., *J. Mol. Spectros.*, **22**, 272 (1967).
3.35 Chan, S.I., Borgers, T.B., Russel, J.W., Strauss, H.L. and Gwinn, W.D., *J. Chem. Phys.*, **44**, 1103 (1966).
3.36 Harris, D.O., Harrington, H.W., Luntz, A.C. and Gwinn, W.D., *J. Chem. Phys.*, **44**, 3467 (1966).
3.37 Ueda, T. and Shimanouchi, T., *J. Chem. Phys.*, **49**, 470 (1968).
3.38 Hughes, R.H., *Ann. N.Y. Acad. Sci.* **55**, 872 (1952).
3.39 Morino, Y., *Molecular Spectroscopy*, I.U.P.A.C., 323 (1967).
3.40 Millen, D. and Burden, F.R., *J. Chem. Soc.*, 1212 (1967).
3.41 Mays, J.M., *Ann. N.Y. Acad. Sci.*, **55**, 789 (1952).
3.42 Lamb, W.E. and Retherford, R.C., *Phys. Rev.*, **72**, 241 (1947).

4 Infrared and raman spectroscopy

4.1 INTRODUCTION TO INFRARED AND RAMAN SPECTROSCOPY

The infrared and Raman methods in general yield similar types of information. They both provide a means of studying pure rotational, pure vibrational, and rotation–vibration energy changes in the ground state of simple and even complex molecules; however, for complex molecules the type of information which may be derived is more limited.

The two methods are based on quite different physical principles. Infrared spectroscopy is mainly concerned with the absorption of energy by a molecule, ion, or radical from a continuum (where the absorption occurs somewhere in the region 10 000–200 cm^{-1}) or with the study of the emission of infrared radiation by species in excited states. In contrast, Raman scattering functions by an entirely different mechanism and depends on the collision of a quantum of incident light with a molecule. The molecule can be induced by the collision to undergo a pure rotational, or a vibrational, or a rotation–vibration change. The scattered light quantum now has a different frequency from that of the incident light, and the difference corresponds to the energy change which has taken place within the molecule. The incident light employed is monochromatic and normally lies in the visible region of the spectrum (see p. 225).

Not only do the mechanisms differ for the two techniques, but the criterion as to whether a particular transition will be observed in the infrared or Raman spectrum depends on widely different principles. For example, the occurrence of a vibrational transition in the infrared region is dependent on an overall change of the electric dipole moment during the particular vibration. The Raman criterion, however, depends on a change in the polarizability during the vibration. In addition, the intensity of the infrared spectrum is dependent on the magnitude of the dipole moment change, whereas the Raman intensity is related to how readily polarizable the vibrating atoms and their bonds are.

The infrared and Raman techniques can be used to study gases, liquids, and solids. Both methods have been applied to a wide variety of problems, and both have yielded desirable knowledge such as internuclear distances and vibrational frequencies. They have also been useful in the determination, by the statistical method, of thermodynamic quantities such as entropy and heat capacity; these have been calculated from the spectral moments of inertia and vibrational frequency data. The infrared approach has been much more prolific so far in determining moments of inertia and internuclear distances of simple molecules, but the Raman method is rapidly developing in this direction, now that laser sources are commercially available.

The vibrational frequencies obtained by these methods have been extensively used to fingerprint certain groups in different molecules, but more fundamental than this has been the use of vibrational frequencies in force constant work and in structural determinations. In fact, the latter has been a major line of investigation for both techniques.

For an unsymmetrical molecule every fundamental vibration is concerned with a change of electric dipole moment and therefore satisfies the criterion for absorption in the infrared region; in such a molecule there is also a change in polarizability during the vibrations, and therefore a Raman spectrum is produced. Thus, either method could be employed to obtain vibrational frequencies, although the intensities of bands are likely to be different in the two spectra. In contrast, for molecules with a centre of symmetry, e.g. CO_2, C_2H_2, and SF_6, the rule of Mutual Exclusion[†] applies and the fundamental frequencies which appear in the Raman spectrum do not appear in the infrared spectrum and vice versa; the two methods are then complementary (see p.253).

This chapter is concerned mainly with outlining the different types of problems studied by each method. The procedure adopted has been to divide the chapter into three sections: (A) Infrared spectra; (B) Raman spectra; (C) Correlation of infrared and Raman spectra. This division is one of convenience, and it must not necessarily be supposed that the points common to the Raman and infrared methods are to be found only under Section C. In fact, some of the problems studied by the infrared method in Section A could well be studied by the Raman or even by both together. In a way, Section C is an attempt to stress that the methods often work hand in hand. This is especially true for the structural studies of polyatomic molecules.

† For molecules with a centre of symmetry, fundamental transitions which are active in the infrared are forbidden in the Raman and vice versa.

(A) INFRARED SPECTROSCOPY

4.2 PRACTICAL ASPECTS

4.2.1 Introduction

For practical purposes the infrared region may be quoted as ranging from 4000 to 200 cm^{-1}. Absorptions due to combination or overtone modes (see p. 171) may occur between 10 000 and 4000 cm^{-1}, but the bands are normally very weak and the experimental work done in this region is very limited. The basic components of an infrared spectrophotometer are shown in Fig. 4.1.

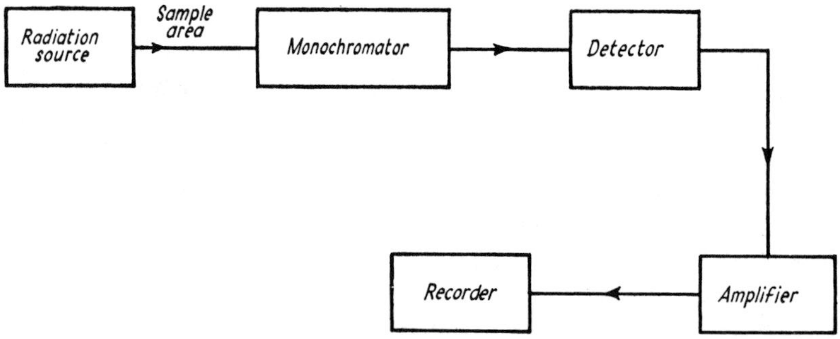

Fig. 4.1 Block diagram of major units in an infrared spectrophotometer.

A source provides radiation over the whole range of the infrared spectrum; the monochromator disperses the light and then selects a narrow wavenumber range; the energy is measured by a detector and the latter transforms it into an electrical signal which is then amplified and registered by a recorder. The individual components will now be discussed in more detail.

4.2.2 Radiation sources

To study infrared absorption spectra an ideal source should emit a continuous spectrum of radiation and the energy output should remain constant. However, all practical sources fall short of the ideal and no one emitter is adequate to cover the necessary range from 4000 to 200 cm^{-1}. The two sources in most general use down to 200 cm^{-1} are the Nernst glower and the Globar, but in addition, high-pressure mercury arcs in silica envelopes show continuous emission below 400 cm^{-1} and they are used in the far-infrared region. Raising the temperature of the source increases the energy output enormously in the high wavenumber region but it has relatively small effects at low wavenumbers. The peak emission also shifts to higher wavenumbers according to the relationship $\tilde{\nu}(cm^{-1}) = 1.96T$ where $\tilde{\nu}$ is the position of the peak emission and T is the absolute temperature (see Fig. 4.2).

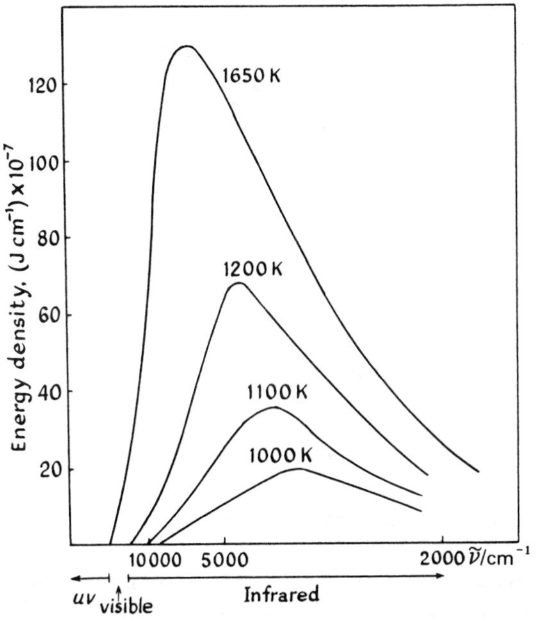

Fig. 4.2 Black-body radiation. Wavenumber distribution of equilibrium energy density with temperature.

The Nernst glower is a rod, 2–5 cm in length and ~ 2 mm in diameter, which consists of 90 parts by weight of zirconium oxide, 7 parts yttrium oxide, and 3 parts erbium oxide held together by a binder. The oxides are insulators at room temperature and electrical preheating is required in order to light the Nernst. When the filament is hot, it conducts electricity and it is then self-maintaining. The amount of energy at each frequency increases with the source temperature and the filament is maintained at incandescence by electrical means. It is normally run at a temperature of ~2000 K. Since the Nernst glower has a negative temperature coefficient of resistance, that is its resistance decreases with increasing temperature, it must be run in series with a number of ballast lamps acting as stabilizers.

The Globar is a rod of silicon carbide (carborundum) which conducts electricity readily at normal temperatures and so does not need to be preheated. It runs at ca. 1500 K and the ends are silver-coated to produce a good electrical contact. The rod is mounted in cups spring-loaded at one end and the voltage across it is regulated in order to achieve a constant radiation source.

Above 1000 cm^{-1} the Nernst glower has the advantage of working at a low power consumption while still having a high intensity. The Globar is most suited to the region below 1000 cm^{-1} but both may be used even down to 200 cm^{-1} although the energy available in that region is rather low, and the slits must be opened wide to allow more energy into the monochromator. The energy emitted by both sources is very low below 400 cm^{-1}.

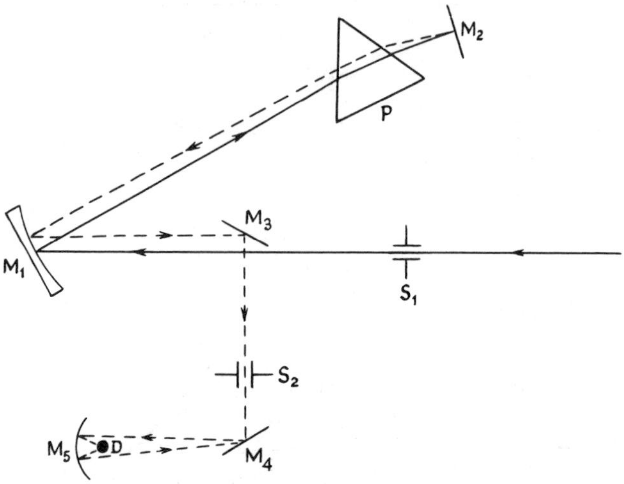

Fig. 4.3 Diagram of a simplified monochromator. M = mirror; P = prism; S = slit.

In small commercial grating infrared spectrophotometers either a Nichrome wire or a platinum filament contained in a ceramic tube is commonly encountered as infrared source for the range 4000–400 cm^{-1}.

Emission spectra are obtained with the sample itself as the source.

4.2.3 Monochromators

A simplified monochromator is illustrated in Fig. 4.3.

The monochromator performs three functions which are basic to the operation of the instrument: (a) it disperses the radiation into its wavenumber components; (b) it restricts the radiation arriving at the detector to a narrow wavenumber band; (c) it maintains the energy incident at the detector at an approximately constant level, when no sample is present, throughout the wavenumber range of the instrument.

The dispersal of the radiation is carried out by a prism or grating while an exit slit (S_2) restricts the radiation passing through it to a narrow band. The mean position of the band corresponds to the wavenumber at which a measurement is being made. Finite slit widths are necessary to enable sufficient energy to reach the detector, and so the radiation emerging from the exit slit is not monochromatic. The effective bandwidth, termed the spectral slit width, varies with the wavenumber setting of the instrument. This is because the energy from the source varies with wavenumber and the slit width is programmed mechanically or electrically, so that with no sample in the beam, the energy focused on the detector is approximately constant for all wavenumber settings. The effective slit width is usually given in cm^{-1} and a spectrophotometer will just resolve a pair of narrow absorption lines if they are separated in wavenumbers by the effective slit width. It follows that the smaller the effective slit width the higher the potential resolution of the spectrophotometer. However, the gain of the amplifier must be correspondingly increased and the scan speed must be decreased because there is less energy reaching the detector. The limit comes when the extra resolution achieved is swamped by the increased noise present on the spectrum. The rotational spacings of the vapour-phase spectra of small molecules can be used to check the resolution of the instrument. The effect of varying both the slit width and scanning speed on the vibrational–rotational spectrum of DCl is shown in Fig. 4.4.

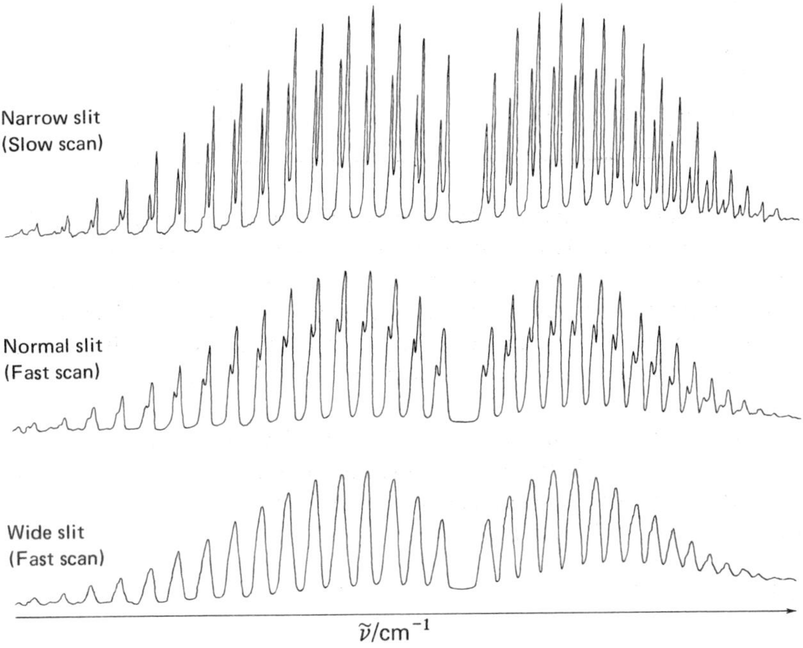

Fig. 4.4 Vibrational–rotational fine structure of DCl. The band is centred on 2090 cm^{-1} and the splitting of each rotational line is due to $^{37}Cl/^{35}Cl$ isotopes in natural abundance.

Glass, because of its opacity to infrared radiation, is quite unsuitable as a lens material. Thus, the wavenumber range of a prism instrument is determined by the availability of suitable materials which are transparent to the infrared region being studied. For example, if a rock-salt prism is used as the dispersing element, the lower wavenumber limit is 650 cm^{-1}, whereas if KBr is used the region under investigation can be extended to 400 cm^{-1}. Some of the chief materials used for prisms and windows of sample cells, together with their useful transmission limits, are listed in Table 4.1.

The majority of near-infrared spectrophotometers have in the past been based

Table 4.1 Lower transmission limit of some window and prism materials; the limits refer to 50 per cent transmittance for a 5 mm thickness

Material	*Lower transmission limit* (cm^{-1})
Glass (Na, Ca silicate)	4000
Quartz (SiO_2)	2700
Silica (SiO_2)	2800
LiF	1500
CaF_2 (fluorite)	1100
NaCl (rock salt)	600
AgCl	400
KBr	350
CsBr	250
CsI	180
High-density polyethylene	10
Diamond	<10

on prisms. Grating instruments were used especially for problems where resolution of the rotational–vibrational structure was essential. However, with the recent introduction of inexpensive diffraction gratings, the trend in modern infrared spectrophotometers is towards small grating instruments, which are rapid and convenient to use and relatively inexpensive to buy. Such instruments are now very common in both undergraduate teaching and chemistry research laboratories.

All infrared monochromators use mirror optics. Fortunately, the reflection from most metallic surfaces is very good in the infrared region, and consequently monochromator mirrors are normally worked in glass and finished with a thin reflective coating of aluminium.

4.2.4 Thermal detectors

The detector must produce an electrical signal which is proportional to the incident radiation intensity over the whole spectral range of the instrument.

The main types are used in the infrared, and (a) the vacuum thermocouple, (b) the bolometer, and (c) the Golay detector. The most desirable features of detectors are the closeness with which they approach the behaviour of a black body, high sensitivity, high speed, and robustness.

The thermocouple most commonly used is constructed from pure bismuth wire soldered to another wire which is an alloy of 95 per cent Bi and 5 per cent Sn. The wire is about 3 mm long with a diameter of 25 μm. Since this is too small for use with a spectrophotometer, a receiving element of blackened gold foil about 0.5 μm thick and 0.2 × 2 mm in area is soldered to the junction of the thermocouple. The couple is mounted in a tube which has a suitably transparent window and the tube is evacuated to a pressure of 10^{-4} mm of mercury. The evacuation reduces thermal losses, but unfortunately it also decreases the speed of response.

Vacuum bolometers have sometimes been employed owing to their short response time, which is of the order of a few milliseconds. The bolometer consists of a fine strip of metal such as platinum (suitably blackened) with dimensions of the order 5×0.5 mm and 1 μm thick. The bolometer is placed in one arm of a modified Wheatstone bridge circuit, and a similar bolometer is placed in one of the other arms. Radiation is only permitted to fall on one of the bolometers. This radiation causes a rise in temperature on the bolometer and its resistance changes, which consequently causes the bridge to become unbalanced. The radiation produces a voltage which has a linear dependence on the intensity of the radiation.

The Golay pneumatic detector consists of a chamber containing a gas of low thermal conductivity which is sealed at one end by a KRS5, KBr, or diamond window through which the radiation reaches a thin absorbing film. The absorbing film has a low thermal capacity and responds readily to infrared radiation, in turn heating the gas with which it is in contact. A rise in temperature of the gas in the chamber produces a corresponding rise in pressure and therefore a distortion of a mirror membrane with which the other end of the gas chamber is sealed. An alternating radiation signal (e.g. 10 Hz) will therefore produce a corresponding deformation of the mirror membrane at 10 Hz which, by a suitable optical system and a photocell, can be converted into an alternating voltage.

Very rapid response detectors, which permit fast rates of scan, are of great research interest, and pyro-electric detectors based on the ferroelectric material triglycine sulphate (TGS) as well as semi-conductor detectors cooled to liquid helium temperature are currently showing promise.

4.2.5 Amplifiers

The function of the detectors described above is to receive the radiant energy and convert it into measurable electrical energy. The resistance of thermocouples and bolometers is small, and amplification of the output is rather difficult. Since it is easier to build a stable high-gain a.c. amplifier, the method of amplification in modern instruments is as follows. The incident radiation is first chopped by a rotating sector at a rate of ca. 10–15 times per second. The resulting energy passes through the system and eventually falls on to the thermosensitive element and produces a pulsating output. This output is fed to an a.c. amplifier via a transformer and then to a pen recorder. Any continuous energy falling on the detector will produce a d.c. output and will not be amplified. This tends to eliminate errors from stray radiation and is a feature of the double-beam instrument described in the next section. If the sample and reference beams are of the same energy, that is no absorption by the sample has taken place, a d.c. voltage is produced by the thermocouple which is not amplified by the a.c. amplifier of the instrument. Only when the intensity of the sample beam is different from that of the reference beam is an unequal (pulsating) signal produced at the detector.

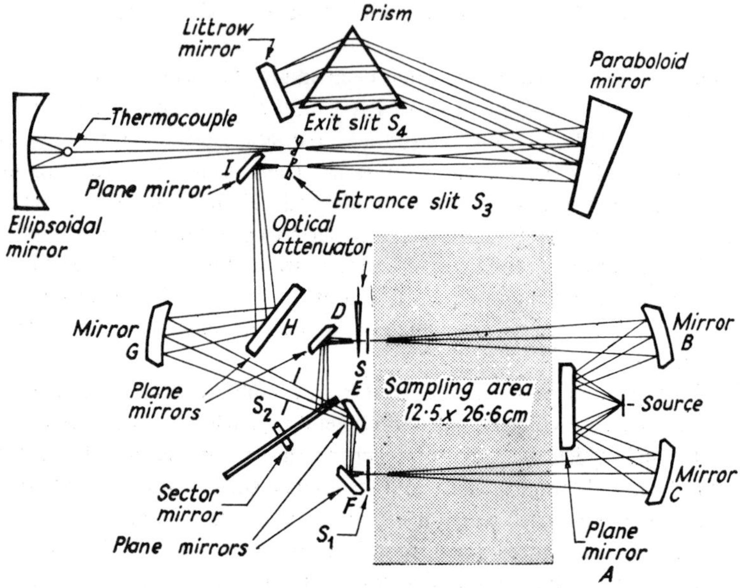

Fig. 4.5 Double-beam infrared spectrophotometer. (Courtesy of Perkin-Elmer Corpn.).

4.2.6 Double-beam spectrophotometers

A double-beam spectrophotometer is illustrated in Fig. 4.5. Radiation from the source is split into two equivalent beams, the sample beam and the reference beam, by the plane mirror A. Mirror C focuses the sample beam through the sampling area on to the slit S_1. Mirror B focuses the reference beam on to the slit S. The sample is placed in the sample area as close to the slit S_1 as possible. Sample and reference beams, after passing through the sampling area, are recombined by a semicircular sector mirror which rotates at 13 Hz. The orientation of the sector mirror is fixed with respect to the three plane mirrors D, E, and F so that it alternately reflects the reference beam and passes the sample beam through the aperture stop S_2. The aperture stop ensures that both beams are the same size and will thus be able to follow identical paths through the remainder of the system. The signal beam, which now consists of alternate pulses of sample and reference radiation, is focused by the mirror G and reflected by plane mirrors H and I through the entrance slit of the monochromator S_3 (this slit is at the focal point of the mirror G). On leaving the slit S_3 the beam diverges until the off-axis paraboloid mirror reflects it as collimated light on the prism. At the prism the component wavenumbers are dispersed and reflected by the Littrow mirror back into the prism, where further dispersion takes place. The dispersed radiation is then focused on the plane of the exit slit S_4 by the paraboloid. The particular wavenumbers which emerge from the slit strike the ellipsoidal mirror which focuses them on the thermocouple.

If the energy in both sample and reference beams is equal, a d.c. voltage is produced at the thermocouple which is not amplified by the a.c. amplifier of the instrument. When the sample absorbs radiations at characteristic wavenumbers, the intensity of the sample beam is reduced. This results in an intermittent signal at the detector which is converted into an alternating voltage and then amplified by the 13 Hz amplifier. The amplified signal is used to drive a servo-motor which moves the optical wedge (or optical attenuator) into or out of the reference beam to equalize the beam intensities. The position of the wedge is thus a measure of the intensity difference in the sample and reference beams, and is a measure of the transmittance of the sample. Since a recorder pen is coupled directly to the wedge, its movements are a record of transmittance.

The spectrum is scanned by rotation of the Littrow mirror about a vertical axis by means of a micrometer screw. Having been previously calibrated, the setting of the screw corresponds to a definite wavenumber position. The chief advantage of a Littrow mounting is the double dispersion and resolution which results from the beam traversing the prism twice.

The double-beam instrument gives the percentage transmission immediately and has the additional advantages that (a) the absorption of water vapour and carbon dioxide in the atmosphere is automatically compensated for by the double beam and hence should be absent from the final trace, and (b) the double beam automatically compensates for any variation in source intensity, detector sensitivity, etc. with wavenumber and with time.

There are drawbacks to the more refined double-beam instrument, but these are mainly technical difficulties, such as optical adjustment for equality of the two beams, amplification of signal, difficulties in detection, and also the noise level due to extraneous, random, internal, and external currents not connected with that being measured.

4.2.7 Diffraction gratings

The diffraction grating is frequently employed instead of the prism as a dispersing element and has the following points to recommend its use. (a) It may be used below 250 cm^{-1} (far-infrared region) where no suitable transparent prism materials are available. In addition, gratings are commonly used in the region between 10 000 and 250 cm^{-1} because, in general, a grating has a greater dispersion and resolving power than a prism. (b) The grating yields a linear spectrum whereas a prismatic spectrum is non-linear. This is labour-saving when the wavenumber positions of a number of lines are to be measured.

A diffraction grating may be regarded as being composed of a large number of parallel, equidistant, and narrow slits side-by-side made by ruling lines on a suitable surface with a diamond point. The surface can either be transparent or opaque to the radiation used. In the former case the grating is a transmission grating and in the latter a reflection grating. The fundamental theory is the same for both types. A reflection grating is superior to a transmission grating, however, in that, if the lines are ruled on a concave mirror, the need for employing focusing and collimating lenses can be eliminated.

If α is the angle of incidence of the light of frequency $\tilde{\nu}\,cm^{-1}$ on to the grating and β is the emergence angle, where both angles are measured with respect to the normal, the diffraction by the grating is taken into account by the formula:

$$\pm\frac{n}{\tilde{\nu}} = \frac{A}{N}(\sin\alpha + \sin\beta) \tag{4.1}$$

where A is the linear aperture of the grating and is the distance from the first ruled line to the last, and N is the number of lines ruled on the grating. n is known as the order of the spectrum and may take values ±1, ±2, ... For the successive values of n, successive images are formed. The image corresponding to $n = \pm 1$ is termed the first order and for $n = \pm 2$ the second order, and so on. Equation (4.1) gives the angles of the diffracted images for given angles of incidence. For a fixed order, images of different frequencies will be formed at different values of β, but it follows from Equation (4.1) that overlapping of spectra of different orders may occur. Thus, for constant values of α and β the value of the right-hand side of the equation is fixed, and therefore $n/\tilde{\nu}$ is constant.

It follows that, for a given value of β, the spectra of 1000 cm^{-1} in the first order, 1500 cm^{-1} in the second order, and 3000 cm^{-1} in the third order would be observed at the same position on a recorder.

The chief disadvantage of a grating is now obvious; it efficiently separates radiation of the desired frequency from radiation of adjacent frequency, but it does not separate it from radiation of double or high multiple frequencies. It is usually possible to separate these overlapping orders by: (a) the use of a foreprism which provides a narrow range of wavenumbers and acts as a monochromator for the grating; (b) the use of certain cut-off filters which, though transparent to the desired wavenumbers, are opaque to the sub-multiples of these wavenumbers. Examples of such filters used in the infrared region are thin films of PbS, Ag_2S, and Te deposited on a material which is by itself transparent.

Equation (4.1) holds for both transmission and reflection gratings; the positive sign on the left-hand side applies when the incident and diffracted rays are on the same side of the grating normal, and the negative sign when the rays are on opposite sides of the normal.

Light distribution from a diffraction grating

In an ideal grating, consisting of alternate equidistant opaque and transparent strips, the distribution of light should be uniform on each side of the normal to the grating, and the intensity of successive orders should decrease in a regular manner. Such, however, is not the case, since the rulings have in addition to finite width a definite shape, usually taking the form of a flat trough, the sides of which make unequal angles with the vertical. This results in an irregular distribution of light on the two sides of the normal and also within the orders on one given side. The efficiency of a grating ruled in the ordinary manner with a diamond point cannot be predicted but must be determined experimentally.

It has already been indicated that one disadvantage of the diffraction grating is that it disperses the incident energy over a large number of orders. This may be overcome in the visible region by increasing the fineness of the groove spacing. As the frequency being studied becomes shorter, however, it is desirable to use coarser gratings; for example in the infrared, gratings are used ranging from about 3000 to 30 lines per cm.

This waste of energy by diffraction into a variety of orders can be serious when, for example, in the infrared the total amount of energy is small in any case. In the production of modern gratings this problem can to some extent be overcome by shaping the tip of the ruling diamond so that the contours of the grooves are such as to concentrate the diffracted energy into a definite direction. This is called *blazing* the grating, and the angle the groove makes with the vertical is known as the *blaze angle*. The effect may be illustrated by considering a plane reflection grating having ca. 1000 lines per cm and a blaze angle of 29° which gives a first-order concentration at 1000 cm^{-1}. In addition, such a grating would be most efficient in the wavenumber range from 2000 to 1600 cm^{-1} in the second order, ~ 3000 cm^{-1} in the third order, and so on. Figure 4.6 demonstrates this by the plot of spectral slit width (in cm^{-1} units) against wavenumbers, where the criterion for the choice of order is that the best resolution is obtained for small slit widths.

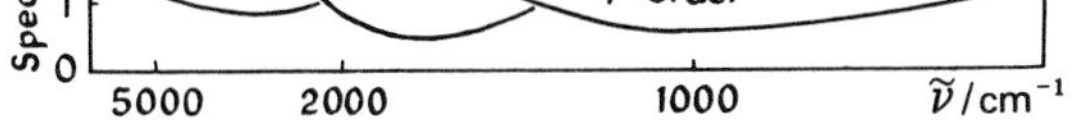

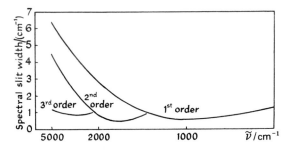

Fig. 4.6 Plot of spectral slit width against wavenumber for three orders of a plane diffraction grating. (Adapted from ref. 4.1. Courtesy of the Institute of Petroleum, London.)

4.2.8 Michelson interferometers

These instruments are especially valuable when energy limitations are very severe. They are designed for use mainly in the 500–10 cm^{-1} region where the stretching and bending modes of bonds involving heavy metal atoms, lattice vibrations, torsional modes, etc. are encountered. The instrumental aspects are discussed in Chapter 5, p.268, and so the details will not be repeated here.

The main advantage of an interferometer over a conventional spectrophotometer is that to obtain high resolution it is no longer necessary to use narrow slits, and so better use can be made of available energy. The resolution of an interferometer is the reciprocal of the distance of travel of the moving mirror and resolutions of up to 0.1 cm^{-1} are readily achieved. Figure 4.7 shows part of the far-infrared spectrum of water vapour recorded on a Beckman-RIIC Model 720 interferometer under routine conditions. Water vapour is often used to calibrate the low-frequency region.

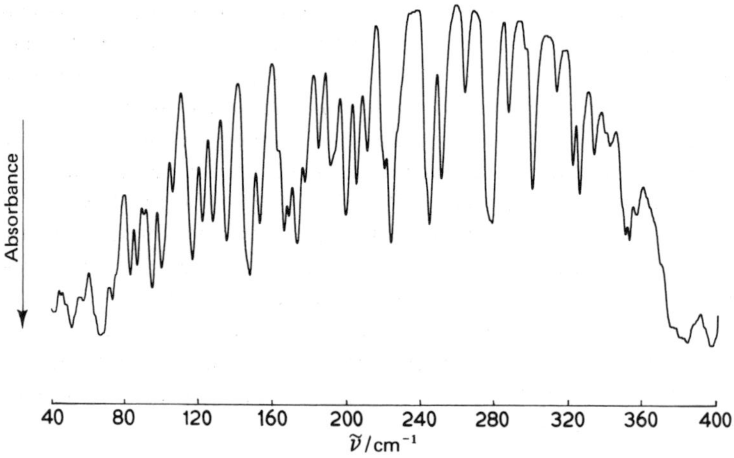

Fig. 4.7 Far-infrared spectrum of water vapour.

Alkali halide plates cannot be used for window materials in the far-infrared because they exhibit strong absorption bands due to lattice vibrations. However, several polymers such as high-density polyethylene show good transmission and they are much easier to handle.

4.2.9 Calibration of spectrophotometers

The spectrophotometer is calibrated by employing substances whose exact absorption positions are known. Naturally, it is essential to use only materials whose absorption bands have been precisely measured. It is desirable to have peaks which are sharp, easily recognized, and found at frequent intervals throughout the range of the spectrophotometer. For greatest accuracy the data given in references 4.2 and 4.3 should be used. A little less accuracy, however, is obtained by using a thin film of polystyrene which has a number of sharp absorption bands whose frequencies are accurately known; a typical infrared spectrum is shown in Fig. 4.8(a), where the minima indicate the positions of maximum absorption. The absorption bands of indene in a 0.05 mm cell also provide a useful series of calibration points and a spectrum is shown in Fig. 4.8(b). The region below 400 cm^{-1} is usually calibrated with water vapour [4.3].

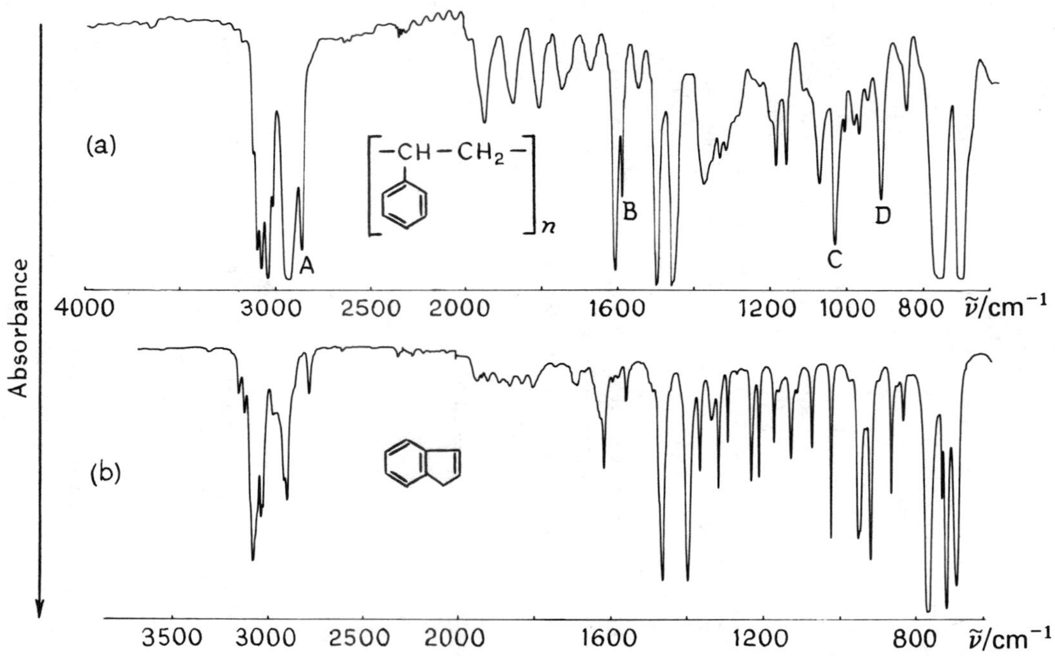

Fig. 4.8 Infrared spectra of (a) polystyrene (solid film 0.05 mm), and (b) indene (liquid film 0.05 mm). (The indene should be sealed in glass ampoules immediately after distillation, and used only from freshly opened ampoules.) Peaks marked: A, 2850.7; B, 1583.1; C, 1028.0; D, 906.7 cm^{-1}.

4.2.10 Infrared spectrum

There is no definite convention regulating the quantities plotted on the two axes, and the following procedures exist.

(i) The wavelength or wavenumber (or occasionally frequency) is plotted against some quantity expressing absorption. In analytical work the wavelength (λ) has been used, whereas in research applications (e.g. structural determinations) wavenumber (cm^{-1}) is a more fundamental and useful unit. Throughout this chapter all band positions have been expressed in wavenumbers, $\tilde{\nu}$ (cm^{-1}), that is, reciprocal wavelength, and following common practice, the term 'frequency' is used in this connection. This is not to be confused with absolute frequency $\nu = c/\lambda$ where c is the velocity of light.

(ii) If I_0 is the intensity of the infrared radiation and I that of the emergent light at the particular frequency concerned, then either the percentage of the infrared light transmitted, that is $100I/I_0$, or the percentage absorbed, $100(I_0 - I)/I_0$, or, absorbance, i.e. $\log_{10}I_0/I$, or even the molar extinction coefficient, i.e. $(1/cl)\log_{10}I_0/I$, may be plotted along the ordinate axis, where c is the concentration (mol dm^{-3}) and l is the length (cm) of the cell through which the radiation travels.

For choice, the optical density is preferable since this is directly proportional to the concentration of the absorbing material. Nearly all the infrared spectrophotometers are limited to the plot of percentage absorption or transmission or optical density. In the case where the percentage transmission is plotted, the frequencies of maximum absorption are indicated by the minima in the curve, while if the percentage absorption is plotted, then the maxima actually represent the maximum absorption.

4.2.11 Preparation and examination of samples

The infrared technique can be used to study the spectra of either pure compounds or mixtures in the gaseous, liquid, and solid states. Alternatively, the compound may be dissolved in a suitable solvent.

(i) *Gases*

Gases are easily handled in the infrared region since the concentration can readily be controlled by varying the pressure. In view of their reduced molecular concentrations, cell path lengths ranging from one centimetre to several metres are used. A typical 10 cm Pyrex gas cell with KBr windows is shown in Fig. 4.9(a). The windows are sealed on with halocarbon wax and the cell is evacuated and tested for leaks before the sample gas is admitted to the required pressure. The small side-arm provides an easy way of condensing gas into the cell. The cells must have windows constructed from material which is transparent in the

Fig. 4.9 (a) Pyrex gas cell with KBr windows.

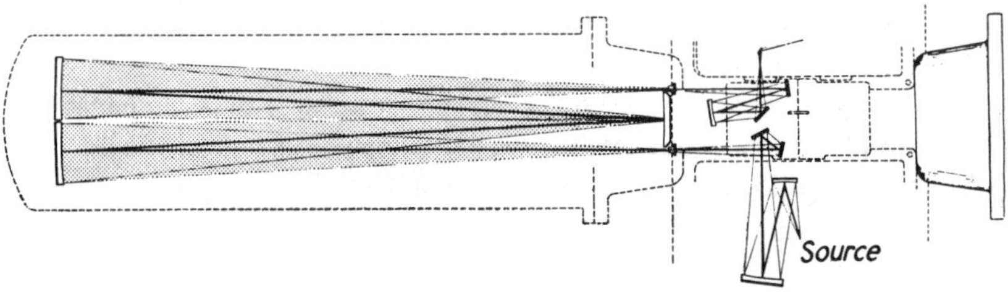

Fig. 4.9 (b) Long-path infrared cell. (Courtesy of Perkin-Elmer Corpn.).

part of the infrared region under investigation (see Table 4.1). Unfortunately, these materials are often readily attacked by chemicals and it is necessary to remove the windows regularly for cleaning. The condition of the windows can easily be checked by running a cell blank.

Special compact multiple reflection cells are available with path lengths of up to 10 m, and for gaseous samples these cells are designed for work between

Fig. 4.9 (c) Demountable cell.

very low pressures and 10 bars. One such long path cell is illustrated in Fig. 4.9(b). This type of cell is particularly useful for the analysis of trace quantities of gas in the atmosphere.

Many compounds do not have sufficient vapour pressure to be examined in the gaseous phase, and in this case the substance may be examined as a pure liquid, or as a solution in a suitable solvent, or as a solid.

(ii) *Spectra of pure liquids*

Pure liquids avoid the interference due to solvent absorptions but complications due to intermolecular interactions may exist to a significant extent (e.g. hydrogen bonding). The liquids are usually examined as very thin films squeezed between the windows of a *demountable cell* such as the one shown in Fig. 4.9(c). These cells are assembled by the operator for each spectrum and then taken apart for cleaning. Since the sample thickness cannot be controlled in cells of this type they are only useful for qualitative or survey studies. For quantitative work it is desirable to use *sealed* cells which have a *fixed path length* in the

Fig. 4.9 (d) Sealed cell.

range 0.005–0.1 mm [see Fig. 4.9(d)]. The space between the windows is determined by a lead or Teflon spacer. These cells are also useful for very volatile liquids, where evaporation is a problem, and for investigating the spectra of solutions.

(iii) *Spectra of solutions*

Solution spectra are examined when the sample absorption is very strong, even using a thin liquid film. The spectra are run in *fixed path length* cells when accurate path lengths are required for quantitative work. It is very important to know the cell thickness accurately, if precise values of extinction coefficients are to be calculated. An accurate value of the cell thickness may be obtained by means of an interference fringe method. The empty cell is calibrated by observing the interference fringe pattern (see Fig. 4.10) created by the reflection of part of the radiation beam from the internal faces of the cell. Thus the

Fig. 4.9 (e) Variable path length cell.

spectrum of an empty cell will look like a sine curve and the thickness of the cell is obtained graphically using the relationship $t(\text{cm}) = n/2\Delta\tilde{\nu}$ where n is the number of complete fringes in the wavenumber interval $\Delta\tilde{\nu}$.

Fixed path length cells must be cleaned by repeated flushing with solvent and the cell blank should be checked. The window surfaces soon deteriorate. Dismantling and repolishing the windows requires skill and patience since it may be necessary in the case of a thin film for the surface to have a high degree of optical flatness to the extent of 10^{-4} mm.

Whenever solution spectra are investigated, it is usual to place a compensating cell, containing an equal thickness of pure solvent, in the reference beam. The absorptions due to solvent bands are thus cancelled out and the recorded spectrum is due only to the compound, provided that the thickness of the two cells

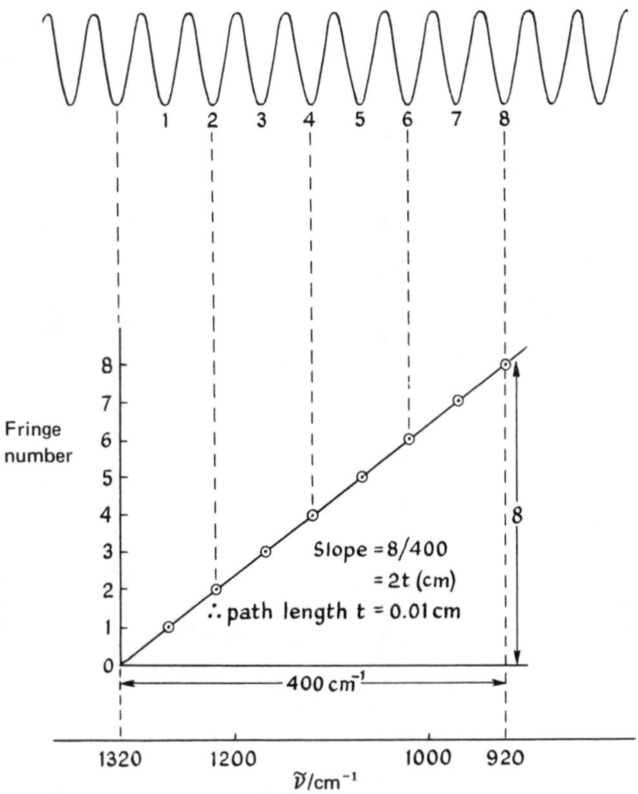

Fig. 4.10 Calibration of a fixed path length cell.

is accurately matched and provided that the solvent absorption is not greater than ~ 70 per cent. When the solvent absorbs very strongly, no light is transmitted by the solution, and regardless of the compound absorptions, the pen draws a straight line in that region of the spectrum. To facilitate the matching of the cell thickness, variable path length cells with micrometer screw arrangements are frequently employed in the reference beam [see Fig. 4.9(e)].

Choice of solvents. The selection of a suitable solvent is governed by two factors: (a) the solvent should not exhibit strong absorption within the wavenumber range of interest; and (b) it should not appreciably interact with the dissolved substance (e.g. hydrogen bonding or complex formation).

Water absorbs strongly over large regions of the infrared spectrum. However, using very thin path lengths (ca. 50 μm) and cell windows of calcium fluoride, silver chloride, or irtran 2, useful information about water-soluble compounds can be obtained for the region 1550–950 cm^{-1}.

One of the best solvents for infrared spectroscopy is CCl_4. Carbon disulphide is also satisfactory but its high volatility makes it difficult to handle in ordinary

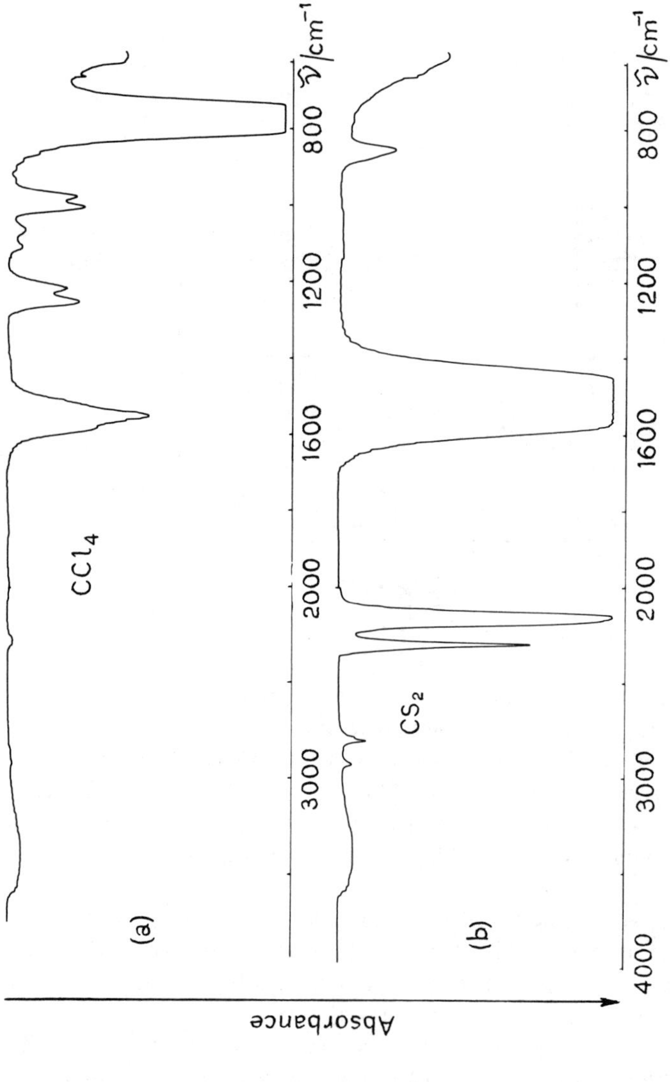

Fig. 4.11 Infrared spectra of two common solvents. (a) CCl_4; liquid film 0.1 mm; (b) CS_2; liquid film 0.1 mm.

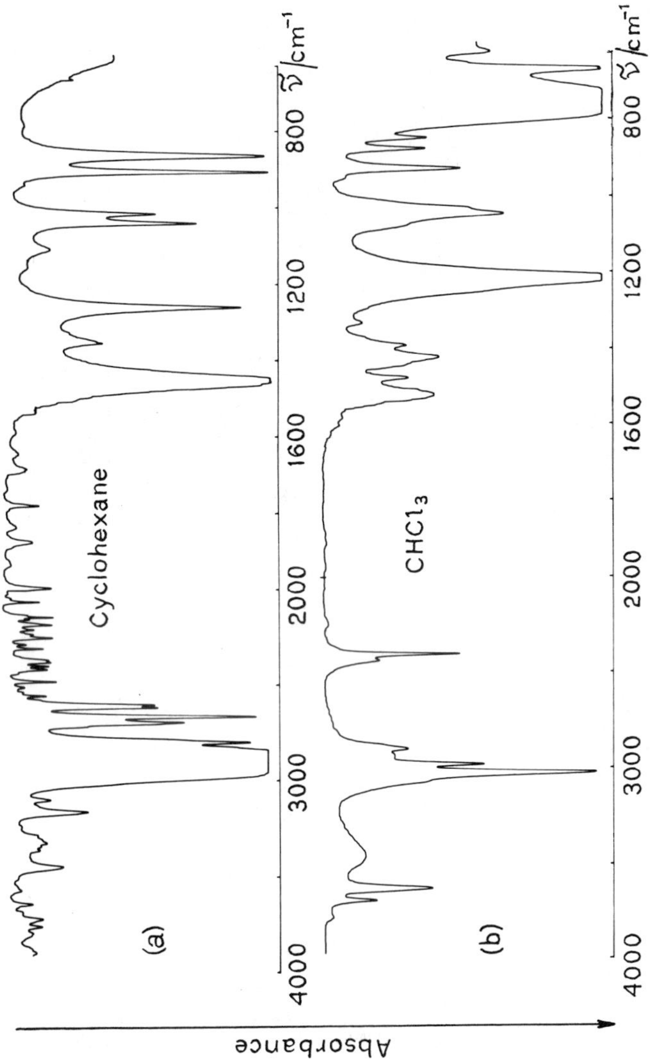

Fig. 4.12 Infrared spectra of (a) cyclohexane; liquid film 0.1 mm; (b) chloroform; liquid film 0.1 mm.

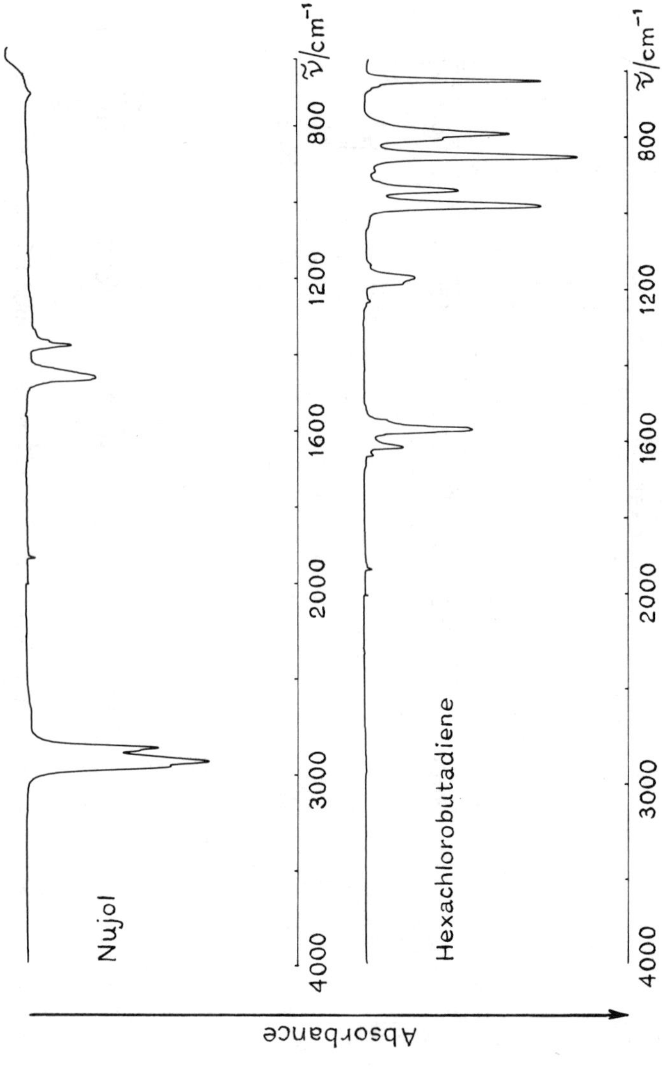

Figs. 4.13 and 4.14 Infrared spectra of two mulling agents (as liquid films).

absorption cells. Figs. 4.11 and 4.12 show the spectra of four of the commonest solvents but others frequently encountered include benzene, methylene chloride, dioxan, acetonitrile, nitromethane, pyridine, acetone, tetrachloroethylene, and dibromo- and tribromo-methane. The examination of solids may be made in one of several ways, some of which are given below.

(iv) *Spectra of solids*

Mull technique. The finely ground solid (~ 2–5 mg) is wetted with two drops of a suitable liquid which has only a few absorption bands in the infrared. Nujol and hexachlorobutadiene have been widely used for this purpose. The suspension is then smeared between two cell windows. The mull spreads out as a very thin film. The function of the liquid is to surround the particles of solid with a medium of high refractive index and so reduce light scatterings. If Nujol is employed, it absorbs at approximately 2900, 1460, 1380, and 725 cm^{-1} but apart from this does not further interfere with the absorption spectrum of the sample (see Fig. 4.13). The CH absorption region may be examined if a fluorocarbon or chlorocarbon is used as the mulling agent (see Fig. 4.14).

Evaporation. A solution or a suspension of a solid is made, and a film of this solid is formed on a suitable transparent window by evaporation of the solvent. The window may be of NaCl for non-aqueous solutions, while AgCl is useful for solutions where water is the solvent.

Melting. A solid may be melted between two plates and then cooled, the resulting crystalline film showing little or no scatter.

For polymers (e.g. nylon), no suitable solvents are available although molten $SbCl_3$ has been employed to dissolve polypeptides. However, if the material has a reasonably low melting point, a high-temperature absorption cell could be used and the substance investigated in the molten state. A further possibility is to examine a very thin film of the substance, the film being cut by means of a microtome.

Pressed disc technique. 1–2 mg of sample is ground up with 200 times its weight of powdered KBr and the mixture is compressed [pressures of ca. 7.5 kbar (ca. 10^5 lb in^{-2} are used] under vacuum to form a clear transparent disc. The advantages of this method over the 'mull' method are that: (a) the concentration and the thickness of the disc are readily determined, thus permitting its use in quantitative analysis; (b) it gives a distribution of particles in the suspending medium, and a smaller particle size is possible; (c) it reduces scatter and sharpens the absorption bands which show more detail; (d) it is possible to store the sample for future reference.

The technique is not limited to KBr; most alkali halides and many other substances can be used (e.g. polythene discs are employed in the far-infrared).

However, unless great care is taken the KBr discs exhibit bands due to the absorption of water. In addition, a serious disadvantage is that the electric field of the positive and negative ions modifies the spectrum. Anomalous spectra due to physical or chemical changes induced by grinding and pressure have been reported also.

In addition to the infrared cells described so far, it is possible to buy commercial cells which are specially designed to enable the samples to be examined at either high or low temperatures. Micro cells for handling small amounts of sample and even low-cost throwaway cells are all readily available.

4.3 PURE ROTATIONAL SPECTRA OF GASEOUS DIATOMIC MOLECULES

The theory of pure rotational energy changes for gaseous molecules will not be discussed in detail because it is described in the chapter concerned with microwave spectroscopy. However, it is worth reminding the reader of the basic equations for both a diatomic rigid rotator and a non-rigid rotator because the expressions will be referred to again in the theory of vibrational–rotational energy levels. (For a more mathematical approach see Appendix A.3.)

The rotational energy of a diatomic rigid rotator is quantized. That is, the energy of two atoms joined together by a rigid bond is limited to certain definite values. The allowed energy levels are given by the expression:

$$E_r = \frac{h^2}{8\pi^2 I} J(J+1) = hcBJ(J+1) \text{ joules}, J = 0, 1, 2, \ldots \quad (4.2)$$

where B is the rotational constant expressed in cm^{-1} which is related to the moment of inertia I by the equation:

$$B = h/8\pi^2 cI \quad (4.3)$$

J is the rotational quantum number of the molecule and governs the rotational angular momentum which has the value of $\sqrt{[J(J+1)]}\, h/2\pi$, c is the velocity of light, and h is Planck's constant. Equation (4.2) expresses the allowed energies in joules; we are more interested, however, in differences between these energies, or more particularly, in the corresponding frequency ($\nu = \Delta E/h$ Hz) or wavenumber ($\tilde{\nu} = \Delta E/hc$ cm^{-1}) of the radiation absorbed as the molecule undergoes a rotational change. Thus in terms of cm^{-1} Equation (4.2) can be rewritten:

$$\frac{E_r}{hc} = BJ(J+1)\,\text{cm}^{-1}, J = 0, 1, 2, 3, \ldots \quad (4.4)$$

We can represent these allowed energy levels diagrammatically as in Fig. 4.15. In principle, there is no limit to the rotational energy the molecule may have, but in practice the point is reached at which the centrifugal force of a rapidly rotating diatomic molecule is greater than the strength of the bond. This point is not reached at normal temperatures. It is clear from Equation (4.2) that when $J = 0$, then $E_r = 0$ and the molecule is not rotating.

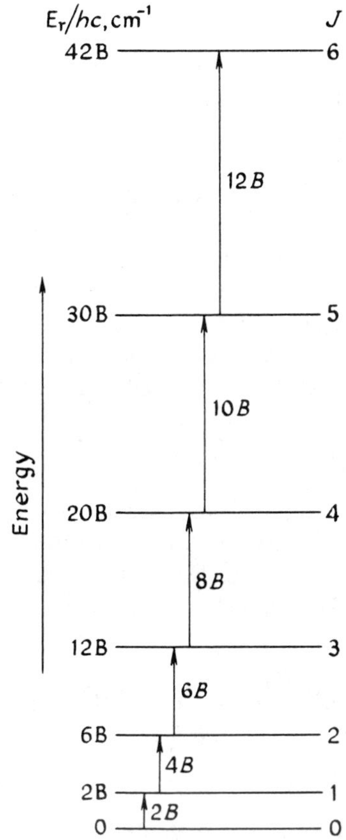

Fig. 4.15 The rotational energy levels and allowed transitions for a rigid diatomic molecule.

Only if the diatomic molecule has a permanent dipole moment can it interact with the infrared radiation and exhibit a pure rotational infrared spectrum. Molecules such as F_2, N_2, and H_2 will be inactive. All heteronuclear diatomic molecules have a permanent electric dipole moment, and hence if a suitable continuous radiation is supplied, the appropriate frequency may be absorbed, and a rotational energy change may take place. The wavenumber of the absorbed radiation $\tilde{\nu}$ can be obtained from the Bohr frequency rule $(E' - E'') = h\nu = hc\tilde{\nu}$ and Equation (4.4):

$$\tilde{\nu} = BJ'(J' + 1) - BJ''(J'' + 1) \tag{4.5}$$

where J'' refers to the lower rotational energy state E_r'', and J' refers to the higher rotational energy state E_r'. The selection rule for changes of J is that $(J' - J'') = +1$ for an absorption transition. On substitution for J' in Equation (4.5) we obtain:

$$\tilde{\nu} = 2B(J'' + 1) \quad J'' = 0, 1, 2, 3, \ldots \tag{4.6}$$

The spectrum of a rigid rotator therefore consists of a series of equidistant lines separated from one another by $2B$ cm^{-1}. The first rotational line (1←0) is to be found at $2B$, that is at $h/4\pi^2 cI$ cm^{-1}, and further absorptions are observed at $4B$, $6B$, $8B$ cm^{-1} etc. (see Fig. 4.15). Thus the region in which the pure rotational spectrum occurs is fixed by the value of B, which in turn depends upon the moment of inertia I.

The concept of a diatomic molecule (or any other molecule) as a rigid rotator is inadequate, since the nuclei must vibrate. Furthermore, the intervals between the main lines are not given exactly by $2B$ because of centrifugal forces in the rotating molecule which cause it to distort and so change its moment of inertia. A more accurate expression for the rotational energy levels of such a molecule is given by:

$$E_r = hcBJ(J+1) - hcDJ^2(J+1)^2 \text{ joules} \tag{4.7}$$

The term $hcDJ^2(J+1)^2$ is only of small magnitude and takes into account the centrifugal distortion in the molecule. D is the centrifugal distortion constant and always has a positive value dependent on the vibrational frequency (ω) of the molecule. D can be related to ω by the equation:

$$D = 4B^3/\omega^2 \qquad (4.8)\dagger$$

Consider a transition $E'' \rightarrow E'$. If we let $(J' - J'') = +1$ and substitute in Equation (4.7), the frequency of the absorbed radiation is given by:

$$\tilde{\nu} = [2B(J''+1) - 4D(J''+1)^3] \text{ cm}^{-1} \tag{4.9}$$

The frequency spacing between successive rotational lines as based on Equation (4.9) is no longer constant but decreases slightly as J increases.

The first infrared study of the pure rotational spectra of the hydrogen halides was made by Czerny [4.4] in 1925. He observed most of the rotational lines of HCl, HF, and HI in the 250–80 cm^{-1} region. Part of an interferometer spectrum of HCl is shown in Fig. 4.16 and the first line is located at approximately 20 cm^{-1}.

The heavier molecules DCl, DBr, and DI have also been studied, although the lowest rotational transitions in these molecules occur below 20 cm^{-1} and are difficult to observe with infrared spectrophotometers.

Czerny found for HCl that the wavenumbers of the rotational lines could be fitted to the equation:

$$\tilde{\nu} = 20.794(J''+1) - 0.00164(J''+1)^3 \text{ cm}^{-1} \tag{4.10}$$

Comparing the coefficients of the linear term in Equations (4.10) and (4.9), the rotational constant B is found to equal 10.397 cm^{-1}. However:

† In the derivation of this equation it is assumed that during the vibration the harmonic oscillator model holds.

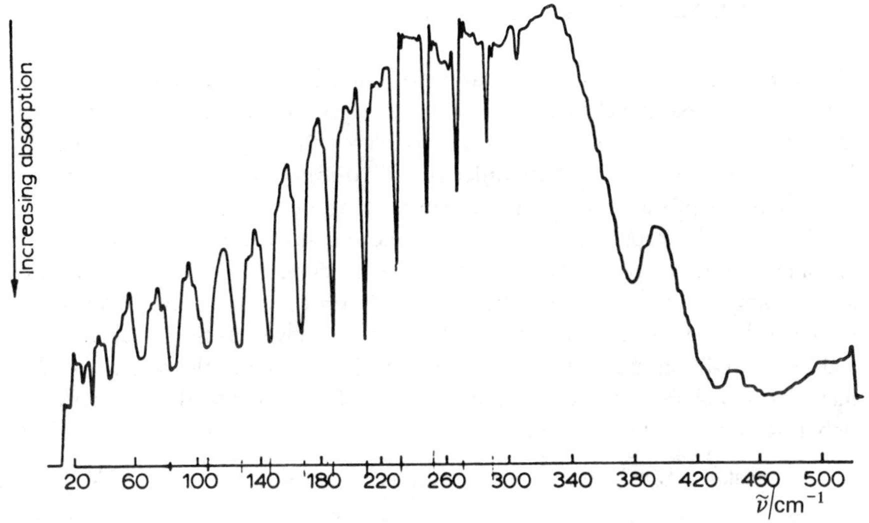

Fig. 4.16 The pure rotational spectrum of gaseous hydrogen chloride recorded using an interferometer. (From ref. 4.5.)

$$B = h/8\pi^2 cI \tag{4.11}$$

and the moment of inertia HCl may be calculated to be 2.72×10^{-47} kg m^2. In addition, since:

$$I = \frac{m_1 m_2}{m_1 + m_2} r^2 = \mu r^2 \tag{4.12}$$

(and the reduced mass μ of HCl is 1.63×10^{-27} kg), the internuclear distance $r_{\text{H-Cl}}$ is found to be 129 pm (1.29 Å). By comparing the coefficients of the term in $(J'' + 1)^3$ we find:

$$4D = 0.00164 \quad \text{and} \quad D = 4.1 \times 10^{-4}\ \text{cm}^{-1} \tag{4.13}$$

Equation (4.8) leads to an approximate value of 3310 cm^{-1} for ω, the vibrational frequency of the HCl molecule. We shall see that a more precise determination leads to a value of $\omega = 2990.6$ cm^{-1} for the $H^{35}Cl$ molecule (see p. 174).

In view of the fact that high-precision data are available from microwave spectroscopy for the pure rotational lines of the hydrogen halides, CO, NO, etc., the lines of these molecules are of considerable value as frequency standards rather than a source of structural data.

Palik [4.6] observed the pure rotational spectrum of DBr, HI, and DI in the 222–60 cm^{-1} region. He found that an equation of the type (4.9) fitted the wavenumbers of the observed lines.

The reader is referred to Chapter 3 for the theory of rotational energy levels for linear, symmetric-, spherical-, and asymmetric-top molecules.

4.4 MOLECULAR VIBRATIONS

The atoms in a molecule are never stationary, whatever the temperature. In fact, even in a solid near the absolute zero temperature the atoms are in constant oscillation about an equilibrium position. However, the amplitude of the oscillation of the atoms is only of the order 10^{-11}–10^{-12} m, while their vibrational frequencies are of the order of the frequency of infrared radiation.

If there are N uncombined atoms free to move in three dimensions, the system would have $3N$ translational degrees of freedom. However, if these atoms are contained in one molecule, there are still $3N$ degrees of freedom of which three degrees are for the translation of the centre of gravity of the molecule, and for a non-linear molecule three degrees for the rotation of the whole which may be resolved about three perpendicular axes. If it is assumed that there are only three types of motion, then:

$$\text{Translational} + \text{Vibrational} + \text{Rotational degrees of freedom} = 3N \tag{4.14}$$

i.e. the vibrational degrees of freedom = $(3N - 6)$.†

For a linear molecule such as Cl_2, CO_2, or C_2H_2 the moment of inertia about the internuclear axis is zero but it has finite moments of inertia about the other two mutually perpendicular axes, and hence there are only two degrees of rotational freedom. Thus, the number of vibrational degrees of freedom for linear molecules is $(3N - 5)$.

To gain some appreciation of the type of vibration which may result from the atoms in a molecule, consider a molecular model, where the nuclei may be visualized as balls and the forces acting between them are represented by spiral springs. It is assumed that the major restoring forces lie along the valence bonds and that the restoring force, for small displacements, would obey Hooke's law. The model chosen is benzene where the carbon balls are twelve times heavier than the hydrogen ones. The springs connecting the carbon and hydrogen balls are now slightly stretched by moving each of the six pairs of weights along the direction of the carbon–hydrogen bonds. The hydrogen balls are moved twelve times further from their equilibrium positions compared with the distance moved by the carbon atoms from their equilibrium positions. When the weights are released simultaneously a vibration sets in where the carbon and hydrogen balls move backwards and forwards along the direction of the bond connecting carbon and hydrogen, that is:

† These values are the number of fundamental vibrational frequencies of the molecule; that is, the number of different 'normal' modes of vibration (see p. 167). In practice, because of degeneracy, the number of fundamental vibrational frequencies which may be measured is often smaller than this.

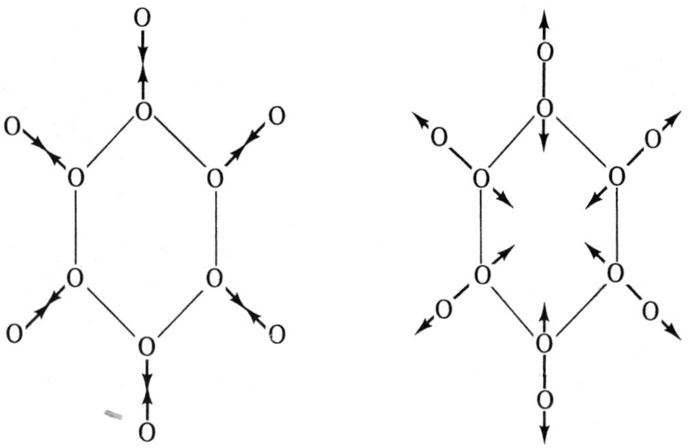

The motion continues along these carbon–hydrogen directions, and the centre of gravity of the system does not alter. This type of vibration is known as the *symmetrical stretching mode of benzene*. Altogether there are theoretically $(3N - 6) = 30$ normal modes of vibration for the benzene molecule. A *normal*

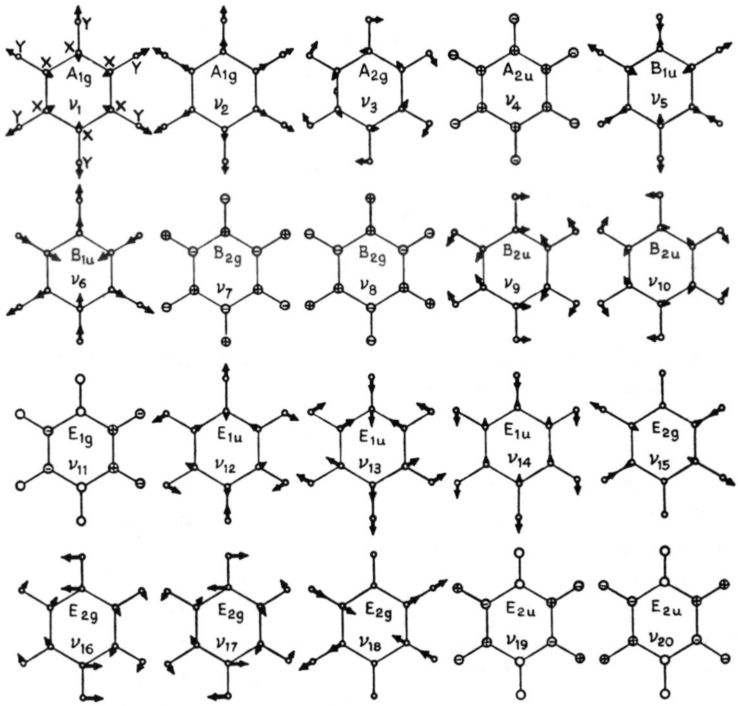

Fig. 4.17 The twenty normal modes of vibration of the benzene molecule. (From ref. 4.7.)

mode of vibration is one where each nucleus executes simple harmonic oscillations about its equilibrium position. All the nuclei move with the same frequency and are in-phase, and the centre of gravity of the molecule remains unaltered. Each normal mode of vibration is treated as the simple harmonic oscillation of what is termed a *normal coordinate*. The normal coordinate is constructed so that it expresses all the individual displacements of the nuclei involved. In practice it is only possible to observe twenty fundamental frequencies for the benzene molecule. This is because ten of the thirty possible modes are doubly degenerate (E), and ten are non-degenerate (A or B). Each degenerate pair of modes gives rise to one frequency. The twenty normal modes of vibration of the benzene molecule are given in Fig. 4.17. A number of these benzene vibrational modes may be analysed by the ball-like nuclei and spring model. Another model may be visualised in which the six carbon balls were displaced slightly above the plane of the ring while the six hydrogen balls were pulled twelve times their displacement below the plane. Then, if all twelve balls were released simultaneously, the resulting motion would be:

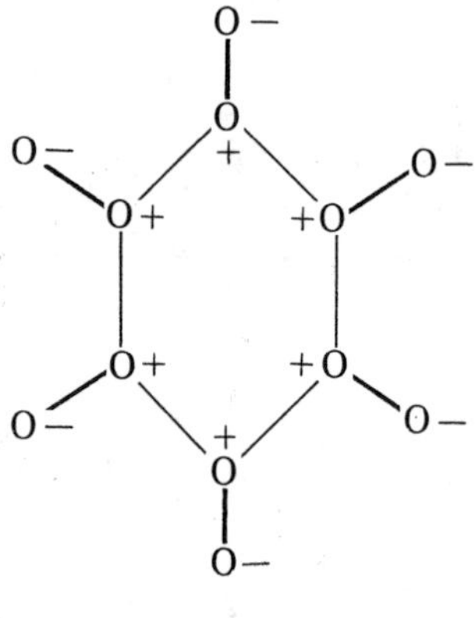

where the positive and negative signs indicate displacements in opposite directions out of the plane of the benzene ring.

In general, if the nuclei were slightly displaced from their equilibrium position and then simultaneously released, the ensuing vibration would have a complex form. The resulting vibration from the displacement could be analysed as a superposition of a limited number of normal vibrations which would have mostly different amplitudes and phases.

To appreciate fully the types of permissible normal modes of vibrations the point group of the molecule must be determined and the symmetry properties of the different modes must be considered. A discussion of molecular symmetry and Group Theory is given in Chapter 2. In Fig. 4.17 the twenty fundamental frequencies of the benzene molecule have been distinguished by the subscripts 1 to 20 on ν and the character species of the modes are included also. The wavenumber positions for eleven of these vibrations are given in Table 4.2. The

Table 4.2 Some of the fundamental frequencies of the benzene molecule

Fundamental modes	$\tilde{\nu}/cm^{-1}$ Raman (liquid benzene)	Infrared (gaseous benzene)
$\tilde{\nu}_{20}$	404	–
$\tilde{\nu}_{18}$	606	–
$\tilde{\nu}_4$	–	671
$\tilde{\nu}_2$	992	–
$\tilde{\nu}_{14}$	1030	1037
$\tilde{\nu}_{17}$	1178	–
$\tilde{\nu}_{13}$	1478	1485
$\tilde{\nu}_{16}$	1585	–
$\tilde{\nu}_{15}$	3047	–
$\tilde{\nu}_1$	3062	–
$\tilde{\nu}_{12}$	–	3099

twenty vibrational modes in Fig. 4.17 are typical for any hexagonal planar molecule (point group D_6h) of the type X_6Y_6. The vibrational modes for the CO_2, C_2H_2, and CH_4 molecules are given on pages 170, 256 and 198 respectively. The frequencies of vibration of the atoms in molecules are of the same magnitude as that of infrared radiation. Provided that the criterion discussed below for infrared absorption is satisfied for a particular vibration, infrared radiation is capable of interacting with the vibrating nuclei and of increasing their vibrational energy.

4.5 ABSORPTION OF VIBRATIONAL ENERGY IN THE INFRARED REGION

4.5.1 Selection rule for infrared absorption

The criterion for the absorption or emission of vibrational energy by a molecule in the infrared region is that a change of electric dipole moment must occur during the normal mode of vibration, that is the vibration should produce temporary displacement of the electrical centre of gravity.

For a diatomic molecule there is only one vibrational mode because $(3N-5) = 3 \times 2 - 5 = 1$. Homonuclear diatomic molecules, X_2, are normally inactive in the infrared region because the molecule has no overall dipole moment initially and the vibration does not produce at any stage a non-zero electric dipole moment. In contrast, heteronuclear diatomic molecules, XY, possess a permanent dipole moment and the vibration of XY about its equilibrium position produces a change in the value of this electric dipole moment. Diatomic molecules of type XY exhibit an infrared spectrum.

In the case of a linear triatomic molecule such as carbon dioxide there are in theory four $(3N-5)$ fundamental vibrational modes. These vibrational modes

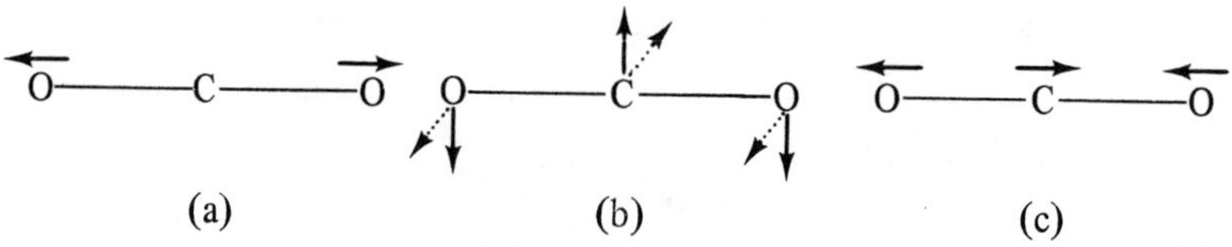

may be represented diagrammatically as follows: (a) is the symmetrical stretching mode of value $\tilde{\nu}_1$; (b) is a deformation mode of value $\tilde{\nu}_2$; during this vibration the molecule ceases to be exactly linear; (c) is the antisymmetrical stretching mode of value $\tilde{\nu}_3$.

The remaining fundamental vibration is accounted for by the fact that the deformation mode is doubly degenerate; that is, if mode (b) as drawn in the plane of the paper were rotated through 90° so that the motion takes place out of the plane of the paper, then the resulting bending mode will occur at the same frequency $\tilde{\nu}_2$. This out-of-plane motion is represented by the dotted lines and thus the expected four frequencies are reduced to three.

To decide which of these vibrations will give rise to infrared bands, one applies the criterion that a change of dipole moment must occur during the vibration. For example:

(i) In vibration (b) the molecule is initially linear but during the vibration it becomes:

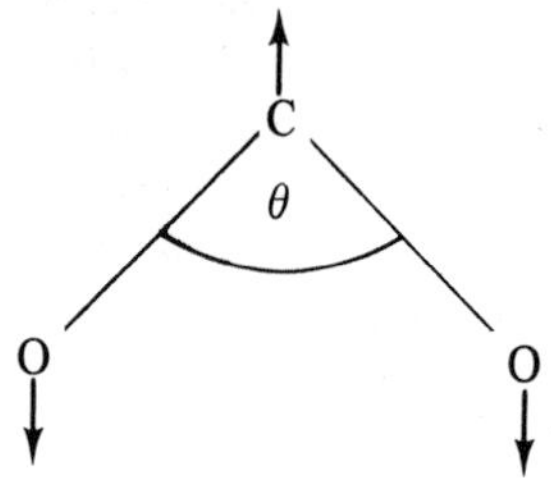

This has a resultant dipole moment of $2\mu_{CO}\cos\theta/2$, where μ_{CO} is the bond moment of one CO link. Hence, the dipole moment varies from 0 through all the $2\mu_{CO}\cos\theta/2$ values, and therefore the criterion for the appearance of an infrared band is satisfied.

(ii) In the case of the symmetrical stretching vibration (a), the molecule never has a resultant dipole moment throughout the vibration because the changes in the bond moments always cancel on vector addition, and therefore no infrared band would be expected.

(iii) For the antisymmetric stretching vibration (c), the two bond moments in each half of the molecule are out-of-phase and thus a non-zero resultant dipole moment is produced. Therefore the molecule is infrared-active when performing this normal mode. It must be stressed that when the antisymmetric vibration alters its vibrational energy by the absorption of the particular frequency $\tilde{\nu}_3$, all the three nuclei are simultaneously involved. The same, of course, holds for $\tilde{\nu}_2$. The motion of a particular fundamental vibrational mode is thus to be treated as a whole and gives rise to definite absorption frequencies. If, during the vibration, a change in dipole moment occurs, an infrared absorption corresponding to that frequency should result. It is not, however, simply a matter of looking at the infrared spectrum and reading off the fundamental vibrational frequencies from the absorption bands, since additional frequencies, which are usually weaker, may also be present. These extra frequencies result from the following.

(i) Integral multiples of the fundamental frequencies; these are known as *overtones*. For example, if $\tilde{\nu}_a$ is the particular fundamental frequency, then the overtones may appear at frequencies of approximately $2\tilde{\nu}_a$, $3\tilde{\nu}_a$, and so on.

(ii) Combinations of the fundamental frequencies and their integral multiples either as their sum or as their difference. Combinations formed from the sum are called *combination bands*. For example, if $\tilde{\nu}_a$ and $\tilde{\nu}_b$ are two fundamental frequencies, possible combination frequencies may occur at $(\tilde{\nu}_a + \tilde{\nu}_b)$ and $(\tilde{\nu}_a + 2\tilde{\nu}_b)$ etc. When the difference between two fundamentals is involved, the bands are called *difference bands* and feasible numerical possibilities are $(\tilde{\nu}_a - \tilde{\nu}_b)$ and $(2\tilde{\nu}_a - \tilde{\nu}_b)$ etc. The activities of overtone and combination bands can be deduced from Group Theory but their intensities are in any case usually weak compared with the intensities of fundamental modes.

4.5.2 Vibrational studies of gaseous diatomic molecules

Simple harmonic vibrations

Only the unsymmetrical type of diatomic molecule, X—Y, may have an infrared spectrum under normal conditions. If the energy values of a vibrating diatomic molecule may be allowed to approximate to that of a harmonic oscillator (i.e. we assume that the bond, like a spring, obeys Hooke's law) then:†

$$E_v = (v + \tfrac{1}{2})hc\omega \text{ joules} \tag{4.15}$$

where v is a vibrational quantum number which can take the values 0, 1, 2, 3, ..., and ω is the vibrational frequency of the oscillator in wavenumbers.

Equation (4.15) gives rise to a set of equally spaced vibrational energy levels at $(1/2)hc\omega$, $(3/2)hc\omega$, $(5/2)hc\omega$, ... etc.

It can be seen that when $v = 0$, $E_v = \frac{1}{2}hc\omega$ joules; this energy is called the *zero-point energy* E_0 and it means that molecules must always vibrate to some extent, even at absolute zero temperature. This is in contrast to the rotational energy E_r which becomes zero when $J = 0$ (see p. 162).

Provided that the vibrational and rotational energies do not interact, the total rotation–vibration energy E_{vr} is given by:

$$E_{vr} = E_v + E_r = (v + \tfrac{1}{2})hc\omega + J(J+1)\frac{h^2}{8\pi^2 I} \tag{4.16}$$

In the infrared region simultaneous changes in both forms of energy are usually

† For a mathematical derivation of Equation 4.15 see Appendix A.4.

studied. Thus, the changes are of the form $v' \leftarrow v''$ and $J' \leftarrow J''$ where v'' and J'' are the values of the vibrational and rotational quantum numbers, respectively, in the lower energy levels from which absorption takes place. The selection rules, which limit the permissible values of the energy absorbed, that is ΔE where:

$$\Delta E = (E_{v'} + E_{r'}) - (E_{v''} + E_{r''}) \tag{4.17}$$

are $\Delta v = (v' - v'') = \pm 1$, if the vibration is simple harmonic,† and $\Delta J = (J' - J'') = 0, \pm 1$, or just ± 1. In the case when $(J' - J'') = 0$ there is no change in the rotational energy accompanying the vibrational energy change:

$$\Delta E = (v' - v'')hc \tag{4.18}$$

and a single absorption line might be expected.

Anharmonic vibrations

It is convenient for the moment to neglect any simultaneous changes in the rotational energies. Equation (4.15) applies only to strictly harmonic vibrations and leads to only one band, because the selection rule $(v' - v'') = \pm 1$ would be rigidly obeyed. In fact, the vibrations of nuclei are never strictly simple harmonic ones. Since the stretching of the oscillator is not perfectly elastic, the vibrations are anharmonic, and the revised equation for the vibrational energy of such an oscillator is given by:

$$E_v = (v + \tfrac{1}{2})hc\omega_e - (v + \tfrac{1}{2})^2 hcx_e\omega_e + (v + \tfrac{1}{2})^3 hcy_e\omega_e + \ldots \text{ joules} \tag{4.19}$$

where x_e and y_e are the so-called *anharmonicity constants*, and ω_e is the hypothetical vibration frequency of negligible amplitude about the equilibrium position of the nuclei. This means that the vibrational energy levels are no longer equally spaced, as in the case of a harmonic oscillator, but they gradually converge as the value of v increases (see Fig. 4.18).

When $v = 0$, the zero-point energy E_0 becomes:

$$E_0 = (1/2)hc\omega_e - (1/4)hcx_e\omega_e + (1/8)hcy_e\omega_e + \ldots \text{ joules} \tag{4.20}$$

When the vibration is anharmonic, there is theoretically no restriction on the change of the vibrational quantum number. Thus, the change Δv may be greater than unity, that is the transitions, $2 \leftarrow 0$ and $3 \leftarrow 0$ may take place, although these higher frequency absorptions are of much lower intensity, and often the amount of absorbing gas has to be increased to observe them. For example, in the case of HCl, if the infrared absorption spectrum is observed with a thin layer of absorbing gas, only the fundamental band is obtained corresponding to a $1 \leftarrow 0$ transition

† The selection rule $\Delta v = \pm 1$ applies for simple harmonic motion only if the dipole moment changes linearly with internuclear distance (see Appendix A.6).

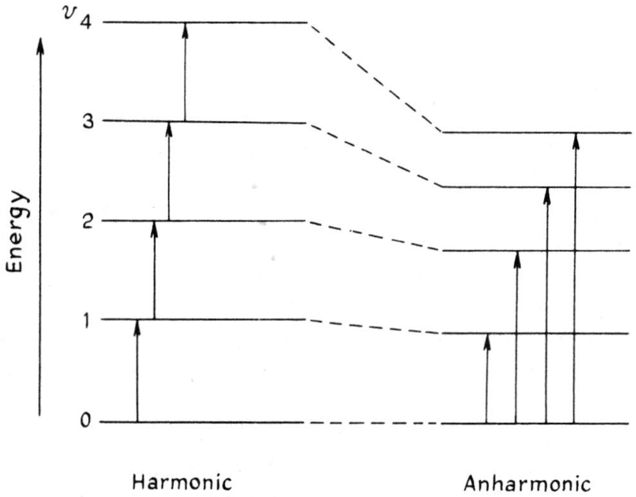

Fig. 4.18 The vibrational energy levels and allowed transitions for harmonic and anharmonic oscillator models. (The convergence of the anharmonic energy levels has been exaggerated.)

at a wavenumber position centred on 2886 cm^{-1}. By studying the absorption with a thicker layer of HCl gas, in addition to the first band, a second band appears corresponding to 2←0 at a wavenumber of approximately 5668 cm^{-1}, that is, at about twice the initial value. This is the *first overtone*. By increasing the path length of HCl gas to several metres at atmospheric pressure, five bands may be obtained whose wavenumbers are approximately two, three, four, and five times that of the 2886 cm^{-1} band, and are, respectively, the first, second, third, and fourth overtones of this vibration. It is to be noted that when all five bands appear the intensities follow the order $1 \leftarrow 0 \gg 2 \leftarrow 0 > 3 \leftarrow 0 > 4 \leftarrow 0 > 5 \leftarrow 0$. If the third and higher terms on the right-hand side of Equation (4.19) are neglected, the energy change in a vibrational transition from the state v'' to the state v' is:

$$E_{v'} - E_{v''} = (v' - v'')hc\omega_e - [v'(v'+1) - v''(v''+1)]hcx_e\omega_e = \Delta E = hc\tilde{\nu} \tag{4.21}$$

where $\tilde{\nu}$ is the wavenumber for the transition between these two vibrational levels. The majority of the molecules are in the $v'' = 0$ state before absorption of energy, and if v'' is made equal to zero in Equation (4.21) then:

$$v' \leftarrow 0, \tilde{\nu} = \omega_e v' - v'(v'+1)x_e\omega_e \tag{4.22}$$

for

$$1 \leftarrow 0, \tilde{\nu}_1 = (1 - 2x_e)\omega_e \tag{4.23}$$

$$2 \leftarrow 0, \tilde{\nu}_2 = (1 - 3x_e)2\omega_e \tag{4.24}$$

$$3 \leftarrow 0, \tilde{\nu}_3 = (1 - 4x_e)3\omega_e \qquad (4.25)$$

Since x_e is usually small (~ 0.01), then from Equations (4.23)–(4.25), $\tilde{\nu}_1 : \tilde{\nu}_2 : \tilde{\nu}_3$ are approximately as 1:2:3. In fact, it was observed for the near-infrared absorption spectrum of HCl gas, that $\tilde{\nu}_1 : \tilde{\nu}_2 : \tilde{\nu}_3 : \tilde{\nu}_4 : \tilde{\nu}_5 = 2886 : 5668 : 8347 : 10\,023 : 13\,397$ cm^{-1}, numbers which are very approximately in the ratio of 1:2:3:4:5. The population of the molecules in the $v = 0$ level is generally much greater than in the $v = 1$ level, and in the case of HCl the ratio of the number of molecules in the first to that in the zero vibrational level at 300 K is 9.77×10^{-7}. Hence, it would be expected that the intensities of transitions from the $v = 1$ level would be very much weaker. The possibility of obtaining such bands is favoured by higher temperatures, and in the case of HCl the ratio at 1000 K is increased to 1.57×10^{-2}. In general, studies are usually made on transitions from $v'' = 0$ to v', where $v' = 1, 2, 3, 4, \ldots$.

If the experimental frequencies are inserted in equations of the types (4.23)–(4.25), mean values of x_e and ω_e can be calculated. The reader can easily confirm that $x_e = 0.0174$ and $\omega_e = 2990$ cm^{-1} for HCl gas where $\tilde{\nu}_1 = 2886$ and $\tilde{\nu}_2 = 5668$ cm^{-1}. This procedure is feasible for the infrared transitions owing to the observable overtones. The numerical values of ω_e and x_e are useful:

(i) for expressing the vibrational energy values in terms of v:

$$E_v = (v + \tfrac{1}{2})hc\omega_e - (v + \tfrac{1}{2})^2 hcx_e\omega_e \qquad (4.26)$$

(ii) to obtain a very approximate value for the dissociation energy from $D_e \approx \omega_e/4x_e$ (cf. Vol. 3, Chapter 5). A much better value can be derived from a Birge–Sponer extrapolation (cf. Vol. 3, Chapter 5).

(iii) to calculate the constant α in the Morse equation. The Morse equation is:

$$\text{Potential energy} = D_e[1 - \exp\{-\alpha(r - r_e)\}]^2 \qquad (4.27)$$

where $\alpha = (8\pi^2\mu x_e\omega_e c/h)^{\frac{1}{2}}$, D_e is the spectroscopic dissociation energy (see Vol. 3, Chapter 5) which includes the zero-point energy, μ is the reduced mass, and r_e is the value of r at the minimum of the potential-energy curve.

The value of ω_e itself may be employed:

(i) to calculate force constants from the equation:

$$k = 4\pi^2 c^2 \omega_e^2 \mu \qquad (4.28)$$

where μ is the reduced mass and k the force constant.

(ii) in conjunction with other data for the calculation of thermodynamic quantities such as entropy and heat capacity (cf. Vol. 2, Chapter 7).

4.5.3 Vibrational studies of triatomic molecules

For a diatomic molecule executing simple harmonic motion with a vibrational frequency of ω, the vibrational energy is given by Equation (4.15).

For a triatomic molecule, however, there are three fundamental frequencies ω_1, ω_2, and ω_3, and for a simple harmonic oscillator the vibrational energy of the molecule would be given by:

$$E_v = \left(v_1 + \frac{1}{2}\right)hc\omega_1 + \left(v_2 + \frac{d_2}{2}\right)hc\omega_2 + \left(v_3 + \frac{d_2}{2}\right)hc\omega_3 \quad (4.29)$$

where the change in vibrational energy, $E_{v'} - E_{v''}$, is governed by the (v_1', v_2', v_3') and the (v_1'', v_2'', v_3'') values in the upper and lower levels, respectively; d_2 is the degeneracy of the bending mode and it has a value of 1 for a non-linear molecule and 2 for a linear molecule.

If the vibrational motion was strictly a simple harmonic one, the changes in the vibrational quantum numbers for each vibration would be restricted to:

$$\begin{aligned} v_1' - v_1'' &= 1 = \Delta v_1 \\ v_2' - v_2'' &= 1 = \Delta v_2 \\ v_3' - v_3'' &= 1 = \Delta v_3 \end{aligned} \quad (4.30)$$

In addition, each vibrational change would occur independently and combination frequencies such as:

$$(\omega_1 + \omega_2); \quad (\omega_1 + \omega_3); \quad (\omega_2 + \omega_3); \quad (\omega_1 + \omega_2 + \omega_3) \quad (4.31)$$

would be forbidden. However, in practice the motion is not strictly a simple harmonic one, and combination frequencies may be obtained. Since the motion is anharmonic, Δv_1, Δv_2, and Δv_3 are not limited to unit changes, and in fact may be small integers. The frequencies of the bands composed of these combination or overtone frequencies are given approximately by the general formula:

$$\tilde{\nu} = \Delta v_1 \omega_1 + \Delta v_2 \omega_2 + \Delta v_3 \omega_3 \quad (4.32)$$

where Δv_1, Δv_2, and Δv_3 are small integers (or zero). When two of the Δv values are zero (e.g. $\Delta v_2 = \Delta v_3 = 0$), and $\Delta v_1 > 1$, overtones of the fundamental vibrational frequency ω_1 are obtained, and when $\Delta v_1 = 2$ this frequency is the *first overtone*, while when $\Delta v_1 = 3$ this is the *second overtone.*

Thus, in practice, equations more complex than (4.29) have to be employed in order to represent the experimentally determined vibrational data most accurately.

For non-linear triatomic molecules, such as water, where the fundamental frequencies are ω_1, ω_2, and ω_3, the experimental data may be fitted to the equation:

$$\frac{E_{v'}}{hc} = x_0 + x_1 v_1 + x_2 v_2 + x_3 v_3 + x_{11} v_1^2 + x_{22} v_2^2 + x_{33} v_3^2 + x_{12} v_1 v_2 + x_{13} v_1 v_3 + x_{23} v_2 v_3 \quad (4.33)$$

The unknowns $x_1, x_3, x_{11}, x_{22}, x_{33}, x_{12}, x_{13}$, and x_{23} can be determined from the data on nine band centres. The fundamental vibrational frequencies can be calculated from these constants by means of the equations:

$$\omega_1 = x_1 - x_{11} - \tfrac{1}{2}x_{12} - \tfrac{1}{2}x_{13} \tag{4.34}$$

$$\omega_2 = x_2 - x_{22} - \tfrac{1}{2}x_{12} - \tfrac{1}{2}x_{23}$$

$$\omega_3 = x_3 - x_{33} - \tfrac{1}{2}x_{13} - \tfrac{1}{2}x_{23} \tag{4.35}$$

Except in the case of diatomic molecules, it will be assumed in future that the vibrational motion is a simple harmonic one. In fact, this is never strictly true, but, as might be inferred from the example of water which has just been considered, unless this assumption is made the treatment becomes too complex, even for very simple molecules.

Table 4.3 Rotation-vibration bands of H_2O in transitions from the ground state (i.e. the $v_1'' = 0$, $v_2'' = 0$, $v_3'' = 0$ state)

	Values of v in excited state $v_1' v_2' v_3'$	*Wavenumber of band* $\tilde{\nu}/\text{cm}^{-1}$	*Type of frequency*
(a)	1, 0, 0	3652	fundamental
(b)	0, 1, 0	1595	fundamental
(c)	0, 2, 0	3151	first overtone of (b)
(d)	0, 0, 1	3756	fundamental
(e)	0, 0, 3	11 032	second overtone of (d)
(f)	1, 1, 1	8807	combination of (a), (b) and (d)
(g)	0, 1, 1	5332	combination of (b) and (d)
(h)	2, 1, 1	12 151	combination
(i)	3, 1, 1	15 348	combination

The number of rotation–vibration bands which may be observed for such a simple molecule like water is quite considerable, and in Table 4.3 some of the vibrational transitions from the ground state, that is the (0, 0, 0) level to the (v_1', v_2', v_3') levels, are given, and the corresponding band is identified as either a fundamental, an overtone, or a combination frequency. From Table 4.3 the overtone and combination frequency values may be accounted for in terms of the given fundamentals.

4.6 INFRARED ROTATION–VIBRATION SPECTRA OF GASEOUS DIATOMIC MOLECULES

The rotational structure of vibrational infrared bands has been neglected so far but the structure is usually resolved either completely or partially in gas-phase spectra and the theory will now be considered. The main object of resolving rotational structure is to determine internuclear distances, and in relevant cases, bond angles.

If the vibrational and rotational energies may be regarded as additive, then the total rotation–vibration energy E_{vr} of the molecule is given by the equation:

$$E_{vr} = (v + \tfrac{1}{2})hc\omega_e - (v + \tfrac{1}{2})^2 hcx_e\omega_e + \ldots hcB_vJ(J+1) - hcD_vJ^2(J+1)^2 + \ldots \tag{4.36}$$

When a rotation–vibration change takes place between $v' \leftarrow v''$ and $J' \leftarrow J''$ the wavenumber of the resultant energy change ΔE is given by:

$$\tilde{\nu} = \tilde{\nu}_0 + B_{v'}J'(J'+1) - D_{v'}J'^2(J'+1)^2 - B_{v''}J''(J''+1) + D_{v''}J''^2(J''+1)^2 \tag{4.37}$$

Equation (4.37) is formed from the vibrational and rotational energy equations, where $\tilde{\nu}_0$ is the wavenumber of the pure vibrational transition and corresponds to a transition between two vibrational levels, where both J' and $J'' = 0$.

The selection rule for J depends on whether the molecule has electronic angular momentum ($\Lambda h/2\pi$, see Vol. 3, Chapter 1) about the internuclear axis. If $\Lambda = 0$, the selection rule for J is:

$$\Delta J = \pm 1 \tag{4.38}$$

Unlike the situation in pure rotational absorption spectroscopy, both of these possibilities are important. For diatomic molecules Λ normally is zero and two branches are observed. These branches are the P-*branch* when $\Delta J = -1$, and the R-*branch* when $\Delta J = +1$. An exceptional case is that of nitric oxide which has an unpaired electron and $\Lambda = 1$. The infrared spectrum of this molecule is governed by the selection rule:

$$\Delta J = 0, \pm 1 \tag{4.39}$$

and the branch where $\Delta J = 0$ is termed the Q-*branch*.

The absorption curve for the fundamental transition, that is the 1←0 transition, of HCl, is given in Fig. 4.19. The lines in the P- and R-branches may be observed; a gap between these two branches occurs at 2886 cm^{-1}, and this position is termed the band centre or band origin. It will be noted that no Q-branch is observed. If such a branch had been theoretically possible and had actually appeared, the centre of its maximum would have been at the same frequency as that of the band centre. In Fig. 4.20 the type of transition which led to the R- and P-branches for the fundamental transition of HCl is indicated. In addition, some of the transitions which lead to the lines of the R- and P-branches of the first overtone (2←0) are also given.

The wavenumber of the lines in rotation–vibration changes may be obtained from Equation (4.37). The wavenumber of a P-line may be obtained in terms of J'' by substituting $(J'' - 1)$ for J' in this equation, while for an R-line the substitution would be $(J'' + 1)$. The resulting equations would reproduce the wavenumbers of the observed lines within experimental error when the appropriate

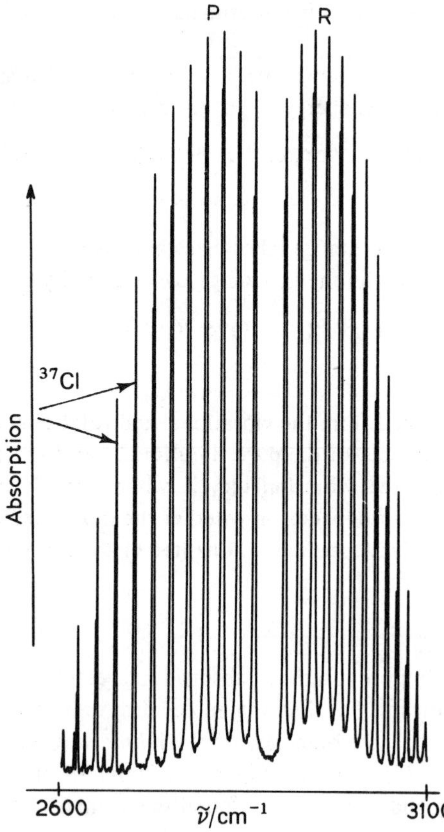

Fig. 4.19 The 1←0 transition of HCl (gas pressure 26 cm Hg in a 10-cm cell), showing the P- and R-branches. The shoulder on the low wavenumber side of each rotational line is due to the ^{37}Cl isotope.

value of J'' was inserted. However, other simpler cases may be considered, where approximations are involved.

Case 1. When both $D_{v'}$ and $D_{v''}$ can be neglected, the resulting equation is:

$$\tilde{\nu} = \tilde{\nu}_0 + B_{v'}J'(J'+1) - B_{v''}J''(J''+1) \tag{4.40}$$

The results for each branch are as follows.

(a) For the R-branch on substitution of $(J''+1)$ for J' in Equation (4.40):

$$\tilde{\nu}_R = \tilde{\nu}_0 + 2B_{v'} + (3B_{v'} - B_{v''})J + (B_{v'} - B_{v''})J^2 \tag{4.41}$$

where $\tilde{\nu}_R$ is the wavenumber of the line whose rotational quantum has the value J. $J'' = J$, and it may take the values 0, 1, 2, 3, ...

(b) For the P-branch on substitution of $(J''-1)$ for J'.

$$\tilde{\nu}_P = \tilde{\nu}_0 - (B_{v'} + B_{v''})J + (B_{v'} - B_{v''})J^2 \tag{4.42}$$

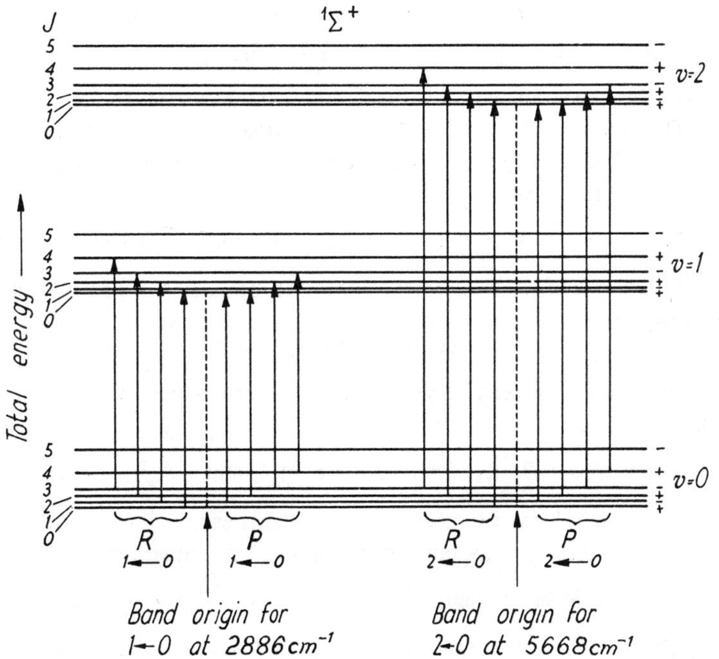

Fig. 4.20 Infrared rotation–vibration transitions for the fundamental and first overtone bands of HCl with a ${}^1\Sigma^+$ ground electronic state.

where $\tilde{\nu}_P$ is the wavenumber of the line whose rotational quantum number has the value J. $J'' = J$, and it may take the values $1, 2, 3, \ldots$.

(c) For the Q-branch on substitution of J'' for J':

$$\tilde{\nu}_Q = \tilde{\nu}_0 + (B_{v'} - B_{v''})J + (B_{v'} - B_{v''})J^2 \qquad (4.43)$$

where $\tilde{\nu}_Q$ is the wavenumber of the line whose rotational quantum number has the value J. $J'' = J$, and J may take the values $1, 2, 3, \ldots$.

It is possible to represent the P- and R-branch lines by means of the formula:

$$\tilde{\nu} = \tilde{\nu}_0 + (B_{v'} + B_{v''})m + (B_{v'} - B_{v''})m^2 \qquad (4.44)$$

where m takes the values $1, 2, 3, \ldots$ for the R-branch (i.e. $m = J'' + 1$), and for the P-branch the values $-1, -2, -3$ (i.e. $m = -J''$). Since m does not equal zero for either the R- or P-branches, a line is missing for this value of m. This missing line whose wavenumber $\tilde{\nu} = \tilde{\nu}_0$ is the band centre and corresponds to the forbidden transition between $J' = 0$ and $J'' = 0$.

Case 2. Equation (4.44) usually gives a good fit to the experimentally observed wavenumbers although it often does not represent them exactly. An exact fit may be obtained by replacing that equation by the equation:

$$\tilde{\nu} = \tilde{\nu}_0 + (B_{v'} + B_{v''})m + (B_{v'} - B_{v''} - D_{v'} + D_{v''})m^2 - 2(D_{v'} + D_{v''})m^3 - (D_{v'} - D_{v''})m^4 \quad (4.45)$$

which takes into account the centrifugal stretching. However, $D_{v'}$ is almost equal to $D_{v''}$, and the $(D_{v'} - D_{v''})m^4$ term is normally very small, and even for exact analyses is often neglected. If $D_{v'}$ is approximately equal to $D_{v''}$ then Equation (4.45) reduces to:

$$\tilde{\nu} = \tilde{\nu}_0 + (B_{v'} + B_{v''})m + (B_{v'} - B_{v''})m^2 - 4D_{v''}m^3 \quad (4.46)$$

and this equation is of the form of the empirical formula (4.47) employed in such analyses:

$$\tilde{\nu} = \tilde{\nu}_0 + am + bm^2 + cm^3 \quad (4.47)$$

The constants a, b, and c can be evaluated by analysis of the rotational structure of the band. Since:

$$a = (B_{v'} + B_{v''}), \quad b = (B_{v'} - B_{v''}), \quad \text{and} \quad c = -4D_{v''} \quad (4.48)$$

it follows that the values of the rotational constants can be determined. Once the values of $B_{v'}$ and $B_{v''}$ have been obtained, from the equations:

$$B = h/8\pi^2 cI \quad \text{and} \quad I = \mu r^2 = m_1 m_2 r^2/(m_1 + m_2) \quad (4.49)$$

the appropriate value of the internuclear distance (r) follows immediately. If several absorption bands are examined, the value of B_e can be obtained (see Vol. 3 Chapter 2) and from this the equilibrium internuclear separation r_e follows.

Figure 4.19 shows that the lines in the rotational fine structure are not equally intense, and its effect can be explained in terms of the Boltzmann distribution and the degeneracy of the energy states (see Vol. 2, Chapter 3, p.82). Finally, it can be seen in Fig. 4.19 that the spacings of the rotational lines in the P-branch are noticeably very different from those in the R-branch. This follows from Equation (4.44) because the value of the rotational constant B will vary depending on the particular vibrational state involved. $B \propto 1/r^2$ and r will increase as the molecule vibrates in higher and higher vibrational levels.

4.7 INFRARED ROTATION–VIBRATION SPECTRA OF SIMPLE GASEOUS POLYATOMIC MOLECULES

Simple polyatomic molecules will now be considered whose infrared spectra can be examined in the gaseous phase and where either the band contour or the rotational structure of the bands can be observed. The theory concerned with rotation–vibration spectra of polyatomic molecules overlaps to some extent the theory presented in the Microwave and Far-Infrared chapters. However, additional features associated with vibrational bands such as P-, Q- and R-branches, parallel and perpendicular vibrations, Coriolis forces, and the advent of tunable infrared lasers with very high resolution, make it necessary to restate the important parts of the theory.

The rotational changes associated with vibrational transitions lead to a fine structure of the vibrational bands (cf. diatomic molecules). Since with polyatomic molecules there are three mutually orthogonal moments of inertia, which generally have quite different values, the rotational fine structure tends to be rather complex. Even if grating spectrophotometers of high resolving power are employed only small molecules have had their rotational structure completely resolved. For larger molecules the grating spectrum shows band contours with no additional characterizing shape and yields no fundamental information on internuclear distances. An example is shown in the lower curve of Fig. 4.21 where the band contour of the antisymmetric stretching mode, $\tilde{\nu}_3$, of SF_6 gas is shown. However, as the resolution is increased the shape of the fine structure begins to emerge and such a case may be seen in the top spectrum of Fig. 4.21 where part of the fine structure of the $\tilde{\nu}_3$ band of SF_6 has been resolved using a tunable infrared diode laser. Tunable infrared lasers are just becoming commercially available and their very high resolution (3×10^{-6} cm^{-1}) means that the theory of rotation–vibration transitions for polyatomic molecules is of key importance in an understanding of these resolved spectra.

The shape and intensity of the P-, Q-, and R-branches depend on the following factors: (a) whether the electric dipole moment change is parallel or perpendicular to an axis of symmetry; (b) the magnitude of the electric dipole moment change during the vibration; (c) the relative values of the three principal moments of inertia. In Fig. 4.22 the R-, P-, and Q-branches for the perpendicular vibration (see below) of CO_2 are shown. For absorption studies, if the vibrational motion is treated as being a simple harmonic one, then the vibrational transitions are governed by $\Delta v = +1$. Changes in J, however, depend on the vibrational mode. If the change in electric dipole moment is in a direction parallel to the axis of symmetry, the vibration is said to be a *parallel vibration* and is given the symbol $\|$, while if the change in electric dipole occurs perpendicular to the symmetry axis, the vibration is described as a *perpendicular vibration* and is given the shorthand symbol $\perp$. Every band, either fundamental or overtone, may be classified as belonging to either the parallel or perpendicular type.

The best source of information for determining internuclear distances and bond angles in very simple molecules results when the rotational structure of a rotation–vibration band is completely resolvable. The rotation–vibration changes for CO_2 corresponding to the antisymmetric stretching vibration are an example of this. The resolved spectrum is given in Fig. 4.23 and the m values have been assigned to each peak value.† From the spacing between consecutive peaks the internuclear carbon–oxygen distance can be calculated. It will be

† For this very strong parallel $\tilde{\nu}_3$ band of CO_2 at 2349.3 cm^{-1} it will be noted that alternate lines are absent. This arises because CO_2 is a linear symmetric molecule, and since the nuclear spin of both ^{12}C and ^{16}O is zero, the antisymmetric rotational levels are missing entirely (see p. 186). In the initial state only those levels where J is even occurs and in the final state only those are present where J takes odd values.

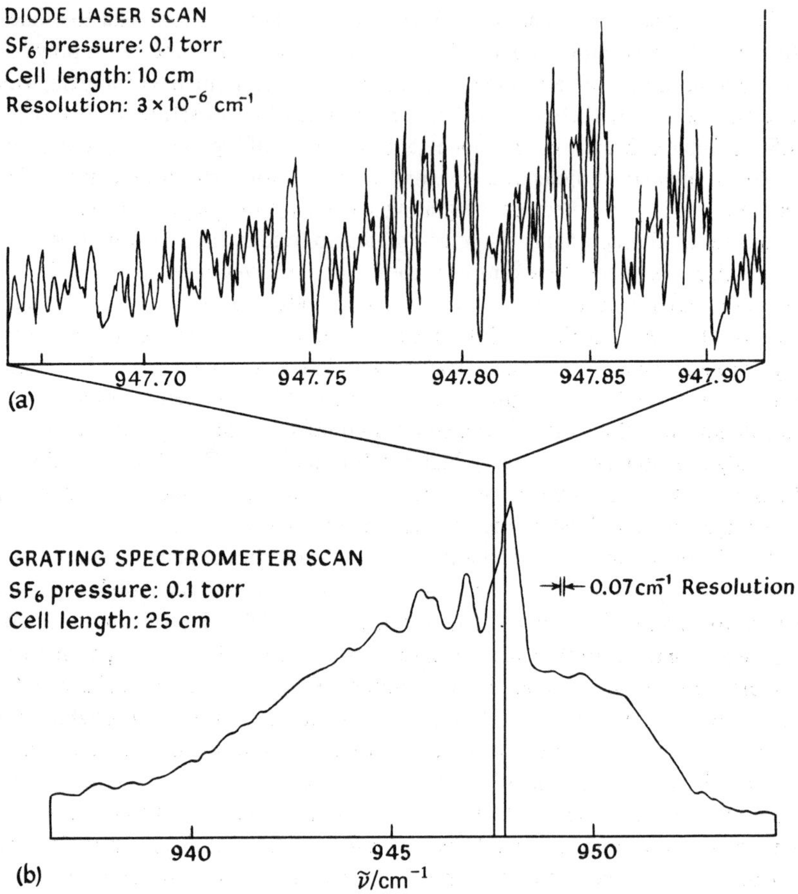

Fig. 4.21 Fine structure of the antisymmetric stretching mode of SF_6 resolved using (a) a diode laser, and (b) a grating spectrometer. (From ref. 4.8.)

noted that for this type of vibrational change there is no Q-branch. This is in harmony with theory, and for this linear molecule it would identify the vibrational change as being of the parallel type. A perpendicular vibration, however, would have a Q-branch as well. This is borne out in Fig. 4.22 for the rotation–vibration band of CO_2 at 667 cm^{-1} which is identified with the fundamental perpendicular bending vibration.

4.7.1 Linear molecules ($I_A = 0, I_B = I_C$)

The energy of a rigid rotator which is linear is given by an expression identical to that for a diatomic molecule, that is:

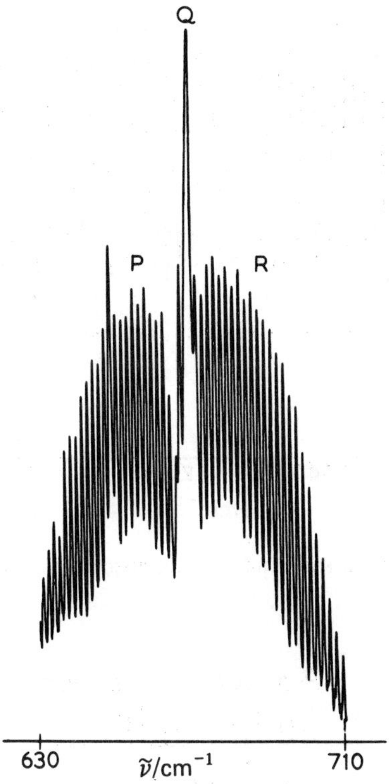

Fig. 4.22 Infrared spectrum of the CO_2 bending mode, showing the PQR structure.

$$E_r = \frac{h^2}{8\pi^2 I} J(J+1) \text{ joules} \tag{4.50}$$

The vibrational energy (assuming simple harmonic motion) is given by an equation of the type:

$$E_v = \sum_i (v_i + d_i/2) hc\omega_i \text{ joules} \tag{4.51}$$

where ω_i is the vibrational frequency of the ith normal mode and d_i is the degree of degeneracy of ω_i. For a triatomic molecule Equation (4.51) becomes Equation (4.29).

It is important to realize that, in contrast to the situation that exists in diatomic molecules, linear symmetric molecules can exhibit rotation–vibration spectra even if the molecule possesses no permanent dipole moment. Examples are the antisymmetric stretching and degenerate bending modes of CO_2 and HC≡CH.

Assuming that the vibrational and rotational energies are additive, then for the total rotation–vibration energy:

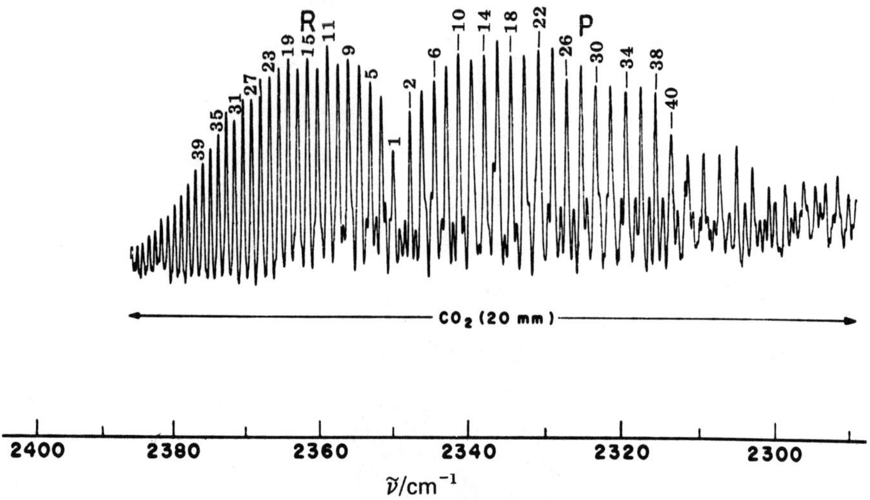

Fig. 4.23 Fine structure in the fundamental absorption band $\tilde{\nu}_3$ of CO_2. (From ref. 4.2.)

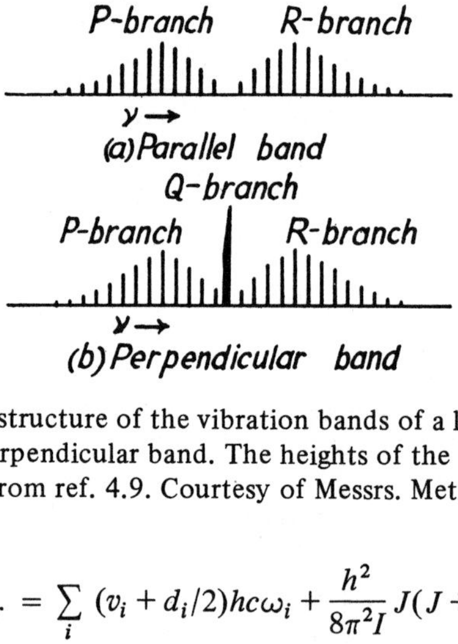

Fig. 4.24 Rotational structure of the vibration bands of a linear molecule: (a) a parallel band; (b) a perpendicular band. The heights of the lines indicate the relative intensities. (From ref. 4.9. Courtesy of Messrs. Methuen and Co. Ltd.)

$$E_{vr} = \sum_i (v_i + d_i/2)hc\omega_i + \frac{h^2}{8\pi^2 I} J(J+1) \tag{4.52}$$

$$E_{vr} = \sum_i (v_i + d_i/2)hc\omega_i + hcB_v J(J+1) \tag{4.53}$$

where B_v is the value of the rotational constant in the vth vibrational level.

The rotational quantum number selection rules for a linear molecule are $\Delta J = \pm 1$ for a parallel vibration and $\Delta J = 0, \pm 1$ for a perpendicular vibration.

When $\Delta J = -1$ for a particular line, this line is a member of the P-branch, while a line for $\Delta J = +1$ would be a member of the R-branch, and for $\Delta J = 0$ the Q-branch.

The rotational structure of typical parallel and perpendicular bands of a linear molecule is shown diagrammatically in Fig. 4.24, where it will be seen that apart from the Q-branch in the perpendicular band the two bands have a similar appearance. In these two bands the intensities of the lines are represented by the heights, and the band contours would be obtained by drawing a line through the uppermost extremities of the lines.

Since the selection rule and the rotational energy part of the E_{vr} equation are the same for linear polyatomic molecules as for diatomic molecules, it follows that the equations for the P-, Q-, and R-branches will have the same form as Equations (4.42), (4.43), and (4.41) respectively if $D_{v'}$ and $D_{v''}$ may be neglected. In this case the equations for a linear polyatomic molecule are:

(i) for $\Delta J = +1$ $\quad R(J) = \tilde{\nu}_0 + 2B_{v'} + (3B_{v'} - B_{v''})J + (B_{v'} - B_{v''})J^2;$
$J = 0, 1, \ldots$ (4.54)

(ii) for $\Delta J = -1$ $\quad P(J) = \tilde{\nu}_0 - (B_{v'} + B_{v''})J + (B_{v'} - B_{v''})J^2;$
$J = 1, 2, 3, \ldots$ (4.55)

(iii) for $\Delta J = 0$ $\quad Q(J) = \tilde{\nu}_0 + (B_{v'} - B_{v''})J + (B_{v'} - B_{v''})J^2;$
$J = 0, 1, 2, \ldots$ (4.56)

where $\tilde{\nu}_R = R(J)$, $\tilde{\nu}_P = P(J)$, and $\tilde{\nu}_Q = Q(J)$. J is the rotational quantum number of the lower rotational state (i.e. J'') and $\tilde{\nu}_0$ is the band origin. $B_{v'}$ is the rotational constant $(h/8\pi^2 cI')$ for the excited state and $B_{v''}$ $(h/8\pi^2 cI'')$ that for the lower state. These formulae therefore take into account the alteration of the internuclear distance in the two vibrational levels. The lines in the R- and P-branches may be expressed in terms of the formula:

$$\tilde{\nu} = \tilde{\nu}_0 + (B_{v'} + B_{v''})m + (B_{v'} - B_{v''})m^2 \qquad (4.57)$$

where for the R-branch $m = (J + 1)$, while for the P-branch $m = -J$. Since $J = 1$ is the lowest value m may have for a P-branch, it follows from Equation (4.57) that one missing line will separate the two branches. This is the case for parallel bands. For perpendicular bands a Q-branch occurs in the vicinity of this missing line. Since both variable terms on the right-hand side of Equation (4.56) involve $(B_{v'} - B_{v''})$, and this value is very small, all the lines in the Q-branch fall very close to the band origin.

When measured accurately, the P and R line separations are not constant (see Fig. 4.23). This was also found to be the case with diatomic molecules and it arises because $B_{v'}$ and $B_{v''}$ do not have the same value. If we let $B_{v'} < B_{v''}$ as is usually the case for parallel bands, the lines of the R-branch will converge towards the high-frequency side and those of the P-branch will diverge towards the low-frequency side following Equations (4.54) and (4.55) for increasing values of J.

Analysis of the rotation–vibration band can lead to values of $B_{v'}$ and $B_{v''}$.

This is done by taking appropriate differences between the frequencies of the P- and R-branch components (see Fig. 4.20). Thus for P and R components that start from the same J value, the relationship $\tilde{\nu}_R(J'') - \tilde{\nu}_P(J'') = 2B_{v'}(2J'' + 1)$ must hold and thus $B_{v'}$ can be evaluated.

Similarly, $\tilde{\nu}_R(J'' - 1) + \tilde{\nu}_P(J'' + 1) = 2B_{v''}(2J'' + 1)$, and from these differences $B_{v''}$ can be obtained.

The dependence of $B_{v'}$, and therefore of I and r, on the vibrational level concerned turns out to be appreciable. The different values of B_v for the ground and excited vibrational states of a linear triatomic molecule can be represented by the equation:

$$B_v = B_e - \sum_i \alpha_i(v_i + d_i/2) \tag{4.58}$$

where α is the vibration–rotation interaction constant and B_e is the rotational constant for a hypothetical vibrationless state; it is the constant for the equilibrium position. B_e and α can be determined using B_0 and B_v for at least one excited vibrational state of each normal mode. For HCN, the B_v values are given by:

$$B_v = 1.4878 - 0.0093(v_1 + \tfrac{1}{2}) + 0.0007(v_2 + 1) - 0.0108(v_3 + \tfrac{1}{2})\,\text{cm}^{-1} \tag{4.59}$$

and the values for B_e and B_0 are 1.4878 and 1.4784 cm^{-1} respectively (see ref. 4.7).

Figure 4.25 shows a typical perpendicular band for the linear molecule and the Q-branch is very prominent. When there are a number of different vibrational modes for a molecule, the rotational constants are governed by the expression (4.58).

In determining precise internuclear distances from rotational constants the B_e values should be used. Since this requires the measurement of rotational constants in excited as well as the ground vibrational state, it is not usually possible to determine B_e values from microwave data. This particular fact limits the accuracy of internuclear distances determined by the latter technique.

Statistical weights of the rotational levels

For linear molecules with a centre of symmetry, i.e. belonging to point group $D_{\infty h}$, the intensity of alternate lines in the P- and R-branches may vary as can be seen in Fig. 4.26, or alternate lines may be completely absent as in CO_2 (see Fig. 4.23). The effect is due to nuclear spin (see Vol. 1) and is an additional factor determining the populations of rotational levels. For CO_2, the like atoms have zero nuclear spin, and all lines corresponding to transitions from the lower state with J'' odd are completely absent. A further consequence is that the gap at the band centre in CO_2 is not $4B$ but $6B$, because the first line of the P-branch is absent. The separation of successive lines is $4B$ instead of $2B$.

For acetylene, which has one pair of identical nuclei (H) of spin $I = \frac{1}{2}$ (the carbon nuclei have no spin and therefore do not have an effect), the odd J levels

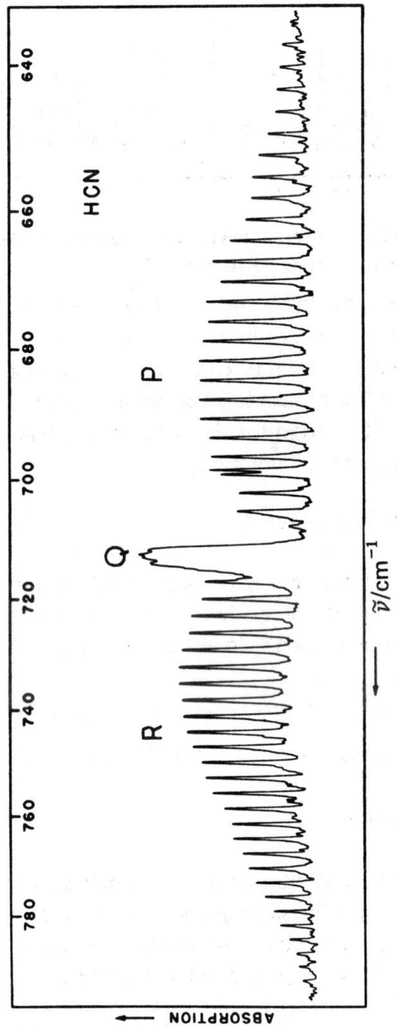

Fig. 4.25 The rotational fine structure of the perpendicular band of HCN. (From ref. 4.2.)

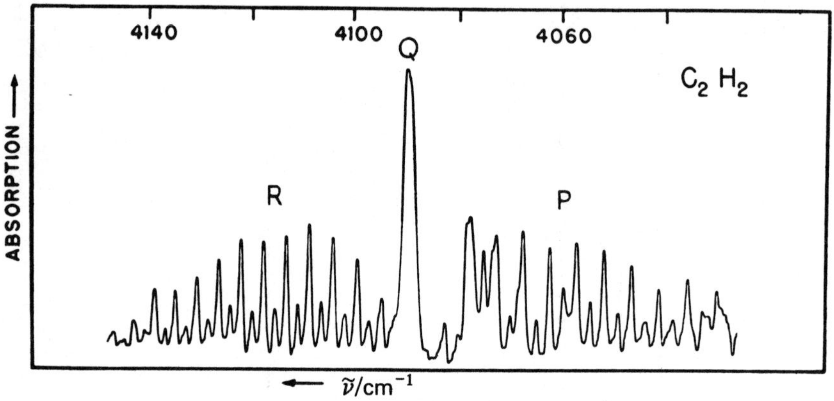

Fig. 4.26 The fine structure of an acetylene combination band showing the intensity variation of alternate lines. (From ref. 4.2.)

in the ground state have three times the statistical weight of the even J levels and this intensity alternation is clearly observed in Fig. 4.26.

The presence or absence of such intensity changes is usually sufficient to show whether or not a linear molecule possesses a centre of symmetry. The absence of an intensity alternation for the fundamental bands of N_2O shows that it cannot possess a centre of symmetry.

Moment of inertia from band maxima

If the rotational fine structure cannot be resolved, an approximate estimate of the moment of inertia I may be made from the spacing of the intensity maxima of the P- or R-branches ($\Delta\tilde{\nu}$) in either the parallel or perpendicular bands by use of the equation:

$$\Delta\tilde{\nu} \approx \sqrt{(8kTB/hc)} = 2.358\sqrt{(TB)}\ \text{cm}^{-1}$$

where k is Boltzmann's constant and T is the absolute temperature.

4.7.2 Symmetric-top molecules

This is the type of molecule where two of the moments of inertia are equal, and the third though finite is not the same as the other two. Typical examples of molecules which have been examined are prolate symmetric tops such as CH_3Cl, with C_{3v} symmetry ($I_A < I_B = I_C$), and oblate symmetric tops such as BF_3, PCl_3, or C_6H_6 (where $I_A = I_B < I_C$).

Two quantum numbers are necessary to define the rotational energy of a symmetric-top molecule. The customary J is employed to govern the total angular momentum of the molecule, but in addition another quantum number, K, is required to fix the angular momentum about the symmetry axis.†

† The total angular momentum of the molecule is $\sqrt{[J(J+1)]}\,h/2\pi$, while the component of the angular momentum along the symmetry axis is $Kh/2\pi$.

The possible values of J are 0, 1, 2, ..., whereas K may take the values 0, ±1, ±2, ..., ±J for each value of J. Thus the numerical value of K is either equal to or less than J, and there are $(2J + 1)$ values of K for each value of J.

The vibrational and rotational energies are normally assumed to be additive, and the vibration a simple harmonic one. The rotation–vibration energy may be expressed by:

$$E_{vr} = \sum (v+\tfrac{1}{2})hc\omega + hc[B_v J(J+1) - D_{vJ}J^2(J+1)^2 + (A_v - B_v)K^2 - D_{vJK}J(J+1)K^2 - D_{vK}K^4] \qquad (4.60)$$

where

$$A_v = h/8\pi^2 cI_A \quad \text{and} \quad B_v = h/8\pi^2 cI_B$$

The constants D_{vJ}, D_{vJK}, and D_{vK} take into account the centrifugal stretching and normally result in very small contributions from the terms in which they are concerned.† More specifically the D_J term results from the stretching due to end-over-end rotation of the molecule whereas the D_K term results from the distortion caused by rotation about the symmetry axis. The D_{JK} term results from the interaction of these two types of motion.

For a rotation–vibration change the ΔJ and ΔK values depend on whether the vibration undergoing the change is parallel or perpendicular to the symmetry axis.

Symmetric-top molecules can be considered according to the way in which they may be analysed to gain information on moments of inertia and internuclear distance. These are: (1) an analysis which neglects centrifugal stretching and ignores the difference between the rotational constants in the lower and excited vibrational states; (2) an analysis which takes into account centrifugal stretching and determines both the $B_{v'}$ and $B_{v''}$ values. In fact, this type of analysis is based on Equation (4.60).

(1) *Analysis of symmetric-top molecules neglecting centrifugal stretching and any alteration in the rotational constants*

When D_J, D_{JK}, and D_K are each placed equal to zero, Equation (4.60) becomes:

$$E_{vr} = \sum (v+\tfrac{1}{2})hc\omega + hc[BJ(J+1) + (A-B)K^2] \qquad (4.61)$$

The type of analysis about to be considered employs Equation (4.61) and neglects any change in the rotational constants A and B in the rotation–vibration transition.

The analysis based on these assumptions may be divided into two types, depending on whether the change in dipole moment during the vibration is parallel or perpendicular to the symmetry axis.

† When it is not necessary to be specific the subscript 'v' is frequently omitted.

Parallel bands. The selection rules are $\Delta J = \pm 1$, and $\Delta K = 0$ when $K = 0$. $\Delta J = 0, \pm 1$; $\Delta K = 0$ if $K \neq 0$. The equations for the P-, R-, and Q-branches are obtained as follows:

P-branch. When the values $K' = K''$ and $J' = (J'' - 1)$ are substituted in Equation (4.61) and the fact that:

$$E'_{vr} - E''_{vr} = \Delta E = hc\tilde{\nu} \tag{4.62}$$

is employed, where $\tilde{\nu}$ is the wavenumber of the particular rotation–vibration transition, then:

$$\tilde{\nu}_P = \tilde{\nu}_0 + B(J'' - 1)(J'') - BJ''(J'' + 1) + (A - B)(K''^2 - K''^2) \tag{4.63}$$

that is

$$\tilde{\nu}_P = \tilde{\nu}_0 - 2BJ'' \tag{4.64}$$

where $\tilde{\nu}_0$ is the band origin, and $J'' = 1, 2, 3, \ldots$.

R-branch. In a similar manner on substituting $K' = K''$ and $J' = (J'' + 1)$, then:

$$\tilde{\nu}_R = \tilde{\nu}_0 + 2B(J'' + 1) \tag{4.65}$$

where $J'' = 0, 1, 2, 3, \ldots$.

Q-branch. In this case $K' = K''$ and $J' = J''$, and:

$$\tilde{\nu}_Q = \tilde{\nu}_0 \tag{4.66}$$

that is, a single line might be expected.

It should be noted that these P-, Q-, and R-branch equations have the same form as those for a linear molecule, where $B_{v'} = B_{v''}$ and $D = 0$. In practice, owing to the rotation–vibration interaction, the Q-branch is once again a band and not a single line. The separation of the rotational lines in the P- and R-branches is approximately equal to $h/4\pi^2 cI_B$ where I_B is the moment of inertia about an axis perpendicular to the symmetry axis. There are, of course, two such directions, but for a prolate symmetric-top $I_B = I_C$. Thus, from parallel bands only one moment of inertia can be obtained; to determine I_A it is necessary to study the more complex perpendicular bands.

Perpendicular bands. The selection rules are $\Delta J = 0, \pm 1$ and $\Delta K = \pm 1$, and for each change in K there will be a band with P-, Q-, and R-branches.

The equations for the P-, R-, and Q-branches for the case where $\Delta K = +1$ are obtained as follows:

P-branch. When the values $K' = (K'' + 1)$ and $J' = (J'' - 1)$ are substituted in Equation (4.61), using the fact that $(E'_{vr} - E''_{vr}) = \Delta E_{vr} = hc\tilde{\nu}$ (where $\tilde{\nu}$ is the wavenumber of the particular rotation–vibration transition), then:

$$\tilde{\nu}_P = \tilde{\nu}_0 + B(J'' - 1)J - BJ''(J'' + 1) + (A - B)(K'' + 1)^2 - (A - B)K''^2 \tag{4.67}$$

that is:

$$\tilde{\nu}_P = \tilde{\nu}_0 - 2BJ'' + (A - B)(2K'' + 1) \ldots \tag{4.68}$$

where ν_0 is the band origin and $J'' = (K'' + 2), (K'' + 3), (K'' + 4), \ldots$.

R-branch. In a similar manner on substitution of $K' = (K'' + 1)$ and $J' = (J'' + 1)$, the equation for lines in the R-branch is found to be:

$$\tilde{\nu}_R = \tilde{\nu}_0 + 2B(J'' + 1) + (A - B)(2K'' + 1) \qquad (4.69)$$

where $J'' = K'', (K'' + 1), (K'' + 2), \ldots$.

Q-branch. In this case $J' = J''$ and $K' = (K'' + 1)$ and the wavenumbers of the Q-lines are independent of the J values; they are given by:

$$\tilde{\nu}_Q = \tilde{\nu}_0 + (A - B)(2K'' + 1) \qquad (4.70)$$

In each of these equations K'' has a fixed value, and the wavenumbers of the lines in a particular P- and R-branch are obtained from the formula by substituting the appropriate J'' values.

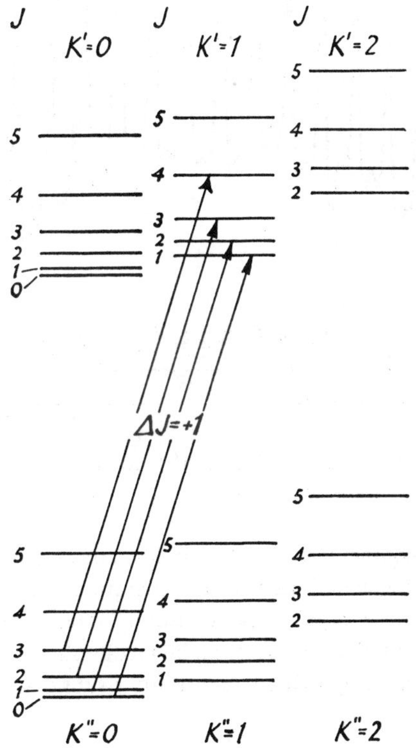

Fig. 4.27 Energy level diagram indicating a few transitions for the R-branch of the first positive band of a prolate symmetric-top molecule.

Each value of K'' gives rise to a P-, Q-, and R-branch of a band. The energy level diagram for a prolate symmetric-top molecule is illustrated in Fig. 4.27. A few typical transitions are given for the R-branch of the case where $K'' = 0$. The

$K' \leftarrow K''$ value $1 \leftarrow 0$ is known as the *first positive subsidiary* and the $2 \leftarrow 1$ is the *second positive subsidiary band*. When $\Delta K = -1$ we have an additional series of P-, Q-, and R-branches, called the *negative subsidiary bands*. Thus the $0 \leftarrow 1$ and $1 \leftarrow 2$ would be known as the *first* and *second negative subsidiary bands*, respectively. The equations relating the wavenumbers of the rotation–vibration lines to the quantum numbers J'' and K'' for the negative subsidiary bands can be obtained in the same way as for the positive subsidiary bands. In this case, however, $K' = (K'' - 1)$ would have to be substituted.

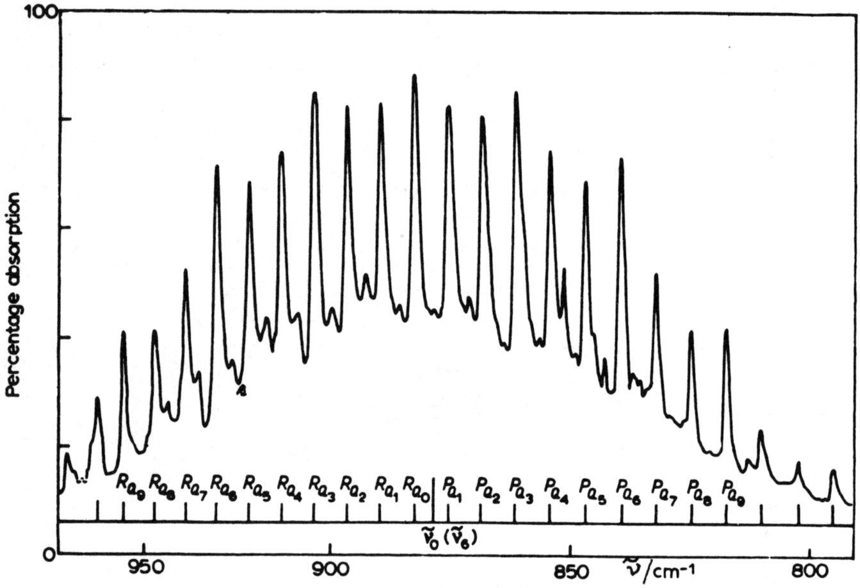

Fig. 4.28 The $\tilde{\nu}_6$ fundamental band of methyl iodide under high resolution. Gas pressure 240 mm Hg in a 10-cm cell. (From ref. 4.10.)

By inserting two consecutive J'' values in formula (4.68), it follows on subtraction of the resulting equations that the separation of consecutive rotational lines in a P-branch is $h/4\pi^2 c I_B$. The same result is obtained for consecutive R-branch lines. From Equation (4.70) it can be readily shown that the wavenumber separation between the Q-branches in consecutive subsidiary bands is

$$\Delta\tilde{\nu} = \frac{h}{4\pi^2 c}\left(\frac{1}{I_A} - \frac{1}{I_B}\right) \tag{4.71}$$

Theoretically, then, I_B can be determined from the spacing of the rotational lines in the P- and R-branches in the parallel bands. From this knowledge of I_B and the measured separation of successive Q-branches, I_A can be found.

In practice, however, it is not always possible to measure the separation between the Q-branches, since the background of the P- and R-branches may be strong enough to obscure the Q-branches unless $I_B \gg I_A$, e.g. in the methyl halides (see Fig. 4.28). A further complicating factor is that the spacing between consecutive Q-branches in different perpendicular bands is not constant but varies markedly. The effect is particularly noticeable in the perpendicular bands of the methyl halides, where the frequency difference between consecutive Q-branches in different bands is very variable. For example, a separation of $7.7\ \mathrm{cm^{-1}}$ is observed for the Q-branches of the CH_3 rocking fundamental ($\tilde{\nu}_6$) of methyl iodide. This is very different from the separation measured in the other two perpendicular bands $\tilde{\nu}_5$, δCH_3 ($11.7\ \mathrm{cm^{-1}}$) and $\tilde{\nu}_4$, $\tilde{\nu}_{\mathrm{antisym}} CH_3$ ($9.0\ \mathrm{cm^{-1}}$). The reason for this variation in the separation of Q-branches in different bands is due to a coupling of the rotational and vibrational angular momenta giving rise to a force of interaction, called the *Coriolis force*, in the perpendicular vibration, which in effect is an additional internal vibrational angular momentum. Owing to the interaction of this force, the spacing between consecutive Q-branches in perpendicular subsidiary bands is no longer given by Equation (4.71) but by:

$$\Delta\tilde{\nu} = \frac{h}{4\pi^2 c}\left(\frac{1-\zeta_i}{I_A} - \frac{1}{I_B}\right) \tag{4.72}$$

where ζ_i in units of $\hbar$ gives the correction due to the internal angular momentum and is constant for a particular band. This Coriolis force has to be considered for perpendicular bands whenever either the upper or lower vibrational states or both are degenerate. The spacing between consecutive Q-branches, as given by Equation (4.72), differs for each perpendicular band.

Owing to the complexities of the perpendicular bands of polyatomic molecules, it has generally not been found possible to resolve the rotational structure. However, the advent of tunable infrared lasers means that very high resolution infrared spectra can be measured and the above theoretical considerations will become increasingly important (see Fig. 4.21).

Gerhard and Dennison [4.11] have used a theoretical approach to relate the shape of the envelope of the band to the factor β, where:

$$\beta = (I_B/I_A) - 1 \tag{4.73}$$

This has been done for the perpendicular bands of the symmetric-top type of molecule for various values of β. The mathematical procedure is too complex and detailed to be considered here. Briefly, the procedure is to plot a quantity known as the *absorption coefficient* against x where:

$$x = \frac{\tilde{\nu}}{\alpha}\sigma^2, \quad \alpha = \frac{h}{4\pi^2 I_B}, \quad \text{and} \quad \sigma = \frac{h^2}{8\pi^2 I_B kT}$$

This has been done in Fig. 4.29, and the full curved lines represent the total absorption due to the whole of the perpendicular band. The broken lines, however, indicate the absorption due to all the Q-branches of the various

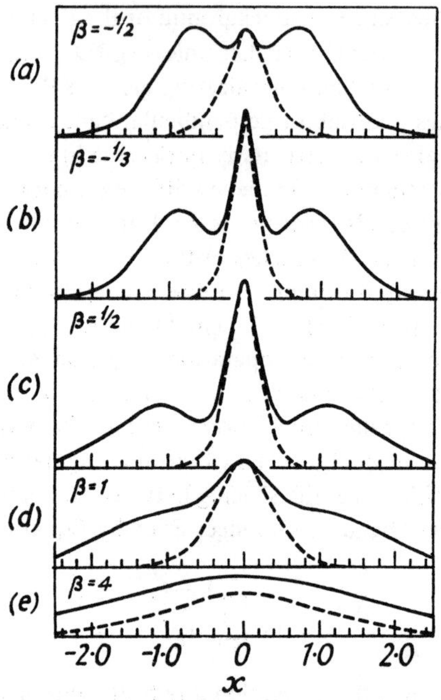

Fig. 4.29 The variation in the contour of the perpendicular band of a symmetric-top molecule in passing from a disc-type molecule where $\beta = -\frac{1}{2}$ to a rod-type molecule where β is large. The full line indicates the total absorption of the whole of the perpendicular band while the broken line indicates the contour of the Q-branches of all the various subsidiary bands. (From ref. 4.11. Courtesy of The Physical Review.)

subsidiary bands. The values of β are as indicated on the figure; for example in Fig. 4.29(a) $\beta = -\frac{1}{2}$, that is $I_A = 2I_B$, and the molecule is of a disc type, while in (e) $\beta = 4$, that is $I_B = 5I_A$, in which case the molecule approximates to a rod. By comparison of such theoretical contours with the experimentally observed ones it is possible to obtain a qualitative estimate of β and hence the general shape of the molecule. One complication in such deductions, however, can be the effect of any internal angular momentum ($\zeta_i h/2\pi$) since the unresolved contour of a band is dependent on the value of ζ_i.

The spacing between maxima in the contours of bands sometimes leads to an approximate value for one moment of inertia. In fact, Gerhard and Dennison have shown that the spacing $\Delta\tilde{\nu}$ between the successive maxima of the P- and R-branches in a parallel band is in certain cases given by:

$$\Delta\tilde{\nu} = \frac{S(\beta)}{\pi c}\sqrt{\left(\frac{kT}{I_B}\right)} \tag{4.74}$$

where k is the Boltzmann constant, while:

$$\beta = (I_B/I_A) - 1 \tag{4.75}$$

The empirical formula which Gerhard and Dennison employed relating $S(\beta)$ to β between $-\frac{1}{2}$ and 100 was:†

$$\log_{10} S(\beta) = \frac{0.721}{(\beta + 4)^{1.13}} \tag{4.76}$$

Thus, if β can be estimated, $S(\beta)$ can be calculated; then if $\Delta\tilde{\nu}$ is determined from the separation of the maxima of the P- and R-branches, I_B can be obtained. If, however, I_B is known from work on parallel bands, then I_B and $\Delta\tilde{\nu}$ can be employed in Equation (4.74) to calculate $S(\beta)$; hence, the value of β follows from Equation (4.76) and the value of I_A from Equation (4.75).

Anyway, it is sometimes possible to decide whether the vibration is of the parallel or perpendicular type from the band contours; e.g. for a symmetric-top, if the band contour can be observed, and it has an intense narrow maximum in the middle of the band, a parallel vibration is involved. However, if a complex structure containing several maxima is observed, then most probably a perpendicular vibration is responsible. Such information may be most valuable in structural determinations.

(2) *Analysis of a symmetric-top molecule involving $B_{v'}$, $A_{v'}$, $B_{v''}$, and $A_{v''}$ values*

For simple symmetric-top molecules such as the methyl halides it is possible to resolve the rotational structure of the bands and to determine the values of the rotational constants A and B in both the ground state and excited vibrational states. The procedure involves the use of Equation (4.60) which takes into account the variation of the rotational constants or the centrifugal stretching.

The structure of a typical perpendicular band of a symmetric-top molecule is illustrated in Fig. 4.30 for K-values from 0 to 5. The ten possible transitions between K' and K'' for these six K values are: 1←0, 0←1, 2←1, 1←2, 3←2, 2←3, 4←3, 3←4, 5←4, and 4←5. The complete band shown at the bottom of the diagram is made up of the superposition of these various positive and negative subsidiary bands resulting from changes in K of ± 1 unit and changes in J of 0, ± 1 unit. Figure 4.30 was constructed by Herzberg [4.7] and is based on the assumptions that $A_{v'} = 5.18$, $A_{v''} = 5.25$, $B_{v'} = 0.84$, and $B_{v''} = 0.85\ \text{cm}^{-1}$. It serves to illustrate the complexity of a perpendicular band and the inadequacies of taking the rotational constants as being the same in the upper and lower vibrational states. The relative intensities of the rotation–vibration lines are indicated by the relative heights of the lines and were calculated for a temperature of 144 K.

The relevant J and K data for the first three subsidiary bands are given in

† In the case of linear molecules $S(\beta)$ is approximately unity.

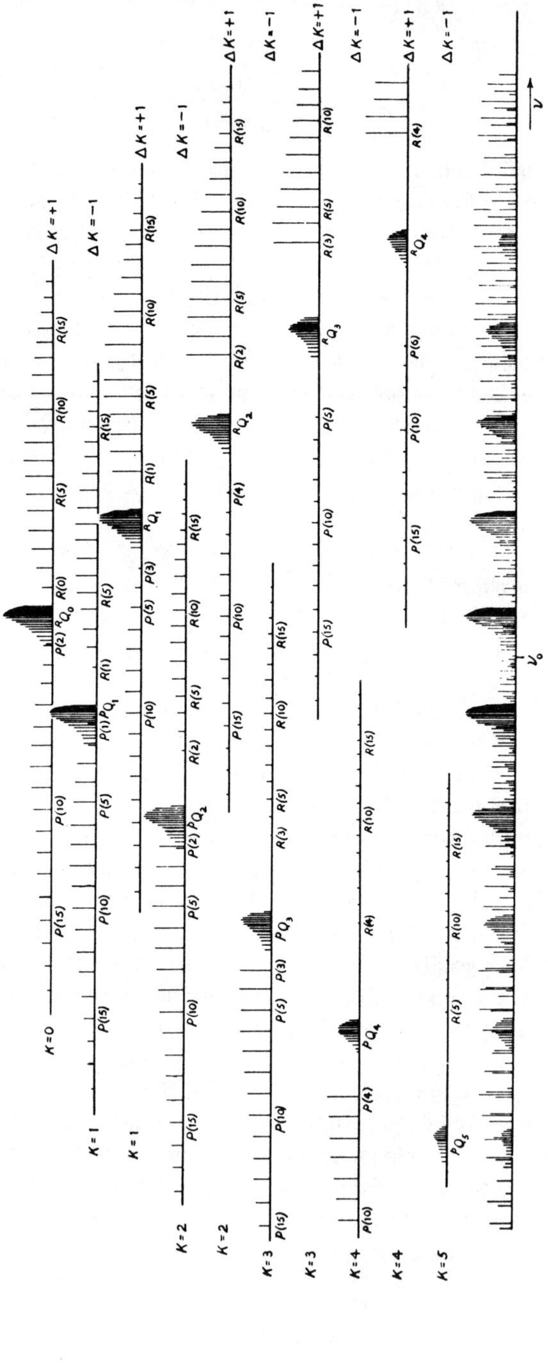

Fig. 4.30 Sub-bands of a perpendicular band, and complete perpendicular band of a symmetric-top molecule. (From ref. 4.7.)

Table 4.4 Data on the sub-bands of a symmetric-top molecule for a perpendicular band

Subsidiary bands	*Changes in J*	*K values*	*J values*	
First positive	P $\Delta J - 1$	$\Delta K = 1$	$J' \geqslant 1$	$P(1)$
	Q ΔJ 0	$K' = 1$	$J'' \geqslant 0$	
	R $\Delta J + 1$	$K'' = 0$		
First negative	P $\Delta J - 1$	$\Delta K = -1$	$J' \geqslant 0$	
	Q ΔJ 0	$K' = 0$		$R(0)$
	R $\Delta J + 1$	$K'' = 1$	$J'' \geqslant 1$	
Second positive	P $\Delta J - 1$	$\Delta K = 1$	$J' \geqslant 2$	$R(0)$
	Q ΔJ 0	$K' = 2$		
	R $\Delta J + 1$	$K'' = 1$	$J'' \geqslant 1$	$P(1) P(2)$
Second negative	P $\Delta J - 1$	$\Delta K = -1$	$J' \geqslant 1$	$R(0) R(1)$
	Q ΔJ 0	$K' = 1$		
	R $\Delta J + 1$	$K'' = 2$	$J'' \geqslant 2$	$P(1)$
Third positive	P $\Delta J - 1$	$\Delta K = 1$	$J' \geqslant 3$	$R(0) R(1)$
	Q ΔJ 0	$K' = 3$		
	R $\Delta J + 1$	$K'' = 2$	$J'' \geqslant 2$	$P(1) P(2) P(3)$
Third negative	P $\Delta J - 1$	$\Delta K = -1$	$J' \geqslant 2$	$R(0) R(1) R(2)$
	Q ΔJ 0	$K' = 2$		
	R $\Delta J + 1$	$K'' = 3$	$J'' \geqslant 3$	$P(1) P(2)$

Table 4.4 together with information about which lines are missing from each of these subsidiary bands.

Types of symmetric-top molecules analysed

The moments of inertia of a number of symmetric-top molecules have been determined by the infrared rotation–vibration technique; these include NH_3, BF_3, C_2H_6, CD_3F, ClO_3F, $CH_3C{\equiv}CCH_3$, and CH_3X, where X = D, F, Cl, Br, and I. Only two different moments of inertia are obtained for these molecules, and it is not always possible to determine all the internuclear distances and bond angles without assuming data from other molecules or resorting to isotopes. For example: (i) if BF_3 is assumed to be planar and symmetrical, the B—F internuclear distance is readily calculated; (ii) in the methyl halides, if the C—H internuclear distance is assumed to be the same as in CH_4, then the C—X internuclear distance and the angles can be evaluated.

4.7.3 Spherical-top molecules ($I_A = I_B = I_C$)

Methane and carbon tetrachloride are typical examples of tetrahedral spherical-top molecules, where the theoretical number of fundamental vibrational modes is (3 x 5 − 6), that is nine modes. However, only four different vibrational modes can be detected by both the infrared and Raman methods together, and these are illustrated in Fig. 4.31. The $\tilde{\nu}_2$ mode is doubly degenerate, while the $\tilde{\nu}_3$ and $\tilde{\nu}_4$ modes are triply degenerate. Thus, the theoretical value of nine fundamental vibrational modes may be accounted for. Only $\tilde{\nu}_3$ and $\tilde{\nu}_4$ are active in the infrared region, because these are the only vibrations which generate an overall change in the dipole moment.

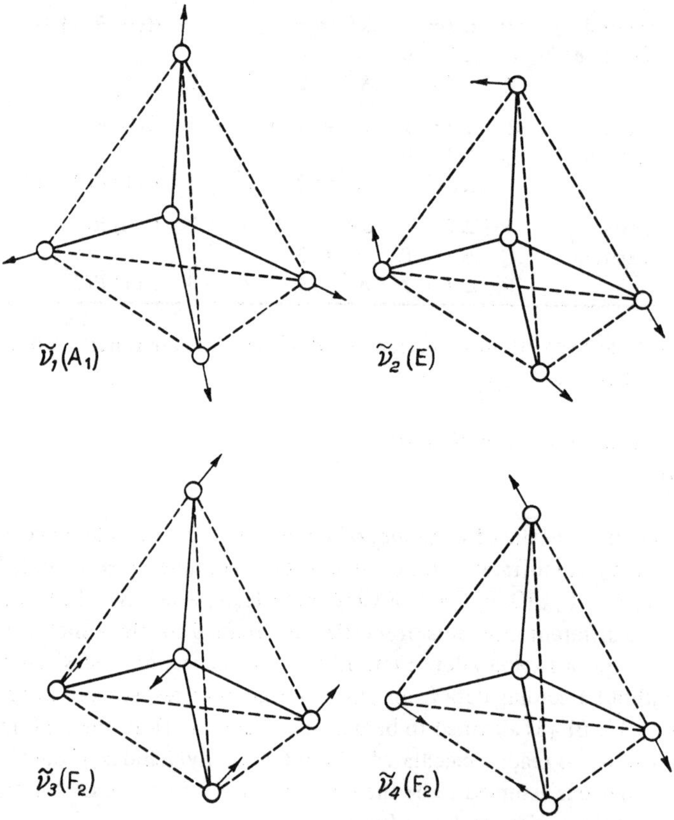

Fig. 4.31 Normal vibrations of a tetrahedral spherical-top molecule.

Disregarding any form of rotation–vibration interaction, the equation governing the rotation–vibration energy is identical with the equation for a linear molecule [Equation (4.53)]. The selection rule is:

$$\Delta J = 0, \pm 1 \tag{4.77}$$

The equations for the wavenumbers of the lines constituting the P- and R-branches are therefore the same as for linear molecules, where the separations of the lines should be approximately given by $h/4\pi^2 cI$ in both the P- and R-branches, but in practice this is not found to hold exactly owing to strong Coriolis interaction; this type of interaction is mentioned under symmetric-top molecules. The Coriolis coupling between rotation and vibration occurs in the upper degenerate (triplet) vibrational state but differs for each excited vibrational state. The derived formulae for the P- and R-branches take this coupling into account by means of the Coriolis coupling constant ζ_i. The P and R may be represented in the form of one series by the formula:

$$\tilde{\nu} = \tilde{\nu}_0 + (B_v' + B_v'' - 2B_v'\zeta_i)m + (B_v' - B_v'')m^2 \tag{4.78}$$

where $m = -J$ for the P-branch and $m = (J+1)$ for the R-branch and the $m = 0$ line is absent. The value of ζ_i differs with each band. The fine structure of the fundamental $\tilde{\nu}_3$ band of CH_4 at 3020 cm^{-1} is given in Fig. 4.32. The numbers written at the top of the lines are m-values. For the $\tilde{\nu}_3$ band of CH_4, B_v' is approximately equal to B_v''. The separation of successive lines is almost constant, and it follows from formula (4.78) that this is $2B_v''(1-\zeta_i)$. Since the value of ζ_i is dependent on the band, the spacing of the rotational lines varies, even though B_v' is approximately equal to B_v''. The moment of inertia cannot be derived from one band alone, since the values of both B_v'' and ζ_i are unknown. It is possible, though, to determine the moment of inertia if the spacing of the rotational lines in both of the fundamental infrared-active bands is known. This becomes feasible owing to rules which can be derived to relate the two ζ_i values. For example, in the case of CH_4, where the two infra-red active fundamental frequencies are $\tilde{\nu}_3$ and $\tilde{\nu}_4$, ζ_3 is related to ζ_4 by the equation:

$$\zeta_3 + \zeta_4 = \tfrac{1}{2} \tag{4.79}$$

Thus, the spacing of the $\tilde{\nu}_3$ band is:

$$2B(1-\zeta_3) \quad \text{that is} \quad 2B(1-\tfrac{1}{2}+\zeta_4)$$

while that of the $\tilde{\nu}_4$ band is:

$$2B(1-\zeta_4)$$

Thus, the sum of the spacings is $3B$. Hence, B, I_B, and therefore the internuclear carbon–hydrogen distance can be determined. For $\tilde{\nu}_3$ and $\tilde{\nu}_4$ of CH_4, $\zeta_3 = 0.05$ and $\zeta_4 = 0.45$.

The major obstacle in the analysis of spherical-top molecules by the infrared technique is the resolution of the rotational structure of the bands. Internuclear distances have been obtained for a number of molecules including CH_4, CD_4, SiH_4, and GeH_4.

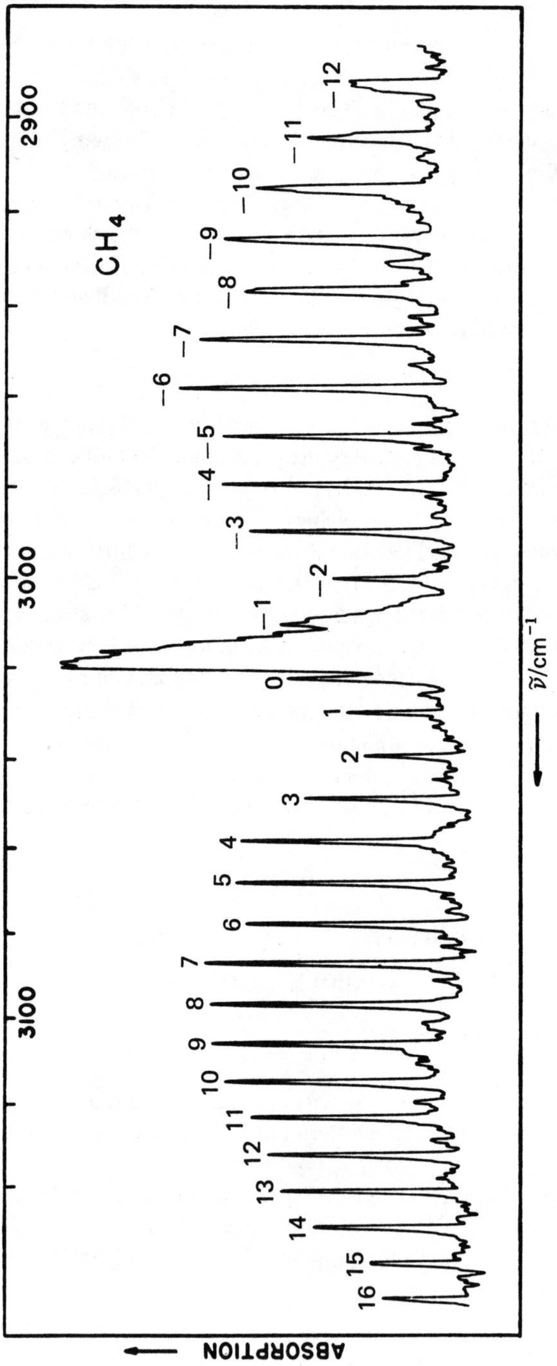

Fig. 4.32 Rotational fine structure of the $\tilde{\nu}_3$ fundamental of CH_4. (From ref. 4.2.)

4.7.4 Asymmetric-top molecules ($I_A \neq I_B \neq I_C$)

Water is an example of this class, in which all three moments of inertia are finite and unequal. For such molecules there is no simple overall relation for the rotational energy, although this does not impose any serious limitations on the possibility of analysis.

One of the simplest rotational energy equations is that due to Wang, and if no rotation–vibration interaction occurs for such a molecule, then:

$$E_{vr} = E_v + \frac{h^2}{16\pi^2}\left(\frac{1}{I_B} + \frac{1}{I_C}\right) J(J+1) + \left[\frac{h^2}{8\pi^2 I_A} - \frac{h^2}{16\pi^2}\left(\frac{1}{I_B} + \frac{1}{I_C}\right)\right] W_\tau \tag{4.80}$$

This equation applies to an asymmetric-top molecule provided that it is a rigid molecule, and the considered vibrational state does not interact with a nearby vibrational state.

The energy levels of such molecules are dependent on the total angular momentum quantum number J, and associated with each J value are $(2J+1)$ different energy levels. These energy levels cannot be interpreted in terms of a quantum number which has physical significance, so an index τ is employed to enable the $(2J+1)$ values to be arranged in the order of their magnitudes.

The selection rule governing changes in J is simply $\Delta J = 0, \pm 1$, but the selection rule for the index τ, according to Dennison [4.12], depends on whether the electric moment of the molecule changes along (a) the greatest axis of inertia, or (b) the least axis of inertia, or (c) the other axis of inertia.

The number of allowed transitions is very large and, in fact, the number of rotational lines in a vibrational band of an asymmetric-top molecule is considerable. Even if the rotational lines can be experimentally resolved, the procedure seems very complex, and at the present time the moments of inertia of only a few asymmetric-top molecules have been determined by the infrared rotation–vibration approach.†

The most complete analysis of a strongly asymmetric-top molecule has been performed on the water molecule; this has been fully analysed for its three moments of inertia in the lowest vibrational level and also in the equilibrium position, and the corresponding O—H internuclear distance and the HOH angle have been calculated. Hydrogen sulphide has been analysed for the three moments of inertia in the lower vibrational state; the internuclear distance H–S and the angle HSH in this state were found to be 133 pm and 92° 16′,

† This is to be contrasted with the microwave technique, where appreciable numbers of asymmetric-top molecules have been analysed.

respectively. Other studies of the rotation–vibration spectra of asymmetric-top molecules include those of D_2O, HDO, and H_2Te.†

The most suitable molecules for examination are those in which two moments of inertia are nearly equal and the molecule approximates to a symmetric-top. The procedure used in the analysis of symmetric-top molecules has to be modified. The three moments of inertia of the following molecules have been determined: C_2H_4, C_2D_4, CH_2O, CHDO, and HCO_2H. In the case of formaldehyde the three moments of inertia in the ground state are 2.977×10^{-47}, 21.65×10^{-47}, and 24.62×10^{-47} kg m^2, where the similarity of the latter two values considerably simplifies the analysis. If the carbon–hydrogen distance is assumed to be the same as that found in ethylene, the ground state values of the C–O distance and the angle HCH are 122 pm and 123°26′, respectively. It is not possible to evaluate the internuclear distances and bond angles of the C_2H_4 molecule solely from its determined moments of inertia, and either C_2D_4 has also to be studied or a value for one angle or an internuclear distances has to be assumed.

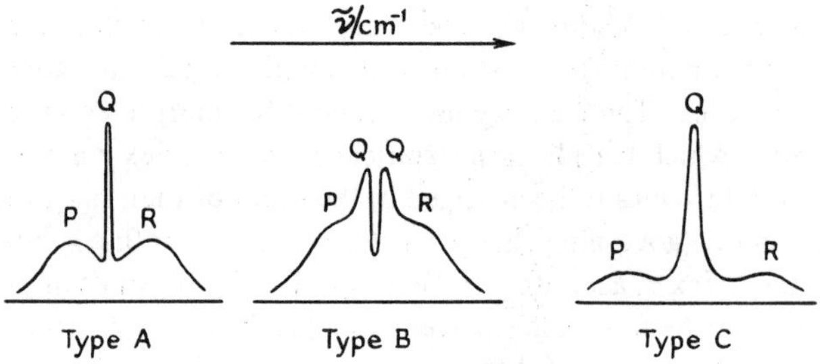

Fig. 4.33 Some typical band contours for an asymmetric-top.

Usually the infrared bands of asymmetric-top molecules remain unresolved. However, useful vibrational analysis information may be gained from studying the band contours, and Badger and Zumwalt [4.13] have represented the A, B, and C type band contours of asymmetric-tops in terms of two parameters $\rho = (A - C)/B$ and $s = (2B - A - C)/(A - C)$. Typical band contours are given in Fig. 4.33. It is not possible to determine moments of inertia from these band contours, but they can be of considerable value in confirming the proposed structure of a molecule or in determining the symmetry class of a particular vibration.

This section on simple molecules examined by the infrared technique is only a brief introduction to the study, and many complicating factors have been

† These are all planar molecules and for such $I_A + I_B = I_C$ would be anticipated from moment of inertia considerations. In practice $(I_A + I_B) - I_C \neq 0$ but shows a linear dependence on the value of the vibrational quantum number.

neglected or only briefly mentioned. For a full treatment Herzberg [4.7] should be consulted.

4.8 INFRARED SPECTRA OF COMPRESSED GASES AND VIOLATION OF SELECTION RULES

In 1949, Crawford, Welsh, and Locke [4.14] reported that H_2, O_2, and N_2 under special conditions had an infrared rotation–vibration absorption spectrum; these are molecules with no permanent electric dipole moment, and from their symmetry the dipole moment would not be expected to change during a normal vibration. The special conditions were a small partial pressure of H_2 in the presence of a high pressure of foreign gas such as He or N_2. Since then numerous studies have been made, and the following are some of the points which have emerged.

(1) Only broadening of the rotational fine structure of the infrared absorption bands occurs for a non-symmetrical diatomic molecule in either the gas or vapour state when studied at a total gas pressure of a few bars.

(2) At pressures between 10 and 100 bars, both in a pure gas and a gaseous mixture, vibrational and rotational transitions may be produced which under normal conditions would otherwise be strictly forbidden, and what are known as 'pressure induced' infrared absorption bands are observed.

(3) In addition, certain gaseous mixtures of $CO_2 + N_2$, $CO_2 + O_2$, and $CO_2 + H_2$ at these high pressures give new frequencies which are equal to the sum of infrared active vibrational frequencies of CO_2 and the vibrational frequency of the diatomic molecule. For example, CO_2 has two active infrared fundamental frequencies, one at 2349 cm^{-1} which corresponds to the antisymmetric stretch and another at 667 cm^{-1} which is the degenerate bending mode. N_2 has its fundamental frequency at 2331 cm^{-1}, but the observed frequencies at high pressures were 2996 and 4670 cm^{-1}; these may be explained as resulting from the combination of the frequencies from the individual molecules, i.e.

$$667 + 2331 = 2998\ cm^{-1}$$
$$2349 + 2331 = 4680\ cm^{-1}$$

The difference between these figures and the observed ones is permissible for combination frequencies. These new bands from mixtures of gases are attributed to 'simultaneous transitions' taking place for component molecules.

(4) The intensity of absorption of the 'pressure induced bands' was directly proportional to the square of the pressure, while for the 'simultaneous transitions' it was directly proportional to the product of the partial pressures of the two components in the mixture. This particular 'square' dependence on the pressure enables this type of transition to be distinguished from another type which arises when certain molecules have weak magnetic dipole or electric quadrupole moments but undergo no change in electric dipole moment during the vibration. In fact, transitions involving molecules with weak magnetic dipoles or

weak quadrupoles are not forbidden by the selection rule. Thus, the criterion of an infrared spectrum contains a corollary that a change of the quadrupole moment or of the magnetic dipole moment may also result in emission or absorption of infrared radiation. The 'pressure induced bands' and the 'simultaneous transitions', on the other hand, definitely break the selection rule for electric dipole radiation.

(5) The pure rotational absorption spectrum of H_2 has been observed also in the infrared by high-pressure studies. The $J' \leftarrow J''$ transitions observed were $3 \leftarrow 1$, $4 \leftarrow 2$, and $5 \leftarrow 3$ where the selection rule $\Delta J = \pm 1$ was disobeyed. Once again the intensity of absorption was proportional to the square of the pressure. In addition, the rotational transitions were also induced by the presence of Ar, N_2, or CO_2 at these high pressures. This violation of the rotational selection rule is not limited to pure rotational transitions; in fact, for the rotation–vibration transition of H_2 from the $v'' = 0$ to the $v' = 1$ level, S$(\Delta J = 2)$, Q$(\Delta J = 0)$, and O $(\Delta J = -2)$ branches are observed and not R and P branches.

It is thought that the bands which result at high pressures when the normally-obeyed selection rules are defied are due to collision pair formation, where both the 'pressure induced' and the 'simultaneous transitions' result from the mutual deformation of the charge distribution of the molecules owing to the mutual interaction at such close proximity. At extremely high pressures the dependence of the intensity of absorption on the square of the pressure is no longer observed, and this is attributed to triple collisions and more involved molecular interactions. In conclusion, it must be stressed that the work on pressure induced infrared studies offers an attractive way of gaining some insight into molecular interaction.

4.9 INFRARED SPECTRA OF LIQUIDS, SOLIDS, AND SOLUTIONS

The preceding sections have shown how an exact description can be obtained for the rotation–vibration spectra of diatomic molecules in the gas phase. The molecules are treated as non-rigid rotators undergoing anharmonic vibrations. Chapter 2 has shown the use of point group theory to derive the number of fundamental modes of vibration and the spectral selection rules for polyatomic molecules in the gas phase. It should be emphasized, however, that the rotational structure of an absorption band is fully resolved only for small molecules. Larger molecules have individual rotational levels which are too close together to be resolved by conventional grating instruments, and overall band contours are observed.

We must now turn our attention to the spectra of non-volatile liquids and solids because an enormous amount of infrared spectroscopy is carried out on involatile compounds. The treatment for detailed rotational structure no longer applies in the condensed phases because molecular collisions are very frequent and the bands do not exhibit fine structure due to molecular rotations (see Fig.

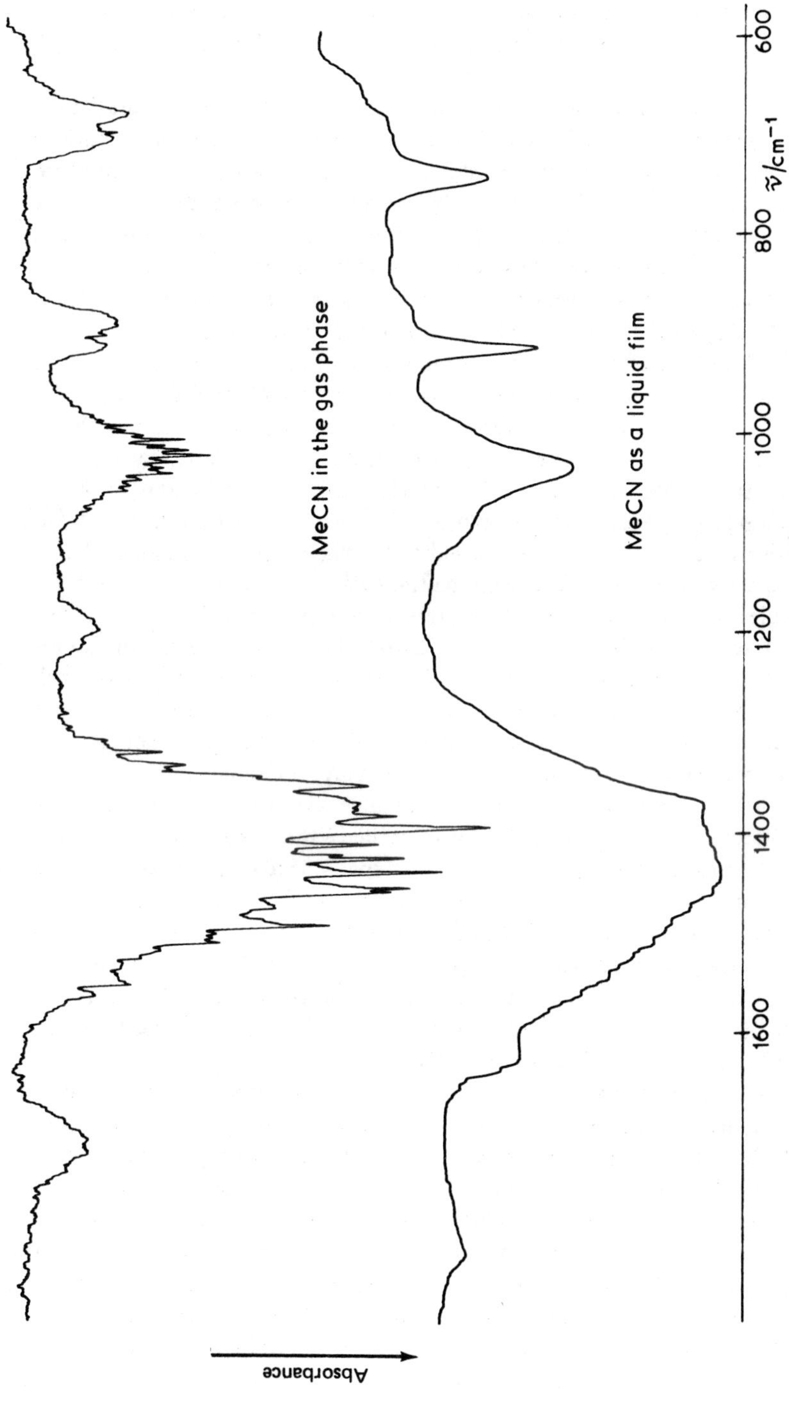

Fig. 4.34 Infrared spectra showing the loss of rotational structure when samples are examined in a condensed phase.

4.34). The observed bands can be described completely in terms of transitions between different vibrational states. This means that bond lengths cannot be calculated from the observed spectra, but group theory can still be applied satisfactorily.

Free rotation is prevented also in most crystalline solids, and so rotational energy changes can be neglected. However, in contrast to liquids and solutions, the vibrational spectra of solids exhibit certain features which are not predicted by group theory if the influence of neighbouring molecules in the surrounding crystalline lattice is neglected. For example, bands observed in a gas-phase spectrum may be split in a solid-phase spectrum. This may be caused either by site group splitting effects (degenerate internal vibrations are split because of the lower local symmetry of the molecule or ion in a crystalline lattice), or correlation field splitting effects (due to interactions with internal vibrations of other molecules or ions within the same unit cell of the crystal) see Fig. 4.35(a). A number of entirely new bands, not seen in the gas-phase spectrum, may also be observed in the low-frequency region ($< 800\ cm^{-1}$). The latter can be assigned to external or lattice modes, and they arise from motions of entire molecules or ions relative to one another within the unit cell. Lattice modes are subdivided into translational modes and rotational or librational modes. Translational lattice vibrations arise from the translational motions of some of the molecules or ions in the unit cell with respect to others in such a way that the unit cell remains stationary [see Fig. 4.35(b)]. Translational lattice modes may be contrasted with acoustic modes which are produced when the translational motions of the molecules or ions in the unit cell all occur in the same direction and hence the whole unit cell moves in a specified direction. An example for the carbonate ion is shown in Fig. 4.35(c). Finally, rotational lattice modes involve the partial rotation of one group of molecular units relative to another. The rotation of each unit is analogous to ordinary rotational motion but the molecular unit does not rotate by 360° [see Fig. 4.35(d)]. The frequency of a translational mode depends upon the total mass of the translating unit while the frequency of a rotational mode will depend upon the moment of inertia for the rotation axis concerned. The two types of mode can be distinguished therefore by isotopic replacement, e.g. for HCl the translational lattice mode occurs at $86\ cm^{-1}$ (cf. DCl = $89\ cm^{-1}$) whereas the rotational lattice mode has been observed at $217\ cm^{-1}$ (cf. DCl = $169\ cm^{-1}$).

A good example to illustrate the complications that can arise due to crystal effects is the vibrational spectrum of solid K_2CrO_4. The tetrahedral chromate ion, CrO_4^{2-}, belongs to point group T_d and group theory predicts 4 normal modes of vibration for the free ion. The spectral activities and observed frequencies (measured in aqueous solution) are given in Table 4.5. Figure 4.36 shows that: (i) the single bands of the free CrO_4^{2-} ion are replaced in most cases by multiplets in the solid-phase spectrum; (ii) bands due to $\tilde{\nu}_1$ and $\tilde{\nu}_2$ appear in the infrared spectrum of solid chromate even though they are expected to be infrared inactive on the basis of the free-ion selection rules; (iii) some very low frequency bands appear below $200\ cm^{-1}$.

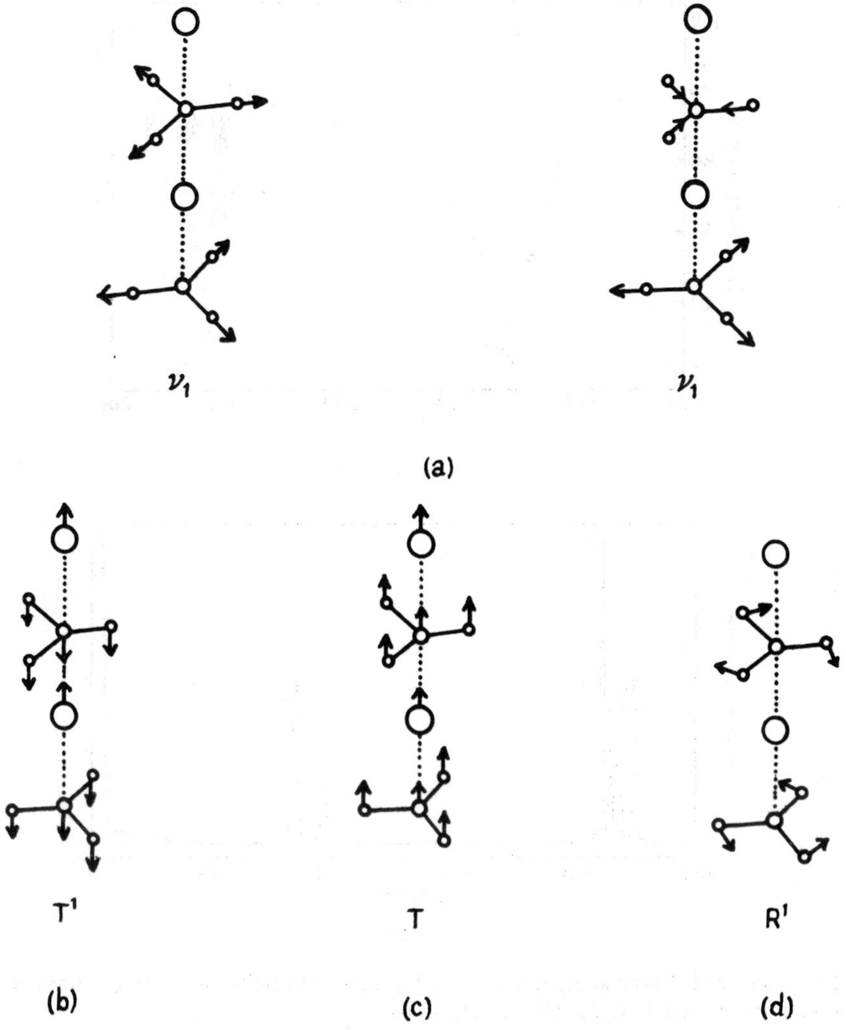

Fig. 4.35 (a) Correlation field splitting effects for the carbonate ion CO_3^{2-}. (b) Translatory lattice mode. (c) Acoustic mode. (d) Rotational lattice mode. (From Sherwood, ref. 4.17.) ○ represents the position of a cation.

These three features of the solid-phase spectra result from the existence of a different symmetry environment for the CrO_4^{2-} ion within the crystal and the presence of significant intermolecular forces which are absent in the gas phase. The most satisfactory method of obtaining a complete assignment is to carry out a *factor group analysis*. The latter assumes that complete vibrational coupling takes place between the normal modes within the *unit cell*. The details of the analysis will not be discussed further in this chapter but the method is described

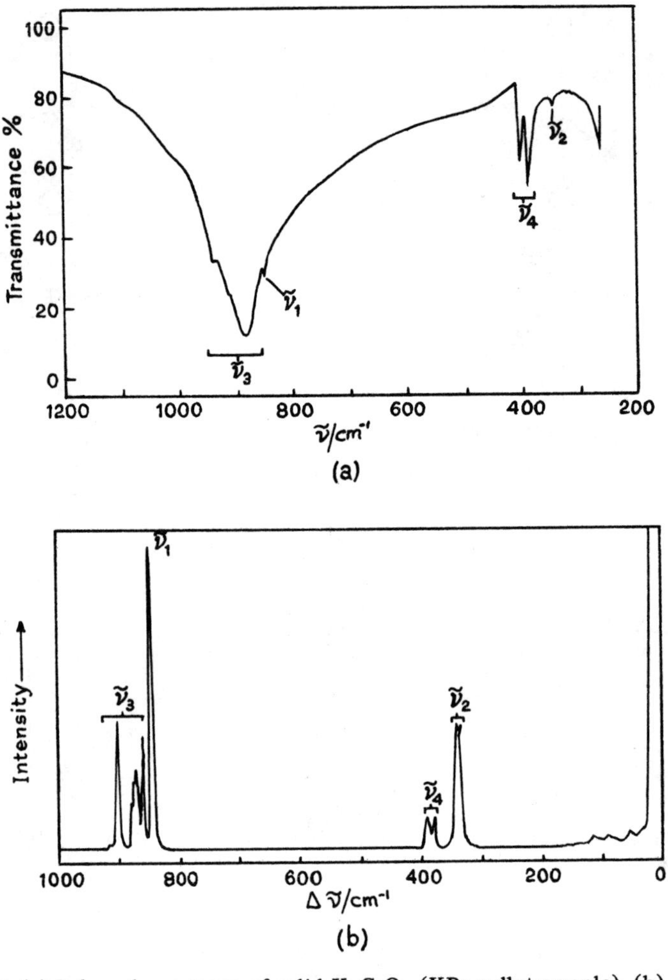

Fig. 4.36 (a) Infrared spectrum of solid K_2CrO_4 (KBr pellet sample). (b) Raman spectrum of solid K_2CrO_4. (From ref. 4.16).

Table 4.5 Spectral activities and observed frequencies for the chromate ion in aqueous solution; Raman data taken from ref. 4.15

Mode	*Activity*	$\tilde{\nu}/cm^{-1}$
$\tilde{\nu}_1$	Raman	847 (pol)
$\tilde{\nu}_2$	Raman	348
$\tilde{\nu}_3$	infrared, Raman	884
$\tilde{\nu}_4$	infrared, Raman	368

Table 4.6 Vibrational frequencies of crystalline K_2CrO_4 [4.16]

Raman frequencies $\Delta\tilde{\nu}/(cm^{-1})$	*Infrared frequencies* $\tilde{\nu}/(cm^{-1})$	*Assignment*	*Number of modes expected using factor group analysis*
918 (B_{2g})	936		
903 (A_g)	910		
881 (B_{2g})	883	$\tilde{\nu}_3$	6 Raman, 5 i.r.
878 (B_{3g})			
876 (B_{1g})			
867 (A_g)	859		
851 (A_g, B_{2g})	850	$\tilde{\nu}_1$	2 Raman, 2 i.r.
396 (A_g, B_{2g})	398		
392 (B_{1g})		$\tilde{\nu}_4$	6 Raman, 5 i.r.
387 (B_{3g})			
386 (A_g, B_{2g})	382		
350 (B_{1g}, B_{2g})			
346 (B_{3g})	342	$\tilde{\nu}_2$	4 Raman, 3 i.r.
345 (A_g)			
157 (A_g)			
138 (B_{3g})			
119 (B_{1g}, B_{3g})			
116 (A_g)			
114 (B_{2g})			
109 (A_g)			
99 (B_{2g})		External modes	24 Raman, 16 i.r.
93 (A_g)			
91 (B_{3g})			
85 (B_{3g})			
83 (A_g)			
67 (B_{1g})			
54 (B_{1g}, B_{3g})			
37 (A_g)			

thoroughly elsewhere [4.17]. To emphasize the complexity of the results compared with free-ion prediction, the factor group analysis for crystalline K_2CrO_4 is given in Table 4.6. The character species (B_{2g}, A_g, etc.) refer to point group D_{2h} and the appropriate group is obtained from a knowledge of the X-ray crystal structure of the compound.

4.10 APPLICATIONS OF INFRARED SPECTROSCOPY

Much of infrared work depends on associating a particular band position with a characteristic group or structural unit in a compound and noting also the maximum intensities of the corresponding absorption band. These points lead into the first topic of qualitative analysis but structural problems and quantitative analysis are related also and will be discussed in turn.

Although molecular vibrations strictly involve motions of all the atoms in the molecule, the energy of a normal mode is sometimes localized almost completely in the stretching or bending of a given bond. An absorption band characteristic of that bond will then be observed. This will be so for terminal groups if the mass of one atom is much less than the other (e.g. M—H vibrations) or if the bond force constant is much higher than those of adjacent bonds (e.g. >C=O, —C≡N etc.). Extensive empirical studies have been carried out, particularly in organic chemistry, and where a correlation between particular bands and molecular groups has been firmly established, the observation of a particular band is strong evidence for the presence of the particular group. The absence of the band also implies the absence of that grouping, but care must be exercised because the band may be weak and hence difficult to observe or it may be masked by a strong neighbouring band. Collections of characteristic group frequencies have been gathered by several authors [4.18] and are widely used in qualitative work.

4.10.1 Vibrational frequencies and qualitative analysis

The exact values of the characteristic group frequencies depend on the molecular environment due to neighbouring groups. The vibrational frequency is partly dependent also on whether the molecule is studied in the solid, liquid, or gaseous phase. The effect of neighbouring substituents is clearly illustrated with the keto group.

In the molecule X(Y)CO, the carbonyl stretching frequency $\tilde{\nu}_{C=O}$ depends on the substitutents X and Y. In simple saturated aliphatic ketones the carbonyl absorption frequency lies within the range 1740–1720 cm^{-1}. However, when an aliphatic C=C is conjugated with the C=O group, the frequency range for simple ketones is 1685–1665 cm^{-1}. Table 4.7 illustrates the effect of X and Y on the frequency. In Table 4.7 the effect of conjugation on the keto vibration frequency may be observed in B, C, and D, while the effect of chelation is illustrated in F.

As will be readily observed, the carbonyl frequency varies with the nature of the substitutent, and therefore, to interpret any absorption frequencies in this range, it is desirable to have all the data for the different types of X(Y)CO. Thus, all the empirical data for these compounds relating frequency to the particular type of structural unit absorbing must be available. In practice the published correlation charts summarizing this information are used.

Considerable experience is required, however, in the application of group frequency charts, and in the case of the ketonic carbonyl vibration, if the frequency range of the carbonyl were quoted to cover all types of ketones, the range would be too great to be useful. For example, the frequencies 1740–1540 cm^{-1} of the quoted ketonic values would include frequencies of the allenes and amines. If, however, a particular frequency, say 1690 cm^{-1}, was

Table 4.7 Characteristic frequencies of ketonic carbonyl vibrations in dilute solution

Ketone		$\tilde{\nu}/(\text{cm}^{-1})$
A. Saturated straight-chain ketones	$-CH_2-CO-CH_2-$	1740–1720
B. α, β-Unsaturated ketones	$-CH{=}CH-CO-$	1685–1665
C. Aryl ketones	C_6H_5-CO-	1700–1680
D. Diaryl ketones	$C_6H_5-CO-C_6H_5$	1670–1660
E. α-Diketones	$-CO-CO-$	1730–1710
F. β-Diketones	$-CO-CH_2-CO-$	1640–1540

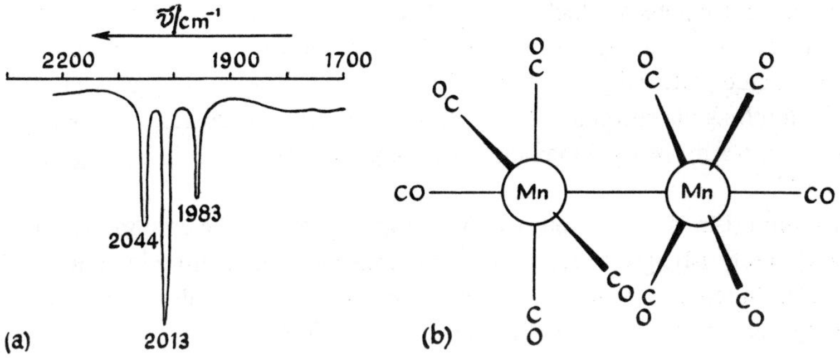

Fig. 4.37 (a) Infrared spectrum (1700–2200 cm^{-1}) of $Mn_2(CO)_{10}$ in cyclohexane solution. (b) Molecular structure of $Mn_2(CO)_{10}$.

obtained, then the presence of an aryl ketone would be indicated. Further confirmatory evidence might be sought; for instance, the intensity of this one band could be compared with that obtained from a similar strength solution of an aryl ketone, and the fact that aryl ketones have another absorption band at 1210–1325 cm^{-1} could be used. Where overlapping between the frequency ranges occurs, e.g. in saturated straight-chain ketones and α-diketones, then the classification of its type may sometimes be made by intensity studies.

The carbonyl group also gives very strong and characteristic frequencies when it occurs in inorganic or organometallic compounds. When CO acts as a bridge between two metal atoms, the $\tilde{\nu}_{CO}$ stretching frequency usually occurs between 1700 and 1850 cm^{-1}. In contrast, terminal carbonyl groups exhibit bands between 1850 and 2150 cm^{-1}. Thus, it can be inferred from the infrared spectrum [see Fig. 4.37(a)] that $Mn_2(CO)_{10}$ contains terminal CO groups only and the structure shown in Fig. 4.37(b) has been confirmed by X-ray crystallography.

The occurrence of characteristic group frequencies enables important conclusions to be drawn in the field of structural isomerism. A well-known example

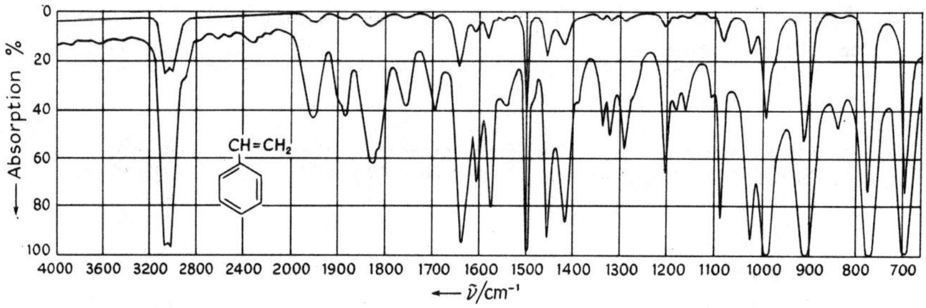

Fig. 4.38 Infrared spectrum of styrene. (From Bellamy, ref. 4.18.)

of this type of application is the analysis of mixtures of *cis*- and *trans*-1,2 disubstituted ethylenes by using the highly characteristic out-of-plane CH deformation frequency which occurs for the *trans*-compound near 970 cm^{-1}

Infrared spectroscopy is used extensively in conformational analysis. For example, an equatorial group X attached to a cyclohexane ring generally shows a C—X stretching vibration at a higher frequency than the corresponding axial group; e.g. for bromocyclohexane, $\tilde{\nu}_{C\text{-}Br}$ (equatorial) = 687 cm^{-1}, $\tilde{\nu}_{C\text{-}Br}$ (axial) = 658 cm^{-1}.

Tautomerism is a common occurrence with heterocyclic compounds, and again characteristic group frequencies can make an important contribution. For example, 2-mercaptobenzothiazole exists primarily in the thione form, as indicated by the strong $\tilde{\nu}_{N-H}$ stretching band at 3410 cm^{-1}

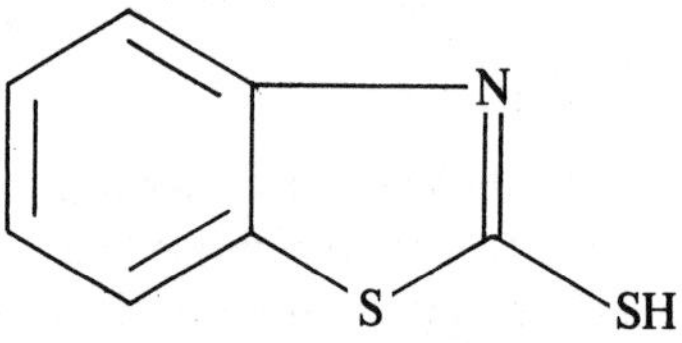

Thiol

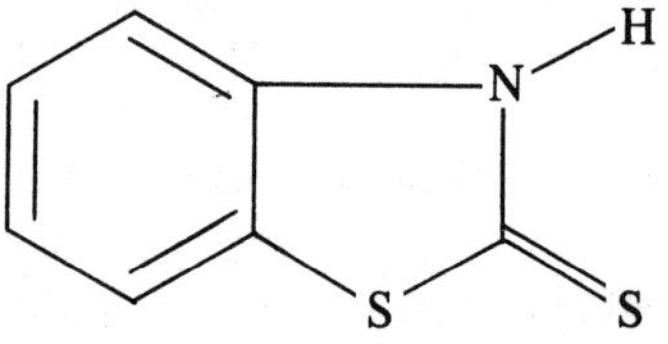

Thione

To illustrate how characteristic frequencies may be used to identify different groups within a molecule, styrene will be considered. Its spectrum is reproduced in Fig. 4.38. The aromatic ring is recognized by the presence of ring carbon—carbon vibrational frequencies in the 1600—1500 cm^{-1} region, since these absorption bands are little affected by substitution. The fact that a mono-substituent is present is indicated by the strong absorption bands at approximately 770—700 cm^{-1}, while the band at approximately 1630 cm^{-1} suggests the C=C aliphatic linkage. From this type of approach the analysis may be made. Absorption in the region 2000—1660 cm^{-1} is particularly characteristic for substituted benzene compounds and usually indicates the number of ring substituents. Emphasis is placed on the relative intensities and the number of bands in

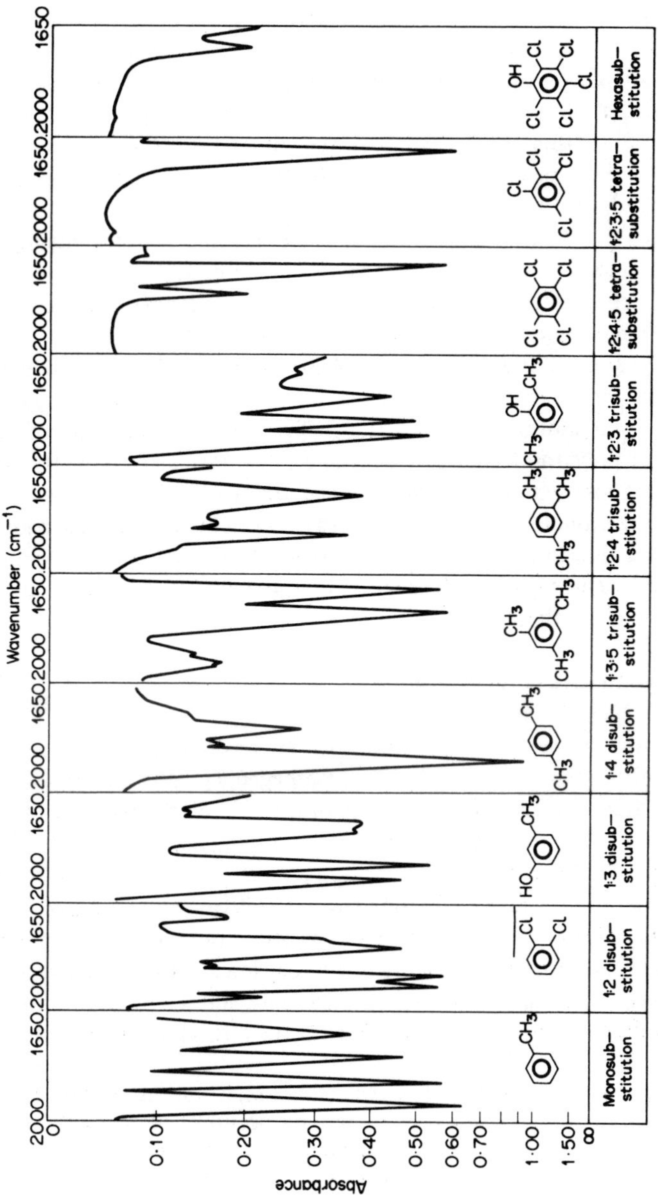

Fig. 4.39 Aromatic substitution patterns in the region 2000–1650 cm^{-1}. (From Bellamy, ref. 4.18.)

deciding the position of the substituent. For example, monosubstituted benzene materials usually give four bands of gradually diminishing intensity towards lower wavenumbers, and although the overall appearance of these bands alters with the type of mono-substituent, varying both in position and relative intensities, the pattern differs so much from the other types obtained by di-, tri-, and tetra-substitutions that it may be readily identified. The 2000–1650 cm^{-1} region patterns of di-, tri-, tetra-, and hexa-substituted benzene compounds are given in Fig. 4.39. Confirmation of the position of the benzene substituents may be obtained by a similar type of study of the 1250–1000 cm^{-1} region or even the 1000–650 cm^{-1} region.

Applications are not nearly so numerous in inorganic chemistry, mainly because bond moments are much more variable than in organic compounds. However, the concept is till useful and Table 4.8 shows some characteristic frequencies of some common inorganic polyatomic ions in the solid state.

Although the qualitative approach involving characteristic group frequencies has had by far the greatest influence on chemistry, the *determination of molecular symmetry* is another important aspect. The number and activity of the vibrations of a polyatomic molecule depend on the molecular symmetry (see Chap. 2). The more symmetrical the molecule, the fewer are the distinct vibrational modes and the fewer the coincidences between infrared and Raman spectra. Thus, vibrational spectra can help to determine molecular structures but, because of the tie-in with Raman spectroscopy, the discussion of this topic will be deferred until Section C.

4.10.2 Quantitative infrared analysis

The quantitative studies in the infrared region are based on the Beer–Lambert law: $\log_{10} I_0/I = \epsilon cl = D$, where D is the absorbance, I_0 is the intensity of incident light of frequency ν, I is the intensity of the light after passing through the absorption cell, c is the concentration of the absorbing material (mol dm^{-3}), l is the thickness of the absorption cell (cm), and ϵ is the extinction coefficient. For a particular absorption band a calibration curve is constructed of $\log_{10} I_0/I$ against c for a number of solutions of known concentration; this curve is then used for estimating the concentration of an unknown solution.† A two- or three-component mixture could be analysed in favourable cases, and the procedure for a three-component mixture is outlined below. If the optical densities of the three

† Direct proportionality exists between c and $\log I_0/I$, and this is known as Beer's law; deviations from this law sometimes occur. The law is only strictly accurate for monochromatic radiation, whereas the exit beam from a spectrophotometer always has a range of frequencies. The range of frequencies depends on: (i) the slit width; (ii) the dispersion of the spectrophotometer; (iii) its resolving power. In addition, the solution must not be too concentrated or exhibit strong molecular interaction; if association or dissociation occurs, wide deviations from Beer's law may take place.

components are D_1, D_2, and D_3, then at a wavenumber position $\tilde{\nu}_1$, where for example $\epsilon_1\tilde{\nu}_1$ is the extinction coefficient of component 1 at wavenumber $\tilde{\nu}_1$:

$$D_{\tilde{\nu}_1} = D_{1\tilde{\nu}_1} + D_{2\tilde{\nu}_1} + D_{3\tilde{\nu}_1} = (\epsilon_{1\tilde{\nu}_1}c_1 + \epsilon_{2\tilde{\nu}_1}c_2 + \epsilon_{3\tilde{\nu}_1}c_3)l \qquad (4.81)$$

and similarly for absorption at wavenumbers ν_2 and ν_3:

$$D_{\tilde{\nu}_2} = D_{1\tilde{\nu}_2} + D_{2\tilde{\nu}_2} + D_{3\tilde{\nu}_2} = (\epsilon_{1\tilde{\nu}_2}c_1 + \epsilon_{2\tilde{\nu}_2}c_2 + \epsilon_{3\tilde{\nu}_2}c_3)l \qquad (4.82)$$

$$D_{\tilde{\nu}_3} = D_{1\tilde{\nu}_3} + D_{2\tilde{\nu}_3} + D_{3\tilde{\nu}_3} = (\epsilon_{1\tilde{\nu}_3}c_1 + \epsilon_{2\tilde{\nu}_3}c_2 + \epsilon_{3\tilde{\nu}_3}c_3)l \qquad (4.83)$$

where $D_{\tilde{\nu}_1}, D_{\tilde{\nu}_2}$, and $D_{\tilde{\nu}_3}$ are determined directly from the experimental absorption pattern. If the extinction coefficients at the given wavenumber position have been previously determined for each of the components in the mixture (since l, the cell thickness is known), c_1, c_2, and c_3 may be determined from the three linear equations. By careful choice of wavenumbers used, the solution of these equations may be simplified. For example, if $\tilde{\nu}_1$ is chosen so that $\epsilon_{2\tilde{\nu}_1}$ is approximately equal to $\epsilon_{3\tilde{\nu}_1}$, which is approximately equal to zero, and if component 1 shows strong absorption at wavenumber $\tilde{\nu}_1$, then the problem is considerably simplified.

This type of method is also used for quantitative analysis in the visible and ultraviolet regions.

The accuracy of quantitative analysis by the infrared method depends on several variable factors, such as band intensities (see p.219), and the presence or absence of interfering substances in the mixture. In favourable circumstances the accuracy can be as high as 0.002 per cent of the amount present. On the other hand, when weak absorption and overlapping bands are being employed, errors of 15 per cent are possible. In mixtures of absorbing substances detection by infrared methods of 0.1 per cent of a particular component is usually quite difficult. The infrared method may be an extremely powerful one for quantitative analysis especially when used for problems for which chemical analysis would prove very difficult or even impossible, e.g. in the estimation of one isomer in the presence of many others.

The following example illustrates the use of the infrared method to solve a chemical problem. Confirmation of the products of mononitration of β-nitronaphthalene was needed and the relative abundance in which they occurred was unknown. Chemical methods had failed to confirm the dinitronaphthalenes which had earlier been reported as products of this reaction. Comparison of the infrared spectrum of the crude nitration mixture with those of pure dinitronaphthalenes revealed that the product consisted of a mixture of 1,6- and 1,7-dinitronaphthalene. After trials with various solvents and examination of the spectrum in the 2000–650 cm^{-1} range, a solution of each component, whose concentration was accurately known, and of the nitration mixture was measured in the 740–770 cm^{-1} range, where characteristic bands of the 1,6- and 1,7-dinitronaphthalene compounds were to be found. The solvent employed was dioxan. The results of these measurements are shown in Fig. 4.40.

Table 4.8 Characteristic frequencies of some common polyatomic ions in the solid state (from ref. 4.19 as adapted in ref. 4.20)

(In some cases only two or three compounds have been studied, so the ranges given must be regarded as approximate, as are the indications of intensities)

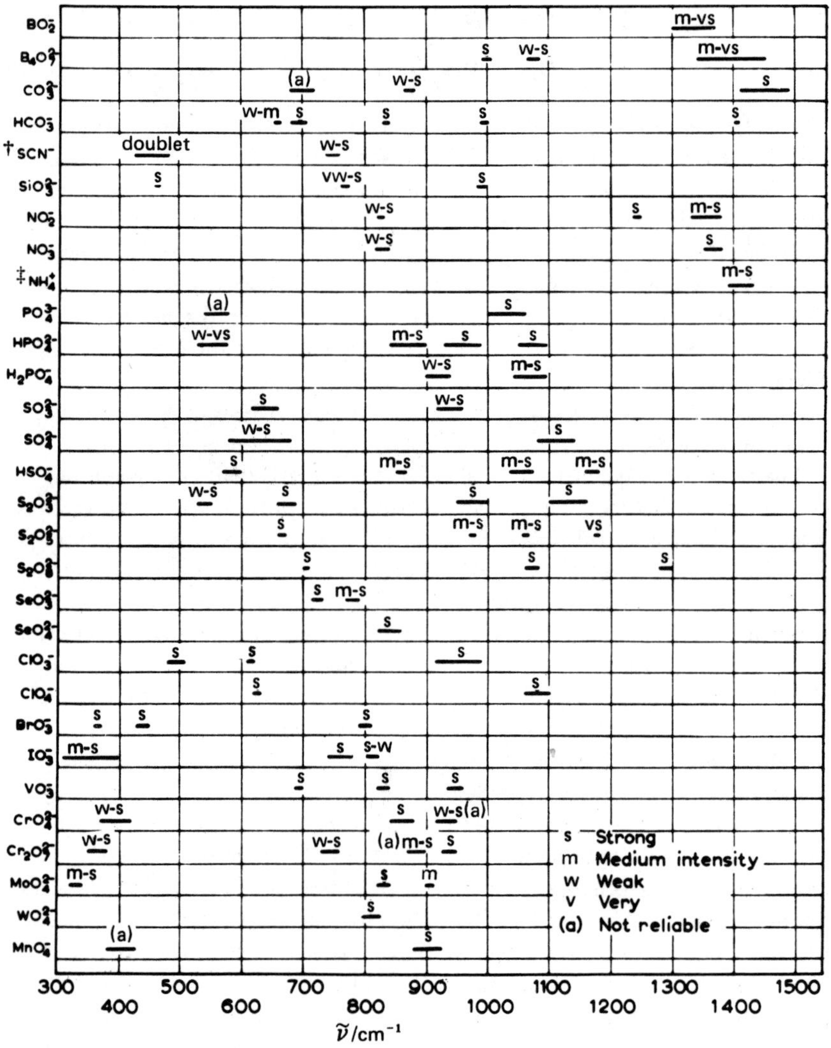

† Bands are also found near 2100 cm^{-1}.
‡ In NH_4F the infrared active vibrations are at about 1490 and 2800 cm^{-1}, other ammonium salts give bands at 3050–3300 cm^{-1}.

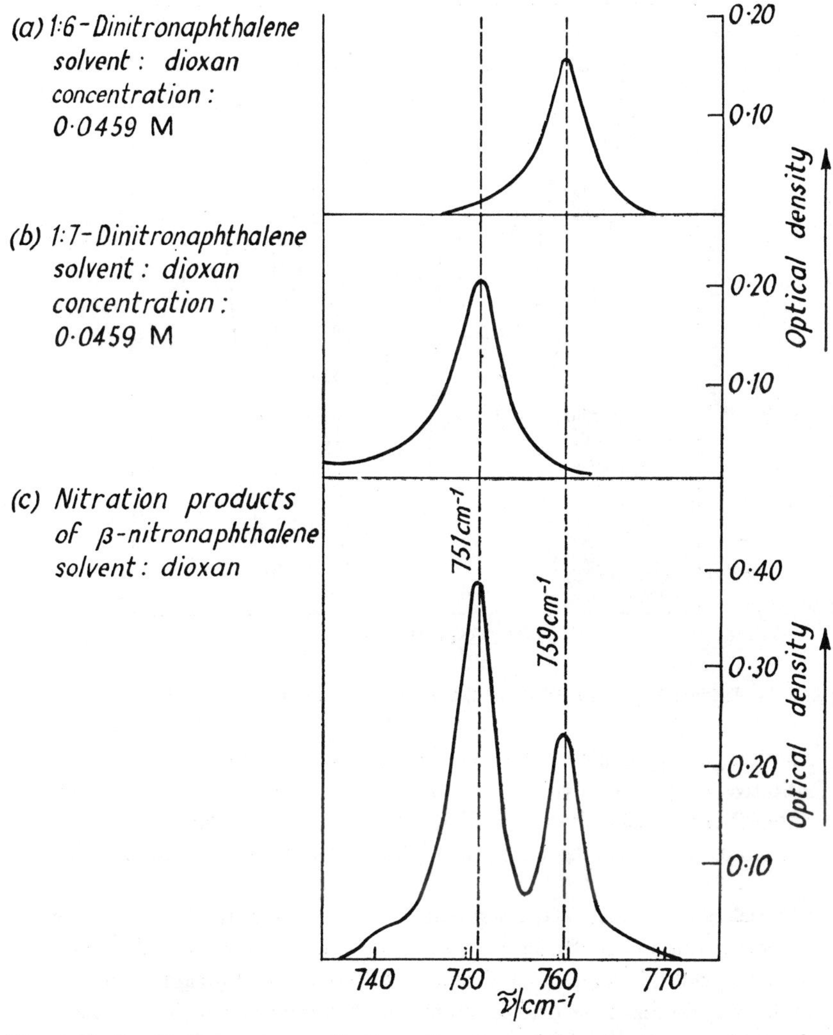

Fig. 4.40 Qualitative and quantitative estimation of the nitration products of β-nitronaphthalene from a study of the bands in the 740–770 cm^{-1} region. (Courtesy of Dr. E.R. Ward and Mr. J. Hawkins.)

The height of the absorption peak in Fig. 4.40(a) is equal to $\epsilon_1 c_1 l$, and thus knowing c_1 and l, ϵ_1 can be calculated. In a similar way ϵ_2 is obtained from Fig. 4.40(b). From the heights of the absorption peaks in Fig. 4.40(c) the concentration of the two components in the mixture can be analysed. If $D_1{}'$ and $D_2{}'$ are the optical density peak values in Fig. 4.40(c) for the bands at 759 and 751 cm^{-1} respectively, then to a good approximation $D_1 = \epsilon_1 c_1 l$ and $D_2 = \epsilon_2 c_2 l$. Hence, the concentrations (c_1 and c_2) of the components in the mixture are determined. The exact approach would be to employ Equations (4.81) and (4.82).

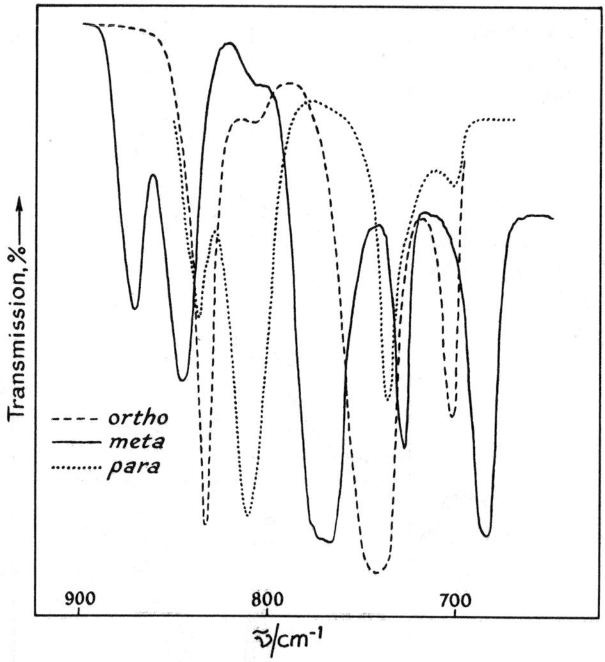

Fig. 4.41 Infrared spectra (900–700 cm^{-1}) of *o*-, *m*-, and *p*-cresol.

Another good example to illustrate the use of infrared spectroscopy for quantitative analysis is provided by the three isomers of cresol. By measuring the intensities of the bands at 813, 775, and 752 cm^{-1} the proportion of *para-*, *meta-*, and *ortho*-cresol respectively in a mixture can be determined (see Fig. 4.41).

In conclusion, it must be stressed that it is extremely difficult to obtain reproducible extinction coefficients for a particular component from different instruments, and this severely handicaps useful exchange of information.

In this respect digital recording, coupled with computer calculations, has considerable potential in this area. Some of the advantages of this more sophisticated approach are: (i) noise can be decreased by mathematical smoothing; (ii) the position of maximum absorption and integrated band intensities are easily calculated; (iii) poorly separated bands can be resolved by mathematical methods; (iv) a large number of samples can be easily analysed in a short time with suitable automation, e.g. fats in milk.

4.10.3 Accurate band intensity considerations

Vibrational absorption band intensity determination

In the past few years much interest has centred around attempts at the measurement of the absolute intensity of vibrational absorption bands. Such measurements, however, are handicapped by the fact that intensity measurements made on one spectrophotometer are rarely reproducible on another. Before mentioning the causes of this discrepancy it will be necessary to introduce several terms. For monochromatic radiation of wavenumber $\tilde{\nu}$ the absorption is governed by the equation:

$$I = I_0 \exp(-K_{\tilde{\nu}}cl) \tag{4.84}$$

where I_0 is the intensity of the incident radiation and I that of the emergent radiation. The length of the absorbing path is l cm and the concentration c mol dm^{-3}. From Equation (4.84) $K_{\tilde{\nu}}$ the absorption coefficient is given by:

$$K_{\tilde{\nu}} = \frac{1}{cl}\log_e(I_0/I)_{\tilde{\nu}} \tag{4.85}$$

The absorption coefficient is related to the molecular extinction coefficient ($\epsilon_{\tilde{\nu}}$) by the equation:

$$K_{\tilde{\nu}} = 2.303\,\epsilon_{\tilde{\nu}} \tag{4.86}$$

An infrared band may be represented by a plot of $\epsilon_{\tilde{\nu}}$ against the wavenumber, and the value of $\epsilon_{\tilde{\nu}}$ at the absorption maximum $\epsilon_{\tilde{\nu}_{max}}$ is quoted to characterize the magnitude of the intensity of the vibrational band. Of more fundamental significance than $\epsilon_{\tilde{\nu}_{max}}$ is the integrated absorption intensity A. This is given by:

$$A = \int K_{\tilde{\nu}}\, d\tilde{\nu} \tag{4.87}$$

$$A = 2.303\int \epsilon_{\tilde{\nu}}\, d\tilde{\nu} \tag{4.88}$$

where $\int \epsilon_{\tilde{\nu}}\, d\tilde{\nu}$ is the area under the absorption curve. The integration extends over the whole of the band examined.

The broadness of a band is defined in terms of the half-bandwidth $\Delta\tilde{\nu}_{\frac{1}{2}}$, that is the width of the band at $0.5\,\epsilon_{\tilde{\nu}_{max}}$. These quantities are illustrated in Fig. 4.42(a).

In Equations (4.84)–(4.88) monochromatic radiation of wavenumber $\tilde{\nu}$ has been assumed. In practice, however, the radiation in a spectrophotometer is never truly monochromatic but contains a range of wavenumbers, the extent of which is partly determined by the effective slit width.

Instead of measuring $K_{\tilde{\nu}}$, $\epsilon_{\tilde{\nu}_{max}}$, and $\Delta\tilde{\nu}_{\frac{1}{2}}$, what is actually obtained is $K_{\tilde{\nu}}^{a}$,

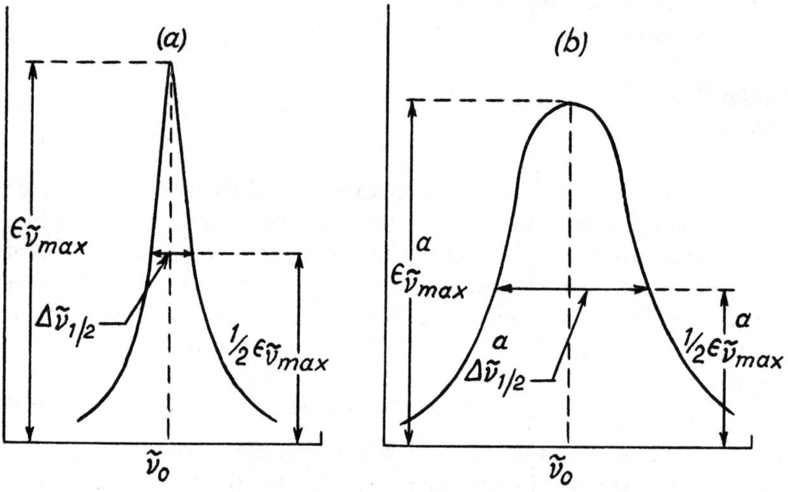

Fig. 4.42 Quantities characterizing the shape and intensity of an infrared absorption band: (a) the true absorption band; (b) the band as modified by the finite spectral slit width. (From ref. 4.21. Courtesy of Interscience Publishers Inc., New York.)

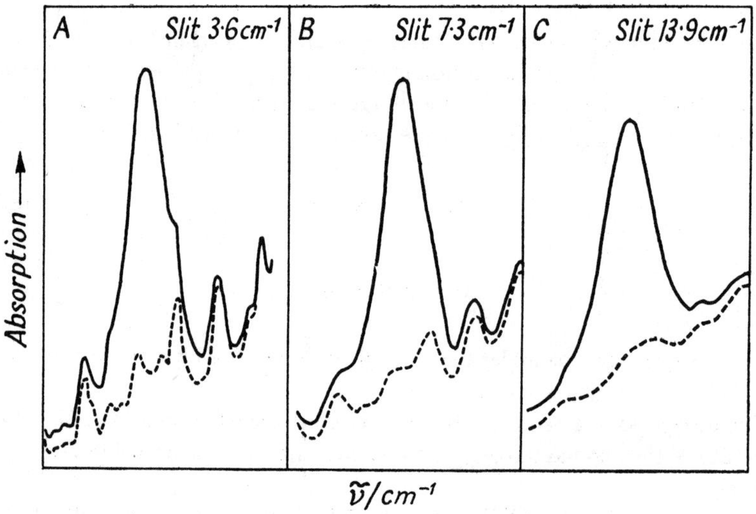

Fig. 4.43 The effect of finite slit width on the band intensity as determined by a single-beam spectrophotometer. Curves A, B, and C were measured on the same CS_2 solution of camphor at increasing slit widths as indicated. The broken curve shows the atmospheric water vapour absorption curve. (After ref. 4.21. Courtesy of Interscience Publishers Inc., New York.)

$\epsilon_{\tilde{\nu}_{max}}{}^{a}$, and $\Delta\tilde{\nu}_{\frac{1}{2}}{}^{a}$, where the superscript 'a' signifies 'apparent' values. Some of these symbols are illustrated in Fig. 4.42(b). An example of the effect of finite slit width on infrared absorption bands is illustrated in Fig. 4.43. As the slit width is increased the band becomes wider and flatter, and the area beneath the curve decreases.

The infrared half-bandwidths for solutions of organic substances are normally between 5 and 20 cm^{-1}. In solids and especially in gases narrower lines are observed. It is now thought that a band does not reveal its true shape until the effective slit width has been reduced to about one-fifth of the true half-band-width, and this in general means very narrow slit widths indeed. Consequently, until such slit widths are employed, A, ϵ, and $\Delta\tilde{\nu}_{\frac{1}{2}}$ cannot be measured directly. Since it is not practicable in the case of conventional prism spectrophotometers to reduce the slit width to these limits, the true band shape cannot be obtained. However, improvements in the band shape can be made by the use of greater resolving power as in suitable grating and double-pass instruments.

Several methods are available for determining A, $\epsilon_{\tilde{\nu}}$, and $\Delta\tilde{\nu}_{\frac{1}{2}}$ from the observed values B, $\epsilon_{\tilde{\nu}}{}^{a}$, and $\Delta\tilde{\nu}_{\frac{1}{2}}{}^{a}$, respectively [4.22], where these symbols are defined as follows:

$$\epsilon_{\tilde{\nu}}{}^{a} = \frac{1}{cl}\log_{10}\left(\frac{T_0}{T}\right)_{\tilde{\nu}} \tag{4.89}$$

$$B = 2.303\int \epsilon_{\tilde{\nu}}{}^{a}\,\mathrm{d}\tilde{\nu} \tag{4.90}$$

$\Delta\tilde{\nu}_{\frac{1}{2}}{}^{a}$ is the apparent half-bandwidth at $0.5\,\epsilon_{\tilde{\nu}}{}^{a}$, and T_0 and T are written for I_0 and I to indicate that truly monochromatic radiation is not being employed for the measurements.

Two methods of determining the true integrated band intensity from the apparent value will now be briefly considered.

Wilson and Wells [4.22] studied the infrared absorption bands of gases broadened by the introduction of a foreign gas at various pressures. They determined the value of $\int \log_e(T_0/T)_{\tilde{\nu}}\,\mathrm{d}\tilde{\nu}$ for a particular band at different pressures and extrapolated these values to zero partial pressure of the absorbing gas; the integrated band intensity was then given by:

$$A = \lim_{p\to 0}\frac{1}{pl}\int \log_e(T_0/T)_{\tilde{\nu}}\,\mathrm{d}\tilde{\nu} \tag{4.91}$$

In the case of a solution the true integrated absorption coefficient A can be obtained by measuring B at several concentrations and extrapolating to zero concentration; alternatively, the concentration can be maintained constant, and the extrapolation carried out to zero path length. These have been used by Ramsey and coworkers [4.23], and the extrapolation was based on the equations:

$$A = \lim_{c \to 0} \frac{1}{cl} \int \log_e \left(\frac{T_0}{T}\right)_{\tilde{\nu}} \mathrm{d}\tilde{\nu} \tag{4.92}$$

$$A = \lim_{l \to 0} \frac{1}{cl} \int \log_e \left(\frac{T_0}{T}\right)_{\tilde{\nu}} \mathrm{d}\tilde{\nu} \tag{4.93}$$

The extrapolation is linear and the line has a small negative slope.

Another procedure is to obtain by graphical integration the area A' beneath the fractional absorption $(I_0 - I)/I_0$ versus wavenumber curve, where:

$$A' = \int \left(\frac{I_0 - I}{I_0}\right)_{\tilde{\nu}} \mathrm{d}\tilde{\nu} \tag{4.94}$$

This area A' is independent of resolving power so that:

$$A' = \int \left(\frac{T_0 - T}{T_0}\right)_{\tilde{\nu}} \mathrm{d}\tilde{\nu} \tag{4.95}$$

and A can be obtained by plotting A'/cl against cl and extrapolating to $cl = 0$. Since, however, the plot of A'/cl against cl is curved, Ramsey [4.24] has employed the reciprocal cl/A' against cl; the plot is nearly linear, and the intercept gives the value $1/A$.

It is most desirable to obtain accurate band intensity measurements for the following reasons.

(1) They can be applied to the quantitative analysis of mixtures so that the data obtained on one spectrophotometer can be accurately used on another.

(2) In ultraviolet absorption spectra of solutions the structural diagnostic procedure is to measure not only the wavelength at which the peak absorption occurs but also its intensity; the latter is often most useful in assisting identification of the particular group absorbing. Such a procedure would be most valuable in the infrared.

(3) The absolute intensity of a vibrational band is related to the change of electric moment during the vibration, and in the case of small molecules it is sometimes possible to derive from the determined absolute intensity the individual bond moment and thus gain some intimate knowledge of molecular electronic structure.

Determination of bond moment from the absolute intensity of an infrared absorption band

The intensity of an infrared absorption band depends on the magnitude of the change in electric dipole moment during the vibrational displacement. For a stretching vibration of, for example, a diatomic vibrating group the integrated intensity of a particular absorption band may be related to $(\partial\mu/\partial r)^2$, where $\partial\mu/\partial r$ is the rate of change of the bond moment with internuclear distance. For a molecule it is necessary to determine first the integrated absorption intensity, A, of the fundamental absorption band, and from this the rate of change of dipole moment with each normal coordinate (see p. 168). That is $\partial\mu/\partial Q$ is calculated where Q is a normal coordinate. The integrated intensity, A, of a band is given by:

$$A = \frac{N\pi}{3c}\left(\frac{\partial\mu}{\partial Q}\right)^2 = \frac{N\pi}{3c}\left(\left|\frac{\partial\mu_x}{\partial Q}\right|^2 + \left|\frac{\partial\mu_y}{\partial Q}\right|^2 + \left|\frac{\partial\mu_z}{\partial Q}\right|^2\right) \qquad (4.96)$$

where x, y, and z represent fixed axes within the molecule, the $\partial\mu/\partial Q$ terms are vector quantities, N is the number of molecules per cubic centimetre of sample, and c is the velocity of light. Q is expressed as some linear combination of the more elementary coordinates, such as bond length and bond angle. To convert $\partial\mu/\partial Q$ into $\partial\mu/\partial r$ it is essential to know the potential-energy function which governs the molecular vibrational frequencies (see Volume 2, Chap. 6). From this it is possible to relate the normal coordinates to the internal coordinates by a set of coefficients which govern the motions of atoms. After this mathematical treatment (normal coordinate treatment) the value of $\partial\mu/\partial r$ is obtained for each of the bonds concerned. In addition to the value of $\partial\mu/\partial r$ the method is capable of yielding individual bond moments in a molecule. One limitation of the method is that, since a square root is involved in the derivation of μ, the sign of the bond moment has to be assumed. For a discussion of absolute intensities for vibrating diatomic molecules see Appendix, p. 345.

4.10.4 Infrared spectra of unstable species

A lot of interest has been shown recently in the identification and structure determination of unstable species such as CH_3, NH_2, OF, $XeCl_2$, etc., mainly because this sort of species may act as intermediates in chemical reactions. There are two basic approaches to the study of such short-lived species: (a) they may be produced in the gas phase and observed very quickly by classical techniques such as flash photolysis; (b) they may be produced and trapped at low temperatures in an inert solid matrix. Method (b) has the advantage that the lifetime of the trapped species is considerably increased, and a substantial number of matrix isolation experiments have been carried out in the last few years.

The essential requirement for matrix isolation studies is that the species should be trapped in a large excess of solid matrix at a low temperature. The reactive species is often produced *in situ* by photolysis of a trapped parent molecule, or it may be quenched rapidly from the gas phase with an excess of inert gas. The noble gases and nitrogen have been widely used as solid matrices, and commercial cryostats enable the matrix to be handled in the temperature range 4–20 K.

At high dilutions ($<$ 1:1000) each reactive species is trapped within a matrix cage. This severely restricts the possiblity of bimolecular collisions, and also it

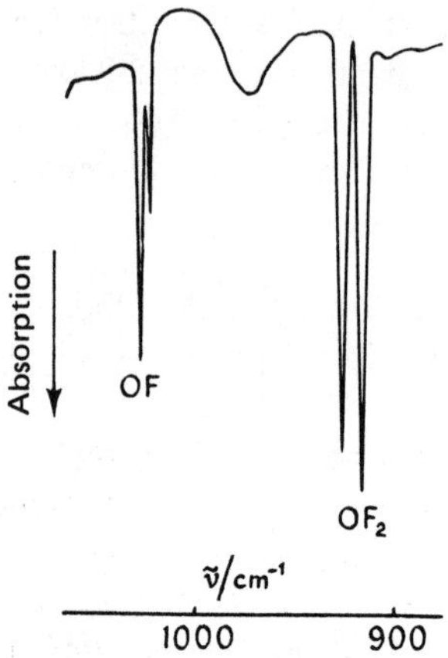

Fig. 4.44 Detection of the OF radical by matrix isolation. (From ref. 4.25.)

may completely prevent the occurrence of molecular rotations. Thus the observed spectra are often very sharp and small isotope effects can be discerned readily. For this reason, matrix studies can be used to obtain high quality infrared spectra of stable molecules. Figure 4.44 shows the detection of the OF radical by matrix isolation. The radical was generated 'in situ' by ultraviolet photolysis of isolated OF_2 molecules. The pair of bands at 1027 and 1031 cm^{-1} (the splitting is due to matrix effects) are assigned to the OF radical and the assignment has been confirmed using $^{18}OF_2$.

Carbon monoxide has been found to be a particularly reactive molecular species at low temperature, and the matrix isolation technique is ideally suited to the characterization of unstable carbonyl intermediates such as HCO, FCO, ClCO, and various metal carbonyls such as Ni(CO), $Ni(CO)_2$, etc.

All four carbonyls of nickel, $Ni(CO)_{1-4}$, have been identified via infrared spectra in argon matrices at 4.2 K. The carbonyls are prepared by the vaporization of the metal atoms followed by co-condensation into a CO/argon (1:500) mixture. The number of predicted *infrared-active* bands corresponding to C—O *stretching* motion for the various possible nickel carbonyl geometries can be summarized as follows: Ni(CO) ($C_{\infty v}$ linear) $1\Sigma^+$; $Ni(CO)_2$ ($D_{\infty h}$ linear) $1\Sigma_u^+$; $Ni(CO)_2$ (C_{2v} angular) $1a_1 + 1b_1$; $Ni(CO)_3$ (D_{3h} trigonal planar) $1e'$; $Ni(CO)_3$ (C_{3v} trigonal pyramid) $1a_1 + 1e$; $Ni(CO)_4$ (T_d tetrahedral) $1t_2$. Thus one might expect between four and six infrared bands in the 2000 cm^{-1} region attributable to $Ni(CO)_{1-4}$. The observed spectrum at 4.2 K is changed by careful warming of the matrix; it results in the growth of some of the $\tilde{\nu}_{CO}$ bands at the expense of others. Experiments with $C^{18}O$ were carried out also, and these provided further evidence which aided the identification of the species present. The spectra show four bands only (2052, 2017, 1967, and 1996 cm^{-1}) which can be identified with simple binary carbonyls and the author [4.26a] has assigned the bands to $Ni(CO)_4$, $Ni(CO)_3$, $Ni(CO)_2$, and Ni(CO) respectively. The author further concludes that either $Ni(CO)_3$ is planar and $Ni(CO)_2$ is linear or the intensities of the a_1 modes for the non-planar and angular geometries are too weak to be observed. For more detailed accounts of the matrix techniques, readers are referred to the reviews given in ref. 4.26b.

(B) RAMAN SPECTROSCOPY

4.11 INTRODUCTION

In contrast to infrared spectroscopy, where we have been concerned with the *absorption* of infrared light, Raman spectroscopy depends on the frequency of the light *scattered* by molecules as they undergo rotations and vibrations. When monochromatic light of frequency ν_0 is directed at a cell containing a dust-free transparent substance, most of the light passes through unaffected. Some of the light, however (~ 0.1 per cent) is *scattered* by the sample molecules in all directions, as shown in Fig. 4.45. The scattered radiation contains photons which have the same frequency ν_0 as the incident light (elastic scattering), but in addition the emergent radiation contains other frequencies (due to inelastic scattering) such as $(\nu_0 - \nu_1)$ and $(\nu_0 + \nu_1)$. This was observed by Raman in 1928 but it had been predicted by Smekal in 1923. The lines of lower frequency than the incident light $(\nu_0 - \nu_1)$ are known as *Stokes lines*, while the high-frequency lines $(\nu_0 + \nu_1)$ are termed *anti-Stokes lines*.

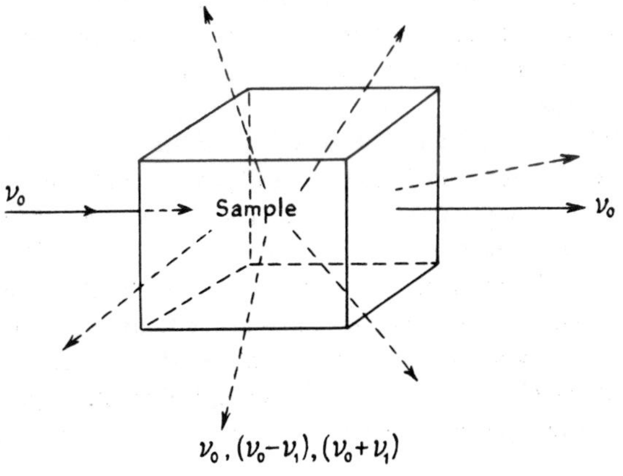

Fig. 4.45 Appearance of scattered light with frequencies ν_0, $(\nu_0 + \nu_1)$, and $(\nu_0 - \nu_1)$.

Normally, an intense monochromatic source of light in the visible region is employed as the incident or exciting radiation. The 4358 Å line from a mercury Toronto arc was commonly used until the 1960s, but this has now been superseded by the highly monochromatic and intense lines available from various gas lasers, e.g. 6328 Å red (helium–neon laser), 4880 Å blue and 5145 Å green (argon laser), 5681 Å yellow and 6471 Å red (krypton laser).

The Raman scattered light is due to rotations and vibrations of the compound under investigation. Since the wavenumber position of the exciting line is approximately 20 000 cm^{-1} (i.e. 5000 Å), the Raman scattered light will have frequencies which are displaced from 20 000 cm^{-1} by amounts which lie somewhere in the range ± 10–4000 cm^{-1}. The exact wavenumber displacements, $\Delta\tilde{\nu}$, will depend upon the rotational and vibrational energies of the particular compound causing the Raman scattering. The same displacement will occur on either side of the exciting line.

The total intensity of all the scattered light is only of the order of 10^{-3}–10^{-5} times the intensity of the incident light and a large part of this light has the same frequency ν_0 as the incident light and is known as Rayleigh scattering. Raman scattered light accounts for $< 10^{-7}$ of the intensity of the incident light, so the Raman effect is very weak indeed. It is therefore essential to use monochromators with very low stray-light characteristics otherwise the weak Raman effect will be 'swamped'. It is also necessary to use very sensitive detectors and efficient optical systems. The relative intensities of the Stokes and anti-Stokes lines are also worth mentioning at this stage. For vibrational transitions, the anti-Stokes (blue shifted) lines are usually considerably weaker than the Stokes (red shifted) lines (see p.234). Since the same frequencies are occurring in both sets of lines, there is no point in most cases in recording spectra on the anti-Stokes side of the exciting line and so most published spectra show the Stokes frequencies only. Figure 4.46 shows the Stokes and anti-Stokes lines of carbon tetrachloride examined as a liquid under low resolution. The detailed analysis of this compound will be dealt with later (see p.257) but the following general points require emphasis.

(1) It should be appreciated that the values of the observed Raman frequencies (in this case the vibrational modes of CCl_4) are completely independent of the laser line that is chosen to excite them. The Raman lines of CCl_4 will be displaced from the frequency of any particular exciting line by 218, 314 cm^{-1}, etc.

(2) The choice of a particular laser line is governed by the requirement that it should be sufficiently intense (i.e. > 10 mW at the sample) and that the molecule itself should not show its normal electronic absorption or fluorescence spectrum in the region of the chosen line. Thus, the 4880 Å blue line is quite unsuitable for the routine investigation of deeply coloured red compounds such as $KMnO_4$.

4.12 EXPERIMENTAL RAMAN SPECTROSCOPY

Raman spectra may be obtained from solids, liquids, gases, and solutions. All Raman spectrometers consist basically of four units – a source, sample optics, a monochromator, and a detector/electronics/recorder system. The sample optics (see Fig. 4.47) must be arranged so that as much scattered light as possible reaches the spectrometer, while the monochromator must have excellent stray-light

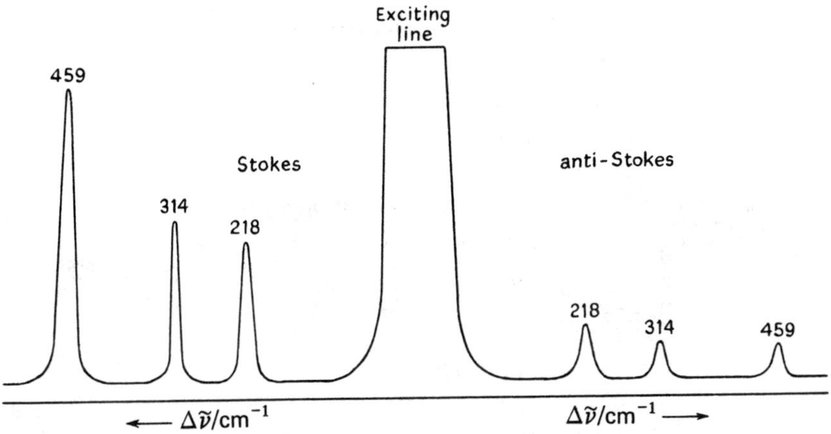

Fig. 4.46 Raman spectrum of CCl_4 liquid showing Stokes and anti-Stokes lines.

characteristics because of the very weak Raman effect. This leads to very sophisticated instrumentation involving double and even triple monochromators. Some general comments about sources and detectors will now be given.

4.12.1 Sources

Low-pressure mercury discharge lamps, which emit an intense blue line at 4358 Å, are frequently employed as radiation sources on older instruments. The discharge, however, produces several other mercury lines (e.g. 4047 Å) and these must be prevented from reaching the sample by surrounding the source with a filter solution (e.g. concentrated sodium nitrate solution). Otherwise the subsidiary lines will also act as exciting lines and a complicated spectrum of overlapping Raman lines would result. The mercury arc lamps have several other serious disadvantages as sources for Raman spectroscopy: (a) the blue line at 4358 Å effectively limits investigations to colourless compounds; even pale yellow liquids absorb this line; (b) comparatively large volumes (~ 5 ml) of either a pure liquid or a very concentrated solution are required in order to obtain spectra of adequate intensity; (c) solids are difficult to illuminate using an arc source, and poor quality spectra are usually obtained.

The advent of stable lasers as exciting sources has completely revolutionised the usefulness of Raman spectroscopy. The majority of lasers in use are filled with noble gases, i.e. He–Ne mixture, Ar, or Kr. They produce highly monochromatic, coherent radiation and the diameter of the beam is usually about 2 mm. The beam is easily focused and the Brewster windows on the laser confer a specific polarization on the radiation. Power outputs ranging from 30 mW to one watt are readily obtainable from various commercial lasers, and all of these properties make lasers almost ideal sources for Raman spectroscopy. An obvious improvement would be a laser which was completely tunable over the whole of

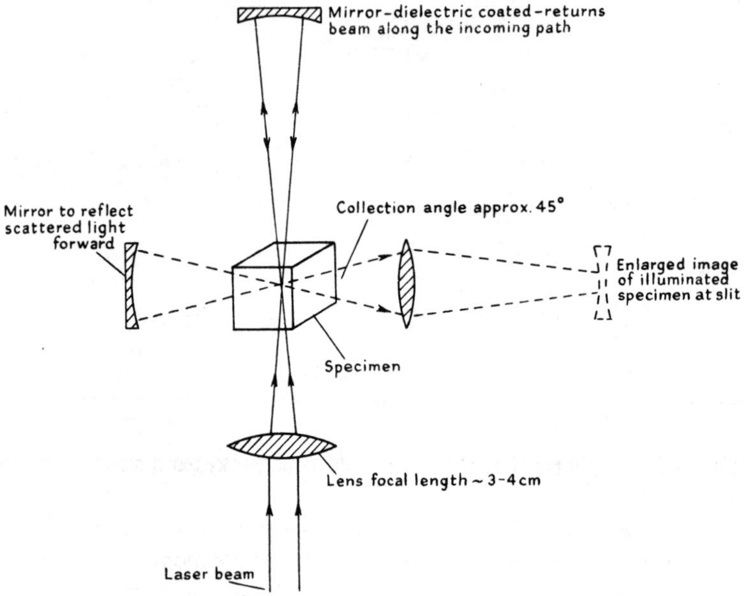

Fig. 4.47 A typical sample optics arrangement. (From ref. 4.27.)

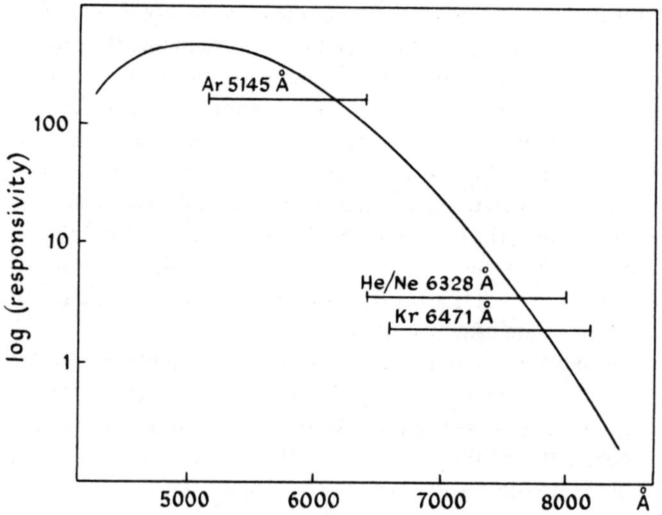

Fig. 4.48 The responsivity of the S20 photocathode surface. (From ref. 4.27.)

the visible region and extending into the infrared and ultraviolet regions; various dye lasers are a significant step in this direction.

4.12.2 Detectors

The early work in Raman spectroscopy used a camera and photographic plates to detect and record the spectrum. Exposure times varied from minutes to hours.

The normal detector for laser work is a multistage photomultiplier. It must be sensitive to a wide range of frequencies and it must have a very low noise character. A typical example is an I.T.T.F.W. 130 photomultiplier which has 16 stages and a cathode surface coated with Na, K, Cs, and Sb (the so-called S20 surface). The responsivity of the S20 photocathode surface is shown in Fig. 4.48 and the drop in sensitivity at the red end of the visible spectrum is apparent. This means that, all else being equal, a C—H stretching mode at $\sim 3000\ cm^{-1}$ will be much easier to detect using the $20\,490\ cm^{-1}$ (4880 Å) green line of an argon laser (it will appear at $20\,490 \pm 3000\ cm^{-1}$) than the red $15\,450\ cm^{-1}$ (6471 Å) line of a krypton laser (it will appear at $15\,450 \pm 3000\ cm^{-1}$).

4.12.3 Calibration

The usual method of checking the calibration of a spectrometer is to record the emission lines from the laser source being used. The standard wavenumber positions for the emission lines of neon, argon, and krypton can be found in ref. 4.2. For routine calibration checks, carbon tetrachloride and indene are frequently used. Table 4.9 gives the frequency shifts $\Delta\tilde{\nu}$ for liquid indene.

Table 4.9 Raman frequencies of liquid indene

$\Delta\tilde{\nu}/cm^{-1}$	*Intensity*	$\Delta\tilde{\nu}/cm^{-1}$	*Intensity*
205.0	w	1286.7	vw
533.7	ms	1361.6	wm
593.0	w	1393.6	wm
730.4	s	1457.6	wm
831.0	mw	1552.7	s
861.0	w	1589.8	w
947.8	w	1610.2	ms
1018.3	s	2892.2	ms
1067.8	mw	2901.2	m
1108.9	mw	3054.7	m
1154.3	vw	3068.2	w
1205.6	s	3112.7	w
1225.6	mw		

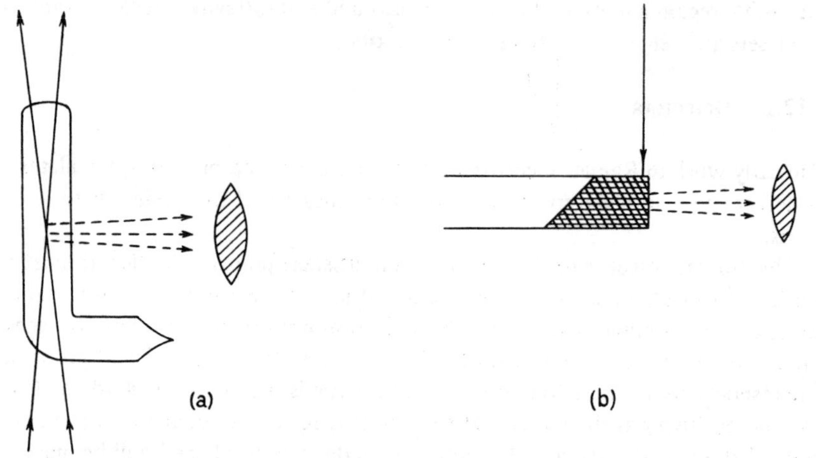

Fig. 4.49 Raman cells for gases, liquids, and solids. (a) Glass-sealed gas or liquid cell. (b) Powder in glass tube.

4.12.4 Examination of samples

The great ease with which samples in all three phases can be examined by laser Raman spectroscopy is one of the main reasons for the resurgence of interest in this technique. Figure 4.49 shows some typical presentations for gases, liquids, and solutions, and it is immediately obvious that samples can be measured (a) directly in reaction vessels, (b) in capillary tubes when only small amounts of liquid are available, (c) under a vacuum or controlled atmosphere, or (d) at various temperatures and pressures. Solids can be examined by holding the sample directly in the beam if the solid is rigid enough or a powder can be packed into a capillary tube.

Single crystals (approximately 1–2 mm long) can be mounted easily in the laser beam. If the crystal is glued on to a glass fibre and then held in a goniometer, as for X-ray work, the unique polarization properties of the laser light can be exploited. Raman spectra can be measured for the crystal in different orientations, and, as we shall see in Section C, this enables the chemist unambiguously to assign the vibrational frequencies of molecules or ions to particular symmetry classes.

4.13 CLASSICAL THEORY OF THE RAMAN EFFECT AND THE SELECTION RULE FOR RAMAN SCATTERING

When a molecule is introduced into an electric field of strength E an electric dipole moment P is induced in the molecule. If α is the polarizability of the molecule, the magnitude of the induced dipole moment is given by:

$$|P| = \alpha |F| \quad (4.97)$$

When electromagnetic radiation of frequency ν_0 falls on the molecule this introduces a varying electric field, E, whose dependence on the time t is given by:

$$E = E^0 \cos 2\pi\nu_0 t \tag{4.98}$$

where E^0 is the amplitude of the electric field. Hence, from Equations (4.97) and (4.98):

$$P = \alpha E^0 \cos 2\pi\nu_0 t \tag{4.99}$$

Thus, the electromagnetic radiation induces a varying electric dipole moment which then permits emission of light identical in frequency with that of the incident radiation. This is *Rayleigh scattering.*

In the equation $|P| = \alpha|E|$, both P and E are vectors, and in the case of an isotropic molecule both their directions are identical. This makes α *scalar*. Howevery, for non-isotropic molecules the application of an electric field in a fixed direction induces a moment in a different direction, and α becomes a *tensor*. In general, molecules are non-isotropic, and the three equations which take account of the unequal polarizability along the different principal axes of the molecule are:

$$P_x = \alpha_{xx}E_x + \alpha_{xy}E_y + \alpha_{xz}E_z \tag{4.100}$$

$$P_y = \alpha_{yx}E_x + \alpha_{yy}E_y + \alpha_{yz}E_z \tag{4.101}$$

$$P_z = \alpha_{zx}E_x + \alpha_{zy}E_y + \alpha_{zz}E_z \tag{4.102}$$

where for example P_z is the induced electric dipole moment in the direction of the z-axis. The meaning of a coefficient such as α_{xy} is that it is the electric moment induced in the x-direction by a unit field E_y polarized along the y-axis. The tensor α is defined by these nine coefficients $\alpha_{xx}, \alpha_{xy}, \ldots, \alpha_{zz}$. However, since $\alpha_{xy} = \alpha_{yx}$, $\alpha_{yz} = \alpha_{zy}$, and $\alpha_{zx} = \alpha_{xz}$, the tensor α is really defined by six coefficients.

These six coefficients together with the coordinates x, y, and z may be expressed in the equation:

$$\alpha_{xx}x^2 + \alpha_{yy}y^2 + \alpha_{zz}z^2 + 2\alpha_{xy}xy + 2\alpha_{yz}yz + 2\alpha_{zx}zx = 1 \tag{4.103}$$

This is the equation of an ellipsoid. Thus, the polarizability of the molecule is divided into three components at right angles along the x, y, and z axes, and the values of these components fix the dimensions of what is termed the polarizability ellipsoid. *If any of the six polarizability tensor components change during a rotation or a vibration, then the theoretical criterion for a Raman spectrum is satisfied.* For very small vibrational amplitudes the polarizability of the molecule is related to the normal vibrational coordinate, q_v by the equation:

$$\alpha = \alpha^0 + \left(\frac{\partial\alpha}{\partial q_v}\right)_0 q_v \tag{4.104}$$

where the attached zero refers to the coordinate values at the equilibrium configuration. An equation such as (4.104) holds for each of the six coefficients which define α.

The dependence of the normal vibrational frequency ν_v on the normal coordinate q_v is given by:

$$q_v = q_0 \cos(2\pi\nu_v t) \tag{4.105}$$

where q_0 is the normal coordinate of the initial position.

By substitution of equations of the type (4.98) into Equation (4.100):

$$P_x = (\alpha_{xx}E_x^0 + \alpha_{xy}E_y^0 + \alpha_{xz}E_z^0)\cos 2\pi\nu_0 t \tag{4.106}$$

On substitution of α from Equation (4.104) and q_v from Equation (4.105) the following expression is obtained:

$$P_x = (\alpha_{xx}^0 E_x^0 + \alpha_{xy}^0 E_y^0 + \alpha_{xz}^0 E_z^0)\cos 2\pi\nu_0 t$$

$$+ \left[\left(\frac{\partial\alpha_{xx}}{\partial q_v}\right) E_x^0 + \left(\frac{\partial\alpha_{xy}}{\partial q_v}\right) E_y^0 + \left(\frac{\partial\alpha_{xz}}{\partial q_v}\right) E_z^0\right] q_0 \cos 2\pi\nu_v t \cos 2\pi\nu_0 t \tag{4.107}$$

Equation (4.107) can be readily transformed into:

$$P_x = (\alpha_{xx}^0 E_x^0 + \alpha_{xy}^0 E_y^0 + \alpha_{xz}^0 E_z^0)\cos 2\pi\nu_0 t + \frac{q_0}{2}\left[\left(\frac{\partial\alpha_{xx}}{\partial q_v}\right)_0 E_x^0\right.$$

$$\left. + \left(\frac{\partial\alpha_{xy}}{\partial q_v}\right)_0 E_y^0 + \left(\frac{\partial\alpha_{xz}}{\partial q_v}\right)_0 E_z^0\right] [\cos 2\pi(\nu_0 - \nu_v)t + \cos 2\pi(\nu_0 + \nu_v)t] \tag{4.108}$$

The first term on the right-hand side of Equation (4.108) contains only one frequency factor ν_0 which is that of the incident radiation. This term is interpreted in terms of Rayleigh scattering. The second term on the right-hand side contains, in addition to the incident frequency ν_0, the frequencies $(\nu_0 \pm \nu_v)$. Thus, the induced dipole moment can also oscillate with the two frequencies:

$$(\nu_0 + \nu_v) \quad \text{and} \quad (\nu_0 - \nu_v)$$

These two frequencies are interpreted as the vibrational Raman frequencies. The $(\nu_0 - \nu_v)$ and $(\nu_0 + \nu_v)$ frequencies are known, respectively, as the *Stokes* and *anti-Stokes lines*. The intensity of the Raman lines for a light source of fixed intensity is determined by the value of $(\partial\alpha/\partial q_v)_0$. In fact, the properties of Raman radiation (e.g. state of polarization, see later) are determined by the tensor $(\partial\alpha/\partial q_v)_0$. Equation (4.104) is important since the α_0 in the first term on the right-hand side determines the properties of Rayleigh radiation, while in the second term $(\partial\alpha/\partial q_v)_0$ determines the properties of Raman radiation.

For a pure rotational change of a diatomic molecule it can be shown that the three frequencies are ν_0, $(\nu_0 - 2\nu_r)$, and $(\nu_0 + 2\nu_r)$, where $2\nu_r$ is the frequency corresponding to the increase of rotational energy. Thus, very simple classical considerations can explain the appearance of both vibrational and rotational changes and of Stokes and anti-Stokes lines. However, once the intensity of Stokes and anti-Stokes lines is considered, this classical theory is most unsatisfactory, since it predicts that the Stokes and anti-Stokes lines should be of equal intensity, whereas, in practice for vibrational changes, the latter are very much less intense than the former. A quantum mechanical approach, however, predicts that the anti-Stokes lines will be much weaker than the Stokes lines for vibrational transitions (see p.234).

As has been indicated already, the criterion for a vibrational Raman spectrum is that one or more of the dimensions of the polarizability ellipsoid must change during the vibration, in order that the vibration may interact with suitable electromagnetic radiation to produce a Raman effect. In the case of a diatomic molecule such as X_2 the polarizability components give an ellipsoid, and during the vibration ←X–X→ the polarizability ellipsoid must change its dimensions. Thus, the condition for a Raman spectrum† is satisfied; it is also satisfied for ←X–Y→. In order to decide which vibrational frequencies will appear as Raman frequencies each bond should be considered as associated with a polarizability ellipsoid, and the criterion for such a change is that the total polarization should change during the vibration. In the case of the parallel vibrations of carbon dioxide (see p.170), for the $\tilde{\nu}_1$ vibrational mode there is reinforcement and a strong Raman spectrum, but for the $\tilde{\nu}_3$ vibrational mode the change in one polarizability ellipsoid tends to be cancelled out by the opposite change in the other, and there is no Raman spectrum.

For a molecule which possesses a centre of symmetry such as CO_2 there is a useful rule known as the *Mutual Exclusion Rule*. This states that, for molecules with a centre of symmetry, fundamental transitions which are active in the infrared are forbidden in the Raman and vice versa.†† For example, in the acetylene molecule (see p.254) the frequencies $\tilde{\nu}_3$ and $\tilde{\nu}_5$ occur in the infrared but not in the Raman, while $\tilde{\nu}_1$, $\tilde{\nu}_2$, and $\tilde{\nu}_4$ appear in the Raman but not in the infrared.

On many occasions both the infrared and Raman methods have to be employed to obtain a complete pattern of the different vibrational frequencies in a molecule; the two methods are often complementary, and when a certain group lacks strong features in the infrared spectrum, it often exhibits intense lines in the Raman spectrum and vice versa.

4.14 QUANTUM THEORY OF THE RAMAN EFFECT

The quantum theory treats monochromatic radiation of frequency ν_0 as a stream of photons having energy $h\nu_0$ where h is Planck's constant.

In Rayleigh scattering, the incident photons collide with a molecule and are

† This should be contrasted with the infrared method where the stretching of a homonuclear diatomic molecule is infrared-inactive under ordinary conditions.

†† It is not to be inferred that the transitions which are forbidden in the infrared appear in the Raman and vice versa. Some transitions may be forbidden to both methods.

scattered without a change of frequency (elastic scattering). In the Raman effect, however, the collision of a photon induces the molecule to undergo a transition and the scattered radiation has a different frequency, and hence energy, from the incident photon (inelastic scattering). Let us take the example of a molecule undergoing a vibrational transition from the ground state ($v'' = 0$) to the first excited state ($v' = 1$). The corresponding frequency will be ν_v and the scattered photon will be *diminished* in energy by an amount $h\nu_v$, i.e. the energy of the scattered photon will be $h(\nu_0 - \nu_v)$. In contrast, if the molecule is already in an excited vibrational state when the photon collides with it, the transition $v' \rightarrow v''$ may be induced, and the photon will be scattered with an *enhanced* energy $h(\nu_0 + \nu_v)$. The photons with diminished energy give rise to Stokes lines while those with an enhanced energy produce the anti-Stokes Raman lines.

At ordinary temperatures the population of molecules in the ground vibrational state is always much greater than in excited vibrational states, and so the intensities of the anti-Stokes lines will always be very much weaker than those of the Stokes lines. This is because the number of molecules in the initial state $v =$ 1 of the anti-Stokes lines is only $\exp(-hc\omega_i/kT)$ times the number of molecules in the initial state $v = 0$ of the Stokes lines (ground state). ω_i is the observed vibrational frequency (in cm^{-1}) and the effect of this Boltzmann distribution can be seen in Fig. 4.46. The above arguments all apply equally well to rotational energy levels except that the higher population of excited rotational states means that anti-Stokes lines will be as intense as the Stokes lines.

4.15 PURE ROTATIONAL RAMAN SPECTRA OF DIATOMIC AND POLYATOMIC MOLECULES

Pure rotational Raman spectra give information about internuclear distances and angles, molecular symmetry, and nuclear statistics. However, in general, it is necessary to use a spectrometer of very high resolving power. In addition, since the lines lie very close to the exciting line, experimental observation becomes difficult owing to the breadth of the corresponding Rayleigh radiation. Low gas pressures to minimize line broadening effects and high-intensity sources are also important requirements.

The criterion for the occurrence of a pure rotational Raman spectrum is that during the rotation the polarizability in a fixed direction should change; in other words, the molecule should have a non-spherical polarization ellipsoid. This is borne out by (1) the absence of such spectra for spherically-symmetrical molecules, e.g. CH_4 and CCl_4, which have spherical polarizability; (2) the fact that linear molecules always have a pure rotational Raman spectrum; this is because the 'polarizability ellipsoid' of a linear molecule is not a sphere, and hence the polarizability in a fixed direction changes during the rotation of the molecule about an axis perpendicular to the internuclear axis.

The attractive feature about pure rotational Raman spectra of diatomic molecules is that the method can be used to obtain internuclear distances for certain symmetrical diatomic molecules which are not accessible by the microwave technique and which were until recently inaccessible in the infrared. By the use of an extreme high-pressure technique, which is outlined on p. 204, it has been possible to study H_2 in the rotation–vibration infrared. The hydrogen molecule, however, had been studied in the pure rotational Raman spectra as early as 1930. In general, the selection rule for pure rotational Raman changes is $\Delta J = 0, \pm 1, \pm 2$. For linear molecules the selection rule is modified to $\Delta J = 0, \pm 2$; the values of the rotational Raman shifts of the linear rigid rotator for the Stokes lines, if centrifugal stretching is neglected, are given by:

$$\Delta\tilde{\nu} = BJ'(J'+1) - BJ''(J''+1) \tag{4.109}$$

Substituting $J' = (J'' + 2)$, then:

$$\Delta\tilde{\nu} = B[(J''+2)(J''+3) - J''(J''+1)]$$

and

$$\Delta\tilde{\nu} = 2B(2J+3) \tag{4.110}$$

where $J'' = J$. This is on the low-frequency side of the exciting line; for the anti-Stokes lines which are on the high-frequency side:

$$\Delta\tilde{\nu} = -2B(2J+3) \tag{4.111}$$

where $\Delta\tilde{\nu}$ is the wavenumber separation of the observed rotational line from the exciting line (e.g. 4880 Å argon laser line), and the values of J are 0, 1, 2, 3, ... Since the convention employed is $\Delta J = (J' - J'')$, that is ΔJ is the difference between the upper and lower values of J, it follows for pure rotational Raman transitions that only ΔJ equal to a positive integer will apply. In the case of a linear molecule, therefore, only an S-branch ($\Delta J = +2$) would be observed. The frequency of the $J = 0$ transitions would correspond to that of the exciting line.

However, since both Stokes and anti-Stokes lines are feasible, there will be two series of rotational Raman lines for this S-branch: one will be on the high-frequency side and the other on the low-frequency side of the exciting line. An energy level diagram is given in Fig. 4.50 to illustrate some of the rotational energy changes in the pure rotational Raman spectrum of a linear molecule.

The representation of the pure rotational transitions by the paths such as ABC and DEF is only a diagrammatic method of showing that the incident light interacts with the molecule and either loses or gains the amount of rotational energy indicated. The points such as B and E should not be taken as representing energy levels of the molecule.

In Table 4.10 the $\Delta\tilde{\nu}$ values are listed for the S-branch for both the Stokes and anti-Stokes lines.

Table 4.10 Raman values $\Delta\tilde{\nu}$ for $\Delta J = 2$ of both Stokes and anti-Stokes lines

	$J = 0$	$J = 1$	$J = 2$	$J = 3$
$\Delta\tilde{\nu}$	$6B$	$10B$	$14B$	$18B$
	$-6B$	$-10B$	$-14B$	$-18B$

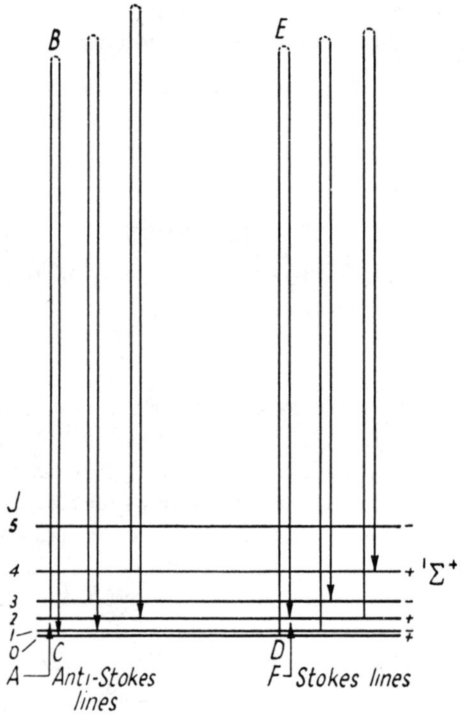

Fig. 4.50 Energy level diagram illustrating some of the rotational energy changes of a linear molecule.

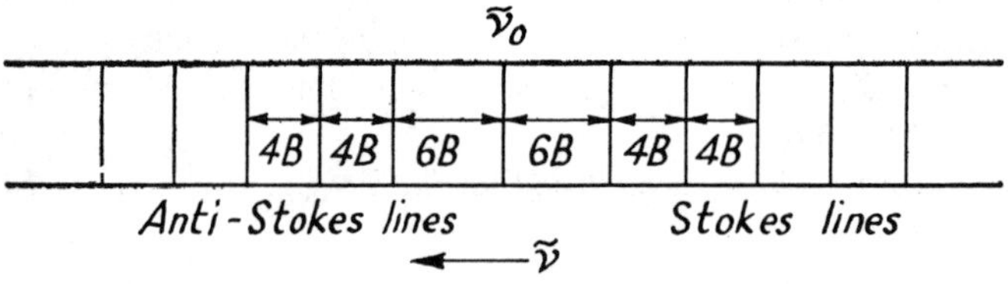

Fig. 4.51 Diagrammatic representation of a rotational Raman spectrum.

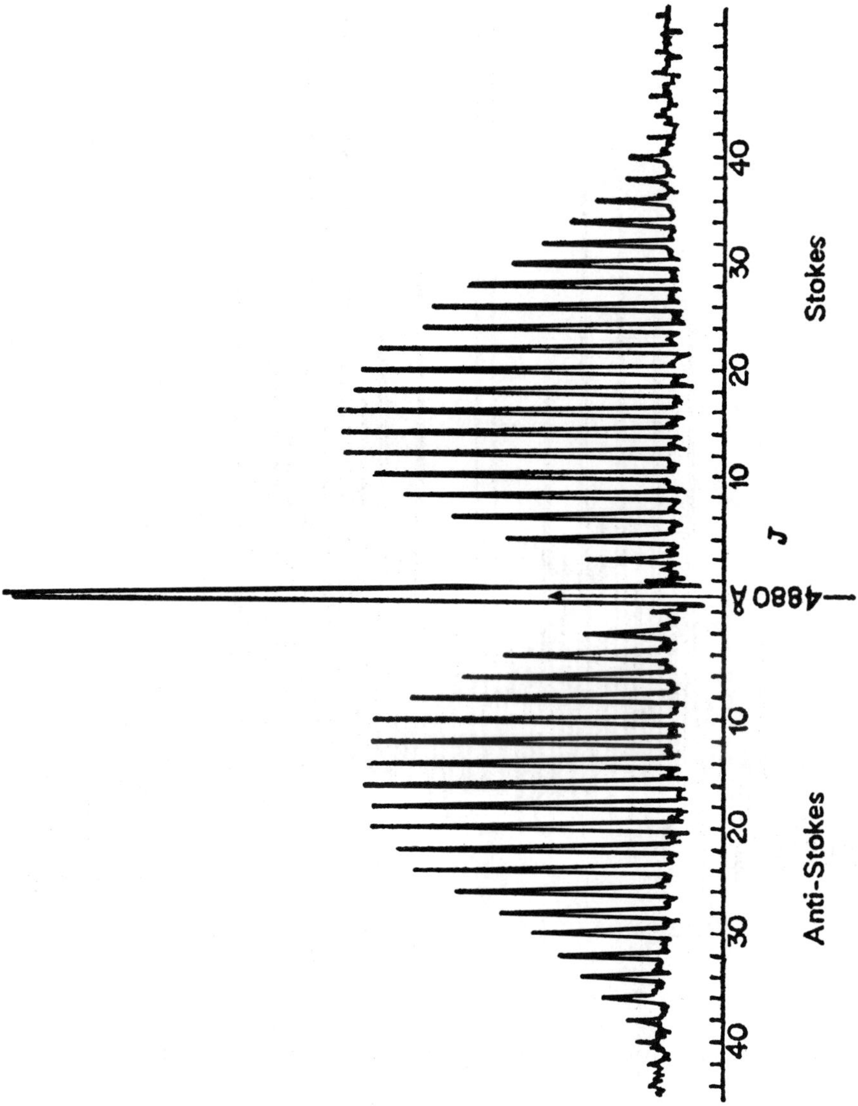

Fig. 4.52 The pure rotational Raman spectrum of CO_2. (From ref. 4.28.)

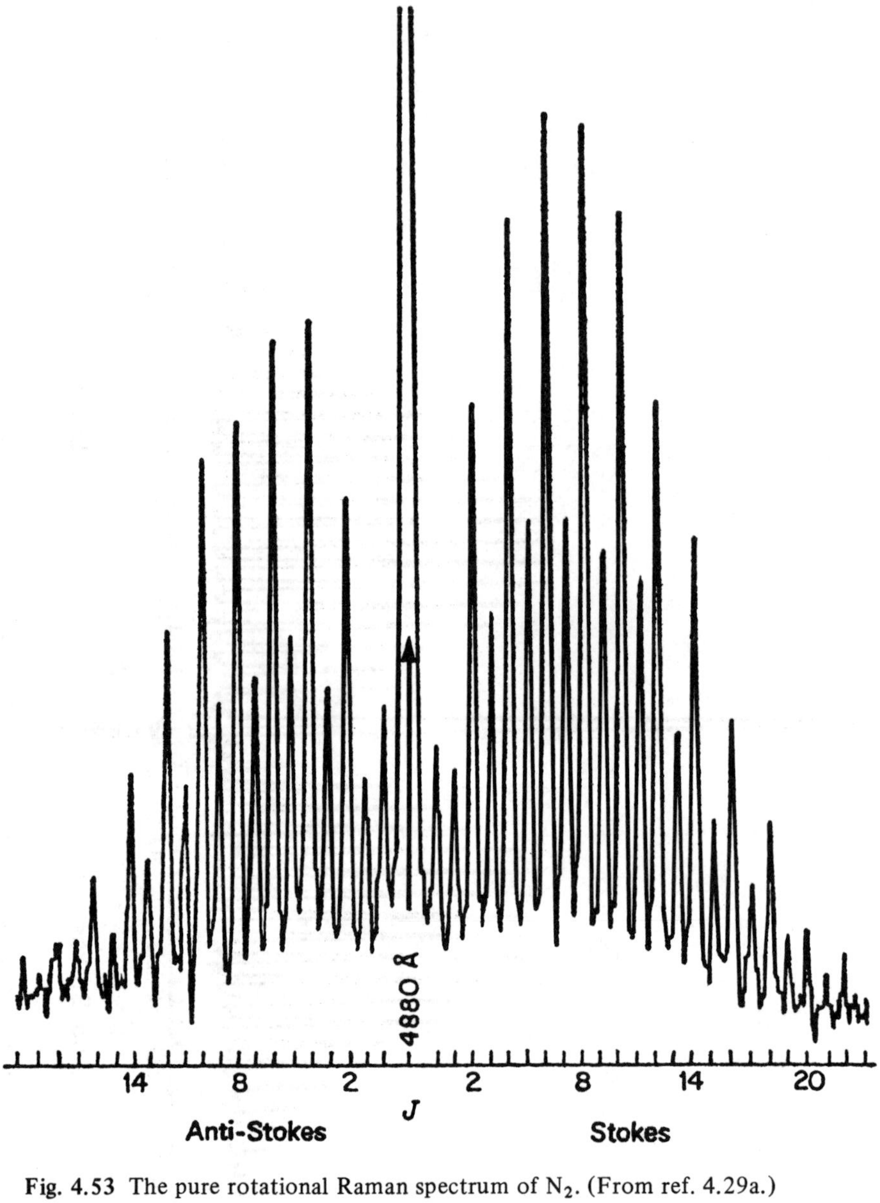

Fig. 4.53 The pure rotational Raman spectrum of N_2. (From ref. 4.29a.)

The difference between consecutive J columns gives the spacing of the rotational lines for both the Stokes and anti-Stokes lines to be $4B$. This value should be compared with that of $2B$ for the infrared. Figure 4.51 is a diagrammatic representation of the rotational Raman spectrum based on Equations (4.110) and (4.111) while Fig. 4.52 is the rotational Raman spectrum of CO_2.

The Raman spectrum of nitrogen ($^{14}N_2$) was studied by Jones and coworkers [4.29b] who obtained and resolved the rotational fine structure using a Fabry–Perot etalon. The gas was examined at a pressure of 1 atmosphere (~100 kPa) inside the cavity of an Ar^+ laser. The Raman spectrum is shown in Fig. 4.53, and the 2:1 intensity alternation arises because ^{14}N has a nuclear spin of 1. Almost as many anti-Stokes as Stokes lines are observed because large numbers of molecules are initially in excited rotational states even at room temperature. From the analysis of the Raman lines the value of B_0 was found to be $1.98950_6 \pm 0.00002_7\ \text{cm}^{-1}$, $D_0 = (5.8 \pm 0.05_5) \times 10^{-6}\ \text{cm}^{-1}$.

Other pure rotational Raman studies of linear molecules have been made on, for example, H_2, O_2, F_2, HCl, CO_2, CS_2, C_2H_2, C_3O_2, and C_2N_2.

In symmetric-top molecules rotations about the symmetry axis yield no change in polarizability during the rotation and hence do not satisfy the criterion for such a spectrum. The rotations analysed in terms of the axes perpendicular to the symmetry axis do lead to a pure rotational Raman spectrum. The rotational energy of a rigid symmetric-top molecule is given by:

$$E_r = J(J+1)\frac{h^2}{8\pi^2 I_B} + \left(\frac{1}{I_A} - \frac{1}{I_B}\right)\frac{h^2K^2}{8\pi^2} \qquad (4.112)$$

where $Kh/2\pi$ is the component of the total angular momentum $\sqrt{[J(J+1)]}\,h/2$ in the direction of the symmetry axis.† The selection rules are $\Delta J = 0, \pm 1, \pm 2$, and $K = 0$. In fact, only the positive values of ΔJ apply (cf. the rotational Raman spectrum of linear molecules). Thus, there should be a set of Stokes and anti-Stokes lines for $\Delta J = +2$ and another for $\Delta J = +1$ [see Fig. 4.54(a) and (b)]. The $\Delta J = 0$ transitions would not be detected since they would correspond to the frequency of the exciting line. Hence, altogether four sets of lines might be expected. Such a spectrum is illustrated diagrammatically in Fig. 4.54, where the following points are to be noted.

(1) The Stokes lines of both the R- and S-branches lie on the low-frequency side of the exciting line, whereas the anti-Stokes lines of the R- and S-branches are on the high-frequency side.

(2) The $\Delta J = +1$ transitions of even J-values are superposed on the lines of the $\Delta J = +2$ transitions.

(3) The wavenumber separation between adjacent lines is approximately $h/4\pi^2 cI_B$, and the first R-branch line is $h/2\pi^2 cI_B$ from the exciting line.

Little work has been done on the pure rotational Raman spectrum of symmetric-top molecules. Ammonia was one of the first molecules to be studied but Stoicheff [4.30] carried out some classic investigations in the 1950s on gases at ordinary pressures. He employed a 21-ft grating spectrograph with an Eagle mounting. The concave grating had an area of 7 in × 3 in ruled with 15 000

† Further terms are added to this equation when centrifugal stretching is taken into account.

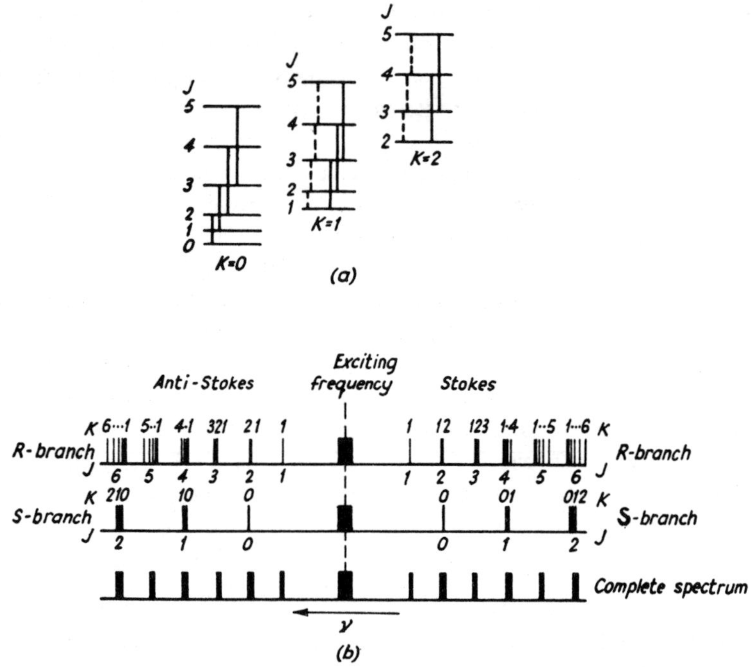

Fig. 4.54 (a) Energy level diagram for a prolate symmetric-top molecule. Full lines indicate the transitions $\Delta J = +2$ (S-branch), broken lines $\Delta J = +1$ (R-branch). (b) Diagrammatic representation of the rotational Raman spectrum of a symmetric-top molecule indicating the fine structure in the R- and S-branches and the appearance of the complete spectrum under low resolution. The $J = 0$ line in the R-branch is absent since $\Delta J = +1$ does not occur for $K = 0$. The different intensities in the Stokes and anti-Stokes branches have been neglected. (Courtesy of Dr. B.P. Stoicheff [4.30].)

lines/in and it was blazed to give maximum intensity in the second order at 5000 Å. The reciprocal linear dispersion was 6.75 $cm^{-1}\,mm^{-1}$. Two high-intensity mercury lamps were used giving a very sharp 4358 Å line, and the time of exposure used ranged from 6 to 24 h. Stoicheff obtained accurate rotational constants for C_6H_6, C_6D_6, and allene and deuterated allenes (C_3D_4, $C_3H_2D_2$).

Although high-resolution Raman spectroscopy in the gas phase has been extended to more complicated molecules such as C_6H_6 and $Zn(CH_3)_2$, the method is still of restricted interest. This is because of the low volatility of many interesting compounds, the experimental difficulties of resolving very closely spaced rotational lines, and because the microwave technique offers greater accuracy; the latter technique, however, is restricted to molecules with a permanent dipole moment.

4.16 RAMAN VIBRATIONAL STUDIES OF DIATOMIC MOLECULES

All diatomic molecules are in theory capable of yielding a vibrational Raman spectrum, since the polarizability must alter during the vibration along the bond. If the diatomic molecule may be treated as a simple harmonic vibrator, then:

$$E_v = (v + \tfrac{1}{2})hc\omega \qquad (4.113)$$

where v is the vibrational quantum number which can take the values 0, 1, 2, ... The selection rule for a transition is $\Delta v = \pm 1$, but in practice only the value $\Delta v = +1$ transitions are observed. When $v'' = 0$ is the initial level, a Stokes line is obtained, whereas when $v' = +1$ is the initial level, an anti-Stokes line results. This is illustrated in Fig. 4.55 where any accompanying rotational energy change is neglected. The representation of the Stokes transitions by the route ABC and the anti-Stokes by DEF is only a diagrammatic way of indicating that the molecule interacts with an incident quantum of light and either diminishes or increases its energy value by the amount ΔE_v, where:

$$\Delta E_v = E_{v'} - E_{v''} = (v' + \tfrac{1}{2})hc\omega - (v'' + \tfrac{1}{2})hc\omega \qquad (4.114)$$

On application of the selection rule $\Delta v = +1$:

$$\Delta E_v = hc\omega = hc\Delta\tilde{\nu} \qquad (4.115)$$

where $\Delta\tilde{\nu}$ is the difference in wavenumber of the observed Raman line from that of the exciting line.

Since the majority of molecules are in the $v = 0$ level at room temperature (in contrast to the situation for pure rotational energy levels), the Stokes lines are obviously the more intense. In fact, the anti-Stokes lines are often so weak that they are difficult to observe; in addition, overtones are normally too weak to be observed. Thus, it is generally not possible to obtain data on the higher vibrational levels, and consequently it is also impossible to determine ω_e or x_e values. This is to be contrasted with the infrared method (see p.172).

4.17 VIBRATIONAL RAMAN SPECTRA OF POLYATOMIC MOLECULES

4.17.1 Introduction

Both the infrared and Raman methods yield highly desirable information in structural work, and for molecules with any degree of symmetry the two methods work hand-in-hand. Since the intensity of an infrared band depends on the magnitude of the electric dipole moment change during the vibration, while the intensity of a Raman band depends on the magnitude of the change in polarizability, it is not surprising that the appearance of a Raman band normally differs from the corresponding infrared band. The wavenumber positions of the two bands are usually in good agreement, however.

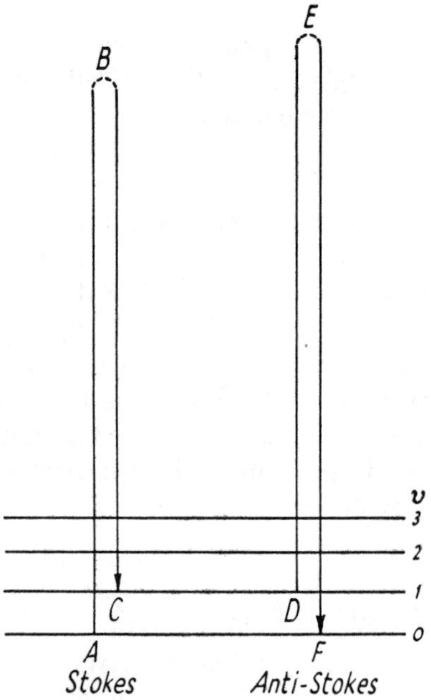

Fig. 4.55 Schematic representation of a vibrational Raman change.

A large number of vibrational frequencies of simple and even relatively complex molecules has been obtained from the vibrational Raman approach. The Raman method is particularly useful for determining low vibrational frequencies which are not readily determined in the infrared. An integral part of the Raman approach to structural problems is the study of whether the lines corresponding to the fundamental frequencies are polarized. From the number of Raman lines observed, and how many of these are polarized, it is possible to decide between certain structural configurations. Any normal mode of vibration which belongs to the totally symmetrical character species of the point group will give rise to a polarized Raman band (see Vol. 2, Chapter 2).

4.17.2 Polarization and depolarization of Raman lines

In the vibrational analysis of simple molecules, the measurement of the degree of polarization of the Raman line gives very valuable information, because it enables that band to be assigned as a totally symmetric mode and it helps to decide the geometrical arrangement of the atoms themselves. The degree of polarization depends on the anisotropic character of the molecules. The meaning

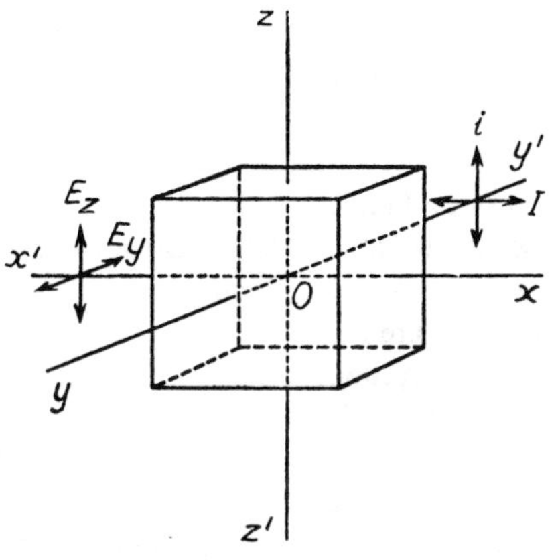

Fig. 4.56 Schematic representation of depolarization of scattered radiation.

of the term depolarization ratio ρ, which measures the degree of depolarization of the line, will now be illustrated. In Fig. 4.56 the three axes x, y, and z are at right angles. If the axis Ox' is the direction of the incident monochromatic radiation, the Raman frequencies are examined in the plane $zy'z'$. If interaction of the unpolarized monochromatic incident light occurs with the molecules in the cell and Raman lines are produced, then these lines may become polarized to different extents. If i is the intensity of the polarized light perpendicular to the plane $x'yxy'$, and I the intensity of the polarized light parallel to this plane, then the ratio of i to I is known as the depolarization ratio (ρ).

It can be shown, from a consideration of the polarizability ellipsoid for the molecule, that in the case where all orientations of the scattering molecules are equally probable (e.g. in a fluid), then for Rayleigh scattering of unpolarized light:

$$\rho = 6\gamma^2/(45\alpha^2 + 7\gamma^2) \tag{4.116}$$

γ measures the anisotropy and may be looked upon as indicating the departure of the ellipsoid from its spherical shape; α is some measure of the overall size of the polarizability ellipsoid. If the ellipsoid reduces to a sphere, then by definition γ is zero, and therefore from Equation (4.116) ρ is also zero, and the Rayleigh line is termed completely polarized. Since α by definition may never be zero, it follows from Equation (4.116) that ρ must always be less than 6/7. In the case of Raman lines Equation (4.116) must be modified, and for a detailed treatment more advanced works should be consulted. Broadly, what emerges is:

(1) Equation (4.116) becomes modified to:

$$\rho = 6(\gamma')^2/[45(\alpha')^2 + 7(\gamma')^2] \tag{4.117}$$

where α' in this case can be zero.

(2) If both γ' and α' are zero, then the Raman line is forbidden.

(3) If γ' is zero and α' is positive, then ρ is zero and the line is completely polarized.

(4) If α' is zero and γ' is positive, then ρ is 6/7, i.e. 0.86, and the line is said to be depolarized.†

(5) If $0 < \rho < 6/7$, then the line is said to be polarized.

These cases are most important in deciding the actual geometrical arrangement of atoms in simple molecules (see later). Obviously, then, a major part of structural Raman work is the experimental determination of the value of ρ.

A few general conclusions are listed below.

(1) Totally symmetrical vibrations result in highly polarized lines, that is, they have a small value of ρ. For example, ←O—C—O→ where in this type of vibration the full symmetry of the molecule is maintained, and the value of ρ is < 0.2. In a similar way the two linear Raman active vibrational modes of acetylene ($\tilde{\nu}_1$ and $\tilde{\nu}_2$, see p.254) are polarized. For symmetrical vibrations of linear molecules the value of ρ often falls between 0.1 and 0.3.

(2) All antisymmetrical vibrational modes are depolarized, e.g.

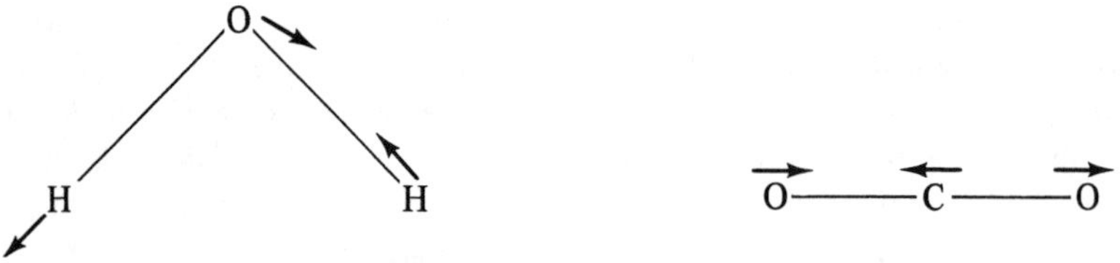

In general, totally symmetrical vibrations can be distinguished by their ρ values from antisymmetric vibrations. The totally symmetrical modes have a value of $\rho < 6/7$, whereas if ρ is equal to 6/7, the vibration is most probably antisymmetrical. A value of about 6/7 itself for ρ does not definitely mean that the vibrations are antisymmetrical, since totally symmetrical vibrations may have a value close to 6/7. It must be emphasized, however, that in simple molecules a value of ρ less than 6/7 is definite proof of a totally symmetrical vibration.

(3) Totally symmetric breathing modes of regular tetrahedral molecules, e.g. P_4 or $SnCl_4$, and octahedral molecules, e.g. SF_6, are completely polarized (i.e. $\rho = 0$).

(4) Raman lines of degenerate vibrational modes are depolarized.

(5) In large unsymmetrical molecules most vibrations are polarized to some degree, although vibrations of symmetrical groups within such molecules can be detected by virtue of their large degree of polarization of the Raman lines.

A typical Raman spectrum of CCl_4 liquid is given in Fig. 4.57. The polarized band at 459 cm^{-1} can be assigned to the totally symmetric stretching mode.

† If the incident light is plane polarized, as in the case of laser radiation, then the maximum value that ρ can have is $\rho = 3/4$.

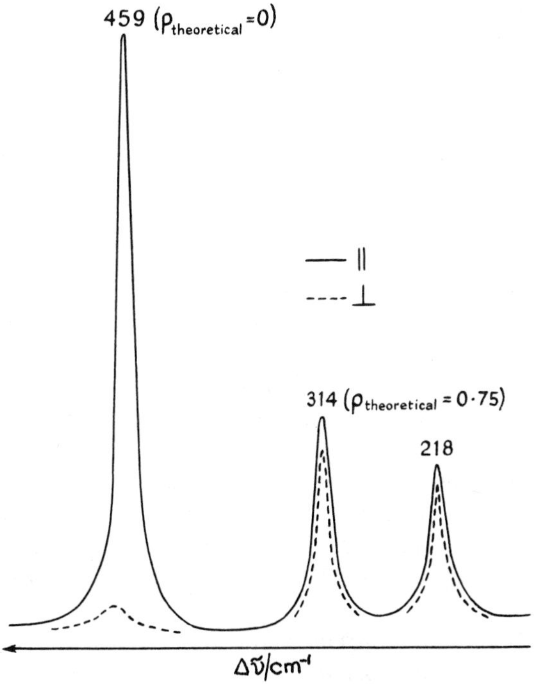

Fig. 4.57 Raman spectrum of CCl_4 liquid showing the polarization of the band at 459 cm^{-1}.

4.17.3 Structural determinations

A few of the different types of molecules whose vibrational frequencies (or some of them) have been determined by Raman spectroscopy are listed below.

(1) Diatomic molecules belonging to point groups $D_{\infty h}$ or $C_{\infty v}$, e.g. H_2 (4161 cm^{-1}), F_2 (892 cm^{-1}), ^{35}ClF (774 cm^{-1}), ^{37}ClF (767 cm^{-1}), and $^{16}O_2$ (1555 cm^{-1}).

(2) Triatomic molecules. For linear symmetric molecules of the type X_3 and XY_2, the selection rules allow the appearance of a single polarized line where the frequency corresponds to the symmetric vibration:

$$\leftarrow X{-}X{-}X \rightarrow \quad \text{or} \quad \leftarrow Y{-}X{-}Y \rightarrow$$

e.g. I_3^- ($\tilde{\nu}_{sym} = 103\ cm^{-1}$), $HgCl_2$ ($\tilde{\nu}_{sym} = 360\ cm^{-1}$). The other vibrational modes are Raman inactive.

In the non-linear symmetric triatomic molecules such as SO_2 all three fundamental frequencies are Raman active and two of the three lines are polarized. Thus, the linear and non-linear symmetrical triatomic molecules ought to be readily distinguished by Raman spectroscopy.

(3) Tetra-atomic molecules. The linear C_2N_2 molecule has been examined,

where three fundamental Raman frequencies are permitted and two of these lines are polarized. The three normal modes and corresponding frequencies for the liquid are:

(a) ←N—C→←C—N→ $\tilde{\nu}_1 = 2322\ cm^{-1}$

(b) N→—C→←C—←N $\tilde{\nu}_2 = 848$

(c) N—C—C—N (↑ on each C, ↓ on each N) $\tilde{\nu}_4 = 506$

where modes (a) and (b) gives rise to polarized lines.

For the planar symmetrical XY_3 molecule (point group D_{3h}) and the pyramidal XY_3 species (point group C_{3v}) it may be possible to distinguish between them when attempting to decide a particular XY_3 structure, since for the planar type three lines are permitted (one polarized) whereas for pyramidal species four lines are observed (two polarized).

(4) Some other important structural determinations are of:

(a) Ions, e.g. BF_4^- tetrahedral, AlH_4^- tetrahedral, and $AuCl_4^-$ square planar. The $AuCl_4^-$ exhibits 3 Raman lines at 347 (polarized), 324 (depolarized), and $171\ cm^{-1}$ (depolarized); this would be expected for a square-planar shape (point group D_{4h}).

(b) Solid PCl_5 at room temperature gives a Raman spectrum which has been interpreted in terms of an ionic structure $PCl_4^+PCl_6^-$. In contrast, the solid, formed by rapid cooling of the vapour, is molecular and contains trigonal bipyramidal molecules. This structure exhibits a Raman spectrum which is significantly different from that of the ionic structure. Solid PBr_5 crystallizes in the form $PBr_4^+Br^-$.

(c) The Raman spectrum of S_8 consists of seven lines, two of which are polarized. This behaviour is consistent with group theoretical predictions for a puckered octagonal ring.

(d) Diborane (point group D_{2h}) has been investigated; the Raman spectrum ruled out an ethane-like model and favoured a bridge structure (I). The bridge

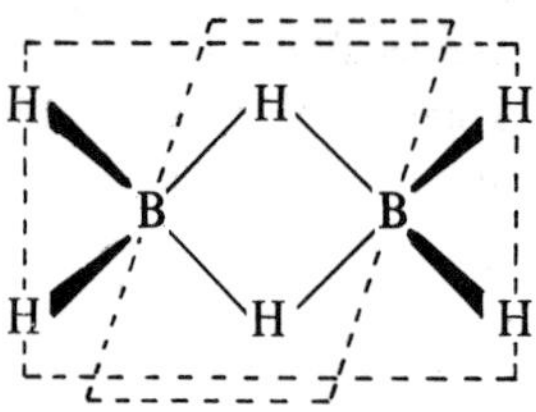

(I)

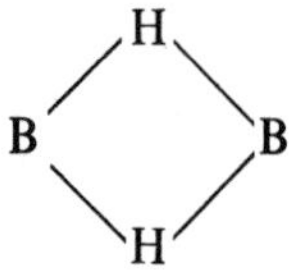

(II)

structure consists of the structure (II) in one plane, while the four outer B—H bonds are also in a plane which is perpendicular to the first one. Again, the employment of the Raman method to distinguish between these two types of structure was based on how many lines would be expected from each structure.

For the bridge type of structure 4 polarized and 5 depolarized lines are predicted from group theory and this has been observed experimentally.

Sufficient examples have now been given to indicate that it is possible to distinguish between certain molecular models of different symmetry from the number of fundamental frequencies observed and the number of these which are polarized. Some other applications are discussed later.

4.18 ROTATION–VIBRATION RAMAN STUDIES

So far only Raman lines have been discussed, and not Raman bands. Further, the difference in frequency between the exciting line and the Raman line has been associated with a vibrational frequency. However, if the change of polarizability during a vibration is not spherical, then it will be influenced by the rotation of the molecule, and the vibrational lines ought to have rotational fine structure associated with them.

For all but the lightest molecules in the gas phase, the spacing of the rotational lines will be very small and great experimental difficulties will be encountered in resolving them. Thus vapour-phase work on most compounds just picks out the rotational band contours. Rotational fine structure is, of course, completely absent from the Raman spectra of non-volatile compounds, and bands due to vibrational changes only are observed. Consider now the case of gases where good resolution can be achieved.

Rotation–vibration Raman spectra of diatomic and polyatomic molecules

In general, the selection rules governing such simultaneous changes in rotational and vibrational energy are $\Delta J = -2, -1, 0, +1, +2$, resulting in O, P, Q, R, and S branches, respectively; the observed changes in v correspond to $\Delta v = \pm 1$. In the case of linear molecules Δv is still ± 1, but $\Delta J = 0, \pm 2$, although sometimes $\Delta J = \pm 1$ is also involved. In Fig. 4.58 a comparison is made of the Raman and infrared rotation–vibration transitions of a linear molecule.†

It may be observed from Fig. 4.58 that, although the mechanism of accomplishing the energy change for the infrared differs from that of the Raman, the effect is similar.

So far the Q-branch has been resolved only for a very limited number of cases, e.g. for H_2 and HCl, which have very small moments of inertia and consequently a larger separation of their rotational energy levels.

Assuming that the vibrational and rotational energies are additive, and that the centrifugal constant D_J may be neglected, and letting $\Delta\tilde{\nu}_0$ be the vibrational

† The higher 'energy levels' represented by dotted lines in Fig. 4.58 do not correspond to any energy states of the system but are merely a diagrammatical way of indicating the energy of the light quantum above the initial state.

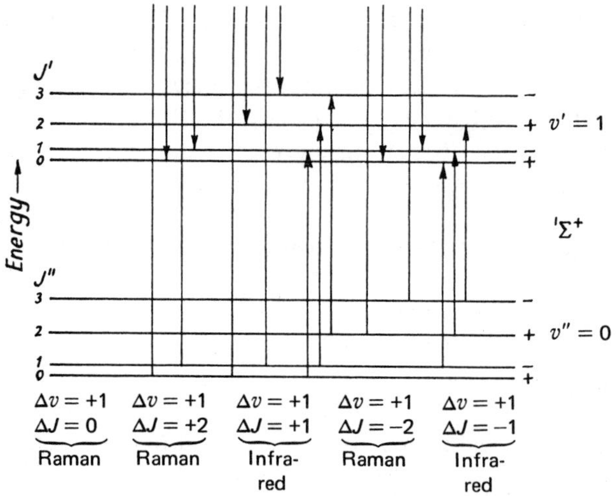

Fig. 4.58 Comparison of some infrared and Raman rotation–vibration transitions.

frequency shift in cm^{-1} units, then $\Delta\tilde{\nu}$ the Raman displacement for the associated rotation–vibration transition of a linear molecule will be given by:

$$\Delta\tilde{\nu} = \Delta\tilde{\nu}_0 + (E_r' - E_r'')/hc = \Delta\tilde{\nu}_0 + B_{v'}J'(J'+1) - B_{v''}J''(J''+1) \qquad (4.118)^\dagger$$

where $\Delta\tilde{\nu} = (\tilde{\nu}_{e.1} - \tilde{\nu})$ or $\Delta\tilde{\nu} = (\tilde{\nu} - \tilde{\nu}_{e.1})$ and $\tilde{\nu}$ is the wavenumber of the Raman line and $\tilde{\nu}_{e.1}$ is the wavenumber of the exciting line. Since nearly all molecules will be in their ground state, i.e. $v'' = 0$, and the usual change in the vibrational quantum number is $\Delta v = +1$, it follows that, in general, $v' = +1$. The equations for the S-, O-, and Q-branches can be obtained from Equation (4.118) by substituting:

$$J' = J'' + 2, \quad J' = J'' - 2, \quad \text{and} \quad J' = J''$$

Thus:

S-branch

$$J' = J'' + 2$$

$$\Delta\tilde{\nu} = \Delta\tilde{\nu}_0 + 6B_{v'} + (5B_{v'} - B_{v''})J'' + (B_{v'} - B_{v''})J''^2 \qquad (4.119)$$

where $J'' = 0, 1, 2, 3$, etc.

O-branch

$$J' = J'' - 2$$

$$\Delta\tilde{\nu} = \Delta\tilde{\nu}_0 + 2B_{v'} - (3B_{v'} + B_{v''})J'' + (B_{v'} - B_{v''})J''^2 \qquad (4.120)$$

where $J'' = 2, 3, 4$, etc.

† For convenience the symbol $\tilde{\nu}_r$ or even $\tilde{\nu}$ is sometimes used instead of $\Delta\tilde{\nu}$.

Q-branch

$$J'' = J'$$

$$\Delta\tilde{\nu} = \Delta\tilde{\nu}_0 + (B_{v'} - B_{v''})J'' + (B_{v'} - B_{v''})J''^2 \qquad (4.121)$$

where $J'' = 0, 1, 2, 3$, etc.†

Molecules such as H_2, CO_2, CS_2, and C_2H_2 have been examined in this way.

The relative intensities of the O-, S-, and Q-branches for linear molecules depend on whether the vibrational angular momenta differ for the two vibrational states concerned. (1) If it is the same, then the Q-branch is much the more intense, and the O- and S-branches do not appear to be even detectable. (2) If it differs, then the intensity of the Q-branch will be low and the O- and S-branches more intense; in addition, R- and P-branches may appear.

Anti-Stokes, Q-, S-, and O-branches are possible, but since the majority of the molecules at ordinary temperatures are in the ground vibrational state, then in the $v = 1$ to $v = 0$ transition, the intensity of the lines would normally be very weak.

In the case of spherical-top molecules, e.g. CH_4, SF_6, CCl_4, for a totally symmetrical vibration, for example:

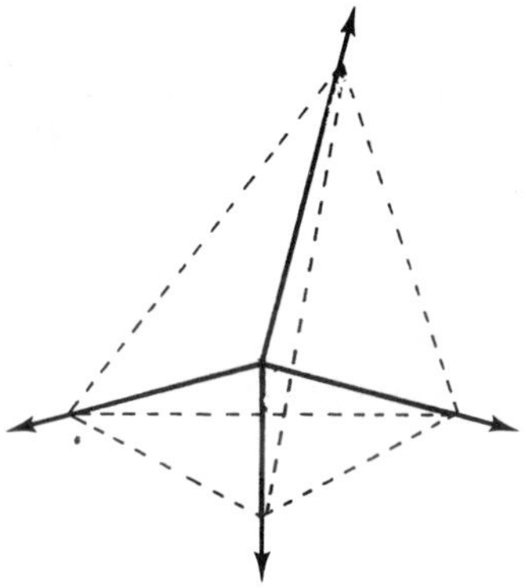

the selection rule is $\Delta J = 0$, and this results in a very sharp Raman line. For antisymmetrical vibrations, since the symmetry is altered, rotational changes accompany the vibrational change.

For symmetric-top molecules, when the top axis is not an axis of symmetry, the second quantum number K is involved, and the selection rules are:

$$\Delta K = 0, \pm 1, \pm 2, \quad \text{and} \quad \Delta J = 0, \pm 1, \pm 2$$

although if the molecule possesses some symmetry, certain values of ΔK do not appear, depending on the type of vibrational transition. The treatment is quite complex and the standard work by Herzberg [4.7] should be consulted. The

† In Equations (4.119), (4.120), and (4.121) it is customary to replace J'' by J.

spectrum of gaseous ammonia is one of the classic examples of a rotation–vibration Raman study.

The difficulties of resolving a rotation–vibration Raman band for an asymmetric-top molecule are formidable and ethylene is the only such molecule to have had the rotational structure of some of its bands resolved and analysed.

As mentioned in the introduction to this section, valueable information can be obtained from rotational band contours. For example, Coriolis constants, ζ, which are concerned with a coupling of the rotational and vibrational energy levels (see p. 193) can be evaluated from the separation of the P and R branch maxima using the relationship:

$$\Delta\tilde{\nu}_{PR} = 4(BkT/hc)^{\frac{1}{2}}(1 - \zeta)$$

where B is the rotational constant. Figure 4.59 shows the rotational contours for SF_6, SeF_6, and TeF_6 molecules.

4.19 APPLICATIONS OF RAMAN SPECTROSCOPY

4.19.1 Qualitative and quantitative analysis

The vibrational frequencies determined from the Raman spectrum of a substance may be used in its identification. The normal spectroscopic procedure of comparing the spectrum of the unknown with that of known compounds is used.

To illustrate the application of the method, benzene and some of its substitution products will be considered. Liquid benzene has two intense Raman bands, one at 3047 and the other at 3062 cm^{-1}, which are due to the fundamental vibrational modes (a) and (b), respectively, together with weak bands at

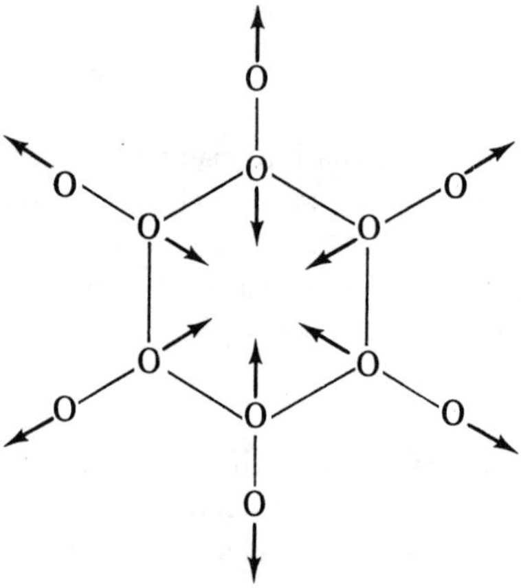

(a)

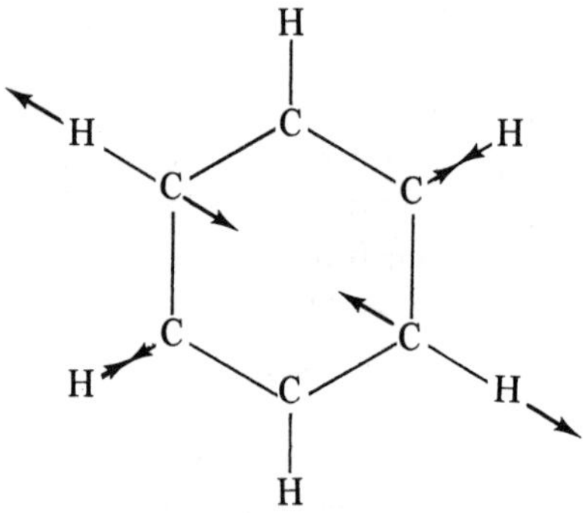

(b)

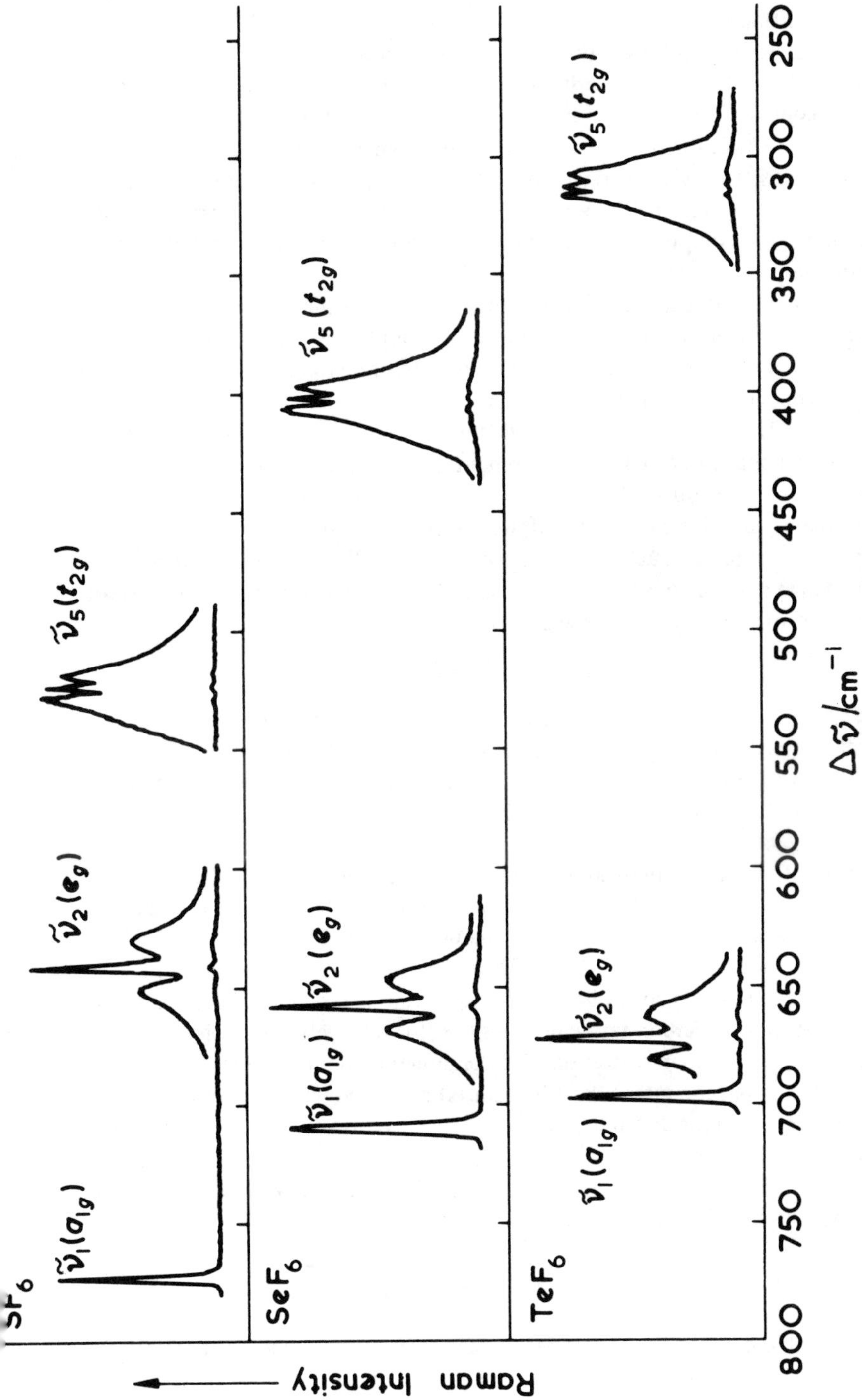

Fig. 4.59 Vapour-phase Raman spectra of sulphur, selenium, and tellurium hexafluorides at 296 K, depicting rotational contours. (From ref. 4.31.)

2925, 2948, 3164, and 3187 cm^{-1}. These bands serve to characterize benzene, but in addition there are a number of other bands which together provide an unmistakable fingerprint for the molecule. The mono-, di-, and tri-substituted benzenes have strong characteristic bands between 3045 and 3070 cm^{-1}. All these frequencies are Raman active. In the case of the di-substituents, it is possible to distinguish the *ortho*-, *meta*-, and *para*-isomers. Only the *meta*-substituent has an intense strongly polarized band at 995 cm^{-1}, while the *para*-substituent has one at about 625 cm^{-1} which is generally absent in *ortho*-derivatives. In addition, the *ortho*-compound usually has a more detailed spectrum than the *para*-compound.

The Raman spectra of a mixture consists simply of the superposed spectra of the components in the mixture. One extensive application of its usefulness has been in the qualitative analysis of mixtures of hydrocarbons. Such analyses have been particularly valuable to oil companies where a standard procedure can be employed. The instruments used for this work scan automatically and produce a record on a chart. Before this development the time required to photograph the Raman effect was longer than the time required by a recording infrared spectrophotometer, and this caused the infrared technique to be much more widely used for the study of qualitative analysis problems. This led to the Raman technique being applied to more special problems, for instance as a guide in preparative work; one example of this was to follow the reaction:

$$SeF_4 + SeO_2 \rightarrow 2SeOF_2$$

The appearance of the $SeOF_2$ spectrum was noted, while the completion of the reaction was detected by the disappearance of the bands due to SeF_4.

The Raman effect is also used in quantitative analysis. Unlike the infrared absorption, the intensity of a Raman line is, in general, directly proportional to the volume concentration of the species. Several experimental difficulties may be encountered in its application to quantitative problems. Two of the most important are the presence of fluorescent materials in the sample to be analysed and the fluctuation in the intensity of the light source. The latter has been largely overcome using laser sources. A definite concentration of a substance with known Raman lines is added before each spectrum is taken; a suitable substance for this purpose is carbon tetrachloride since it is relatively unreactive and it exhibits intense Raman scattering. The component whose concentration is required then has the intensity of its band(s) measured with respect to one of the carbon tetrachloride bands; thus, the concentration of the component can be evaluated.

The quantitative estimation of aromatic and olefinic content of gasolines has been accomplished quite accurately by means of Raman spectra.

4.19.2 Raman studies on inorganic acids

There is a band at 1050 cm^{-1} in the Raman spectrum of aqueous nitric acid which is found also in the spectrum of the alkali nitrates, and this band can be assigned to the symmetric stretching mode of the nitrate ion itself. The 1050 cm^{-1} band is less intense in concentrated nitric acid solution than in solutions of alkali nitrates of the same concentrations. This is taken as evidence of undissociated HNO_3 molecules. By using the above fact and the relation of the intensity of the 1050 cm^{-1} band to the nitrate ion concentration, the dissociation constant for nitric acid was determined to be 23.5 ± 0.5 at 25°.

The maximum concentration of NO_3^- in nitric acid solutions occurs when the acid is about 7 M. Similar studies on perchloric acid indicate that the acid is completely dissociated up to a concentration of 10 M.

4.19.3 Raman studies on inorganic ions and molecules

A very simple but striking Raman study was made by Woodward [4.32] who verified that the mercurous ion in aqueous solution exists as $(Hg{-}Hg)^{2+}$, since one band is observed at 169 cm^{-1} in the vibrational Raman effect. If it had been Hg^+, then obviously no Raman band would be observed. The same method applied to Tl^+ and Ag^+ shows that in both cases there is no appreciable amount of $(M{-}M)^{2+}$ species present.

Many studies have been made on inorganic ions and molecules. Of particular interest is the result when certain metal halides, e.g. $SnCl_4$ and $SnBr_4$, are mixed together; in addition to the Raman bands attributable to the individual halides themselves, new bands, i.e. vibrational frequencies, are observed. These bands have been related to the formation of mixed halides in labile equilibrium with one another and with the original molecules themselves, thus:

$$MX_4 \rightleftharpoons MX_3Y \rightleftharpoons MX_2Y_2 \rightleftharpoons MXY_3 \rightleftharpoons MY_4$$

Other systems which have been shown to involve labile equilibria include $SnBr_4/SnI_4$, $SnCl_4/SnI_4$, $TiCl_4/TiBr_4$, BCl_3/BBr_3, and HgX_2/HgY_2 (X or Y = Cl, Br, I, CN). Measurements of the intensities of the Raman lines have given the value of 2.0 ± 0.2 for the equilibrium constant of $HgCl_2 + HgBr_2 \rightleftharpoons 2HgClBr$ at 288 K. While the original investigations of these equilibria were done in the liquid or solution phase, more recent work has been carried out in the gas phase [4.33] (Fig. 4.60).

(C) CORRELATION OF INFRARED AND RAMAN SPECTRA

4.20 INTRODUCTION

The infrared and Raman methods of investigating molecules have been treated separately earlier in this chapter. It is the object of this section to show that, in fact, the two methods are often complementary; the studies in one field frequently supplement or confirm the information derived from the other.

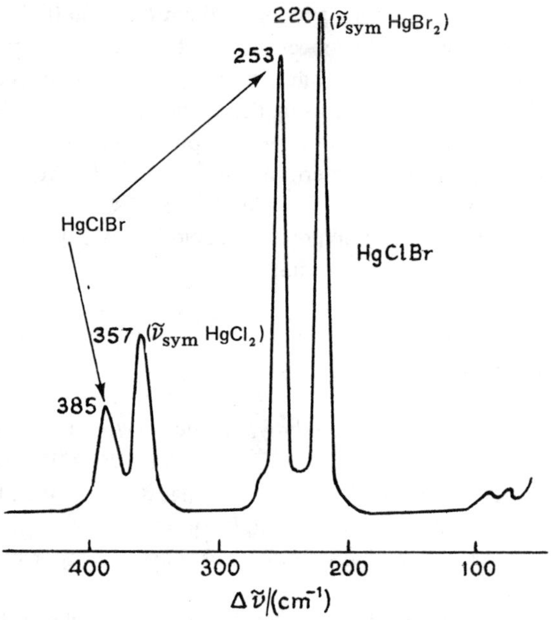

Fig. 4.60 The gas-phase Raman spectrum of an equimolar mixture of $HgCl_2$ and $HgBr_2$ at 730 K. (From ref. 4.33.)

For linear molecules, which have a centre of symmetry, the fundamental vibrations symmetrical about the centre are inactive in the infrared but are active in the Raman. In contrast, the antisymmetrical vibrations are infrared active only, so studies on molecules possessing a centre of symmetry would then be incomplete if both methods were not used.

The volatile noble-gas compounds XeF_2 and KrF_2 are good examples of compounds which obey the Rule of Mutual Exclusion. The Raman spectrum of XeF_2 vapour shows only one band at 515 cm^{-1} ($\tilde{\nu}_{sym}$ XeF_2) while the infrared spectrum exhibits two bands at 558 ($\tilde{\nu}_{antisym}$ XeF_2) and 213 cm^{-1} (δXeF_2). The corresponding fundamentals for KrF_2 are observed at 449 ($\tilde{\nu}_{sym}$ KrF_2), 558 ($\tilde{\nu}_{antisym}$ KrF_2), and 233 cm^{-1} (δKrF_2), and thus both fluorides can be confidently described as linear molecules (point group $D_{\infty h}$) in the vapour phase.

4.20.1 Vibrational analysis of acetylene

The vibrational–rotational spectrum of the acetylene molecule has been studied in great detail. Two of the infrared bands are reproduced in Figs. 4.61 and 4.62 where it is immediately apparent that the 1333 cm^{-1} band has quite a simple structure. It consists of two branches, the P- and R-branches separated by a gap. In addition, the alternating intensities of the individual rotational lines in both the 1333 and the 729 cm^{-1} bands may be readily observed.

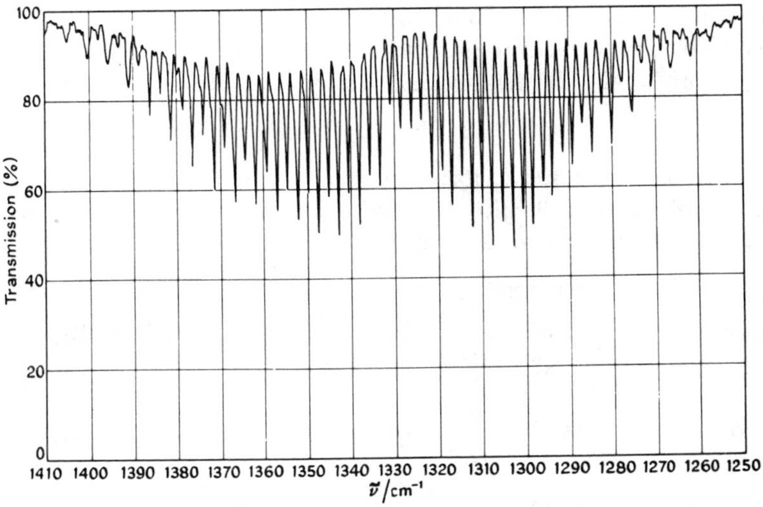

Fig. 4.61 A parallel band of the linear molecule H—C≡C—H which has a centre of symmetry. (From ref. 4.34.)

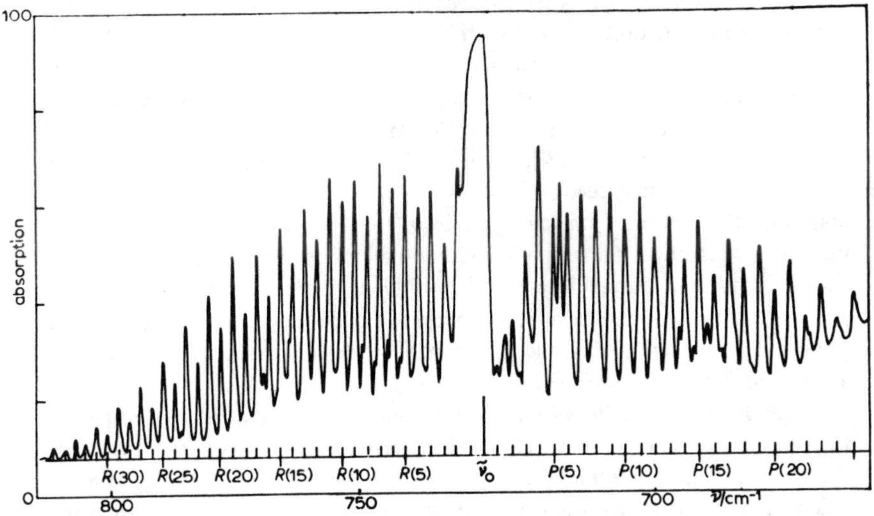

Fig. 4.62 The $\tilde{\nu}_5$ fundamental band of acetylene under high resolution. Pressure 31 mm Hg in a 10-cm cell. (From ref. 4.2.)

Since spherical-top, symmetric-top, and asymmetric-top molecules can never give infrared bands consisting of a single P- and R-branch separated by a gap, it follows, without any detailed study, that the acetylene molecule is linear. Moreover, the presence of the intensity alternation in these two bands proves that the linear acetylene molecule is also symmetrical (point group $D_{\infty h}$). Bearing these two facts in mind, together with the valency of carbon and hydrogen, there can be no doubt that the acetylene molecule has the atoms arranged as follows:

$$H—C{\equiv}C—H$$

A linear molecule which contains four atoms should have seven $(3N-5)$ fundamental modes of vibration. However, two of the modes (π) are doubly degenerate and the five normal vibrations can be represented as below:

$\tilde{\nu}_{sym}$ C—H	3374 cm^{-1}	←H—$\overrightarrow{C}$≡$\overleftarrow{C}$—H→	$\tilde{\nu}_1$	Σ_g^+	∥	R(pol.)
$\tilde{\nu}_{C—C}$	1974	H→—$\overrightarrow{C}$≡$\overleftarrow{C}$—←H	$\tilde{\nu}_2$	Σ_g^+	∥	R(pol.)
$\tilde{\nu}_{antisym}$ C—H	3287	H→←C≡$\overleftarrow{C}$—H→	$\tilde{\nu}_3$	Σ_μ^+	∥	i.r.
δCH	612	H—C≡C—H (↓ H, ↑ C, ↓ C, ↑ H)	$\tilde{\nu}_4$	Π_g	⊥	R(depol.)
δCH	729	H—C≡C—H (↑ H, ↓ C, ↓ C, ↑ H)	$\tilde{\nu}_5$	Π_μ	⊥	i.r.

These vibrations may be accounted for as follows. The acetylene molecule can be regarded as consisting of three units, two C—H and one C≡C group, each of which may vibrate with a frequency related to the force constant of the particular diatomic unit, while the other units are considered fixed. Each C—H group will then vibrate with a frequency related to the ≡C—H stretching or bending force constant. These vibrations may only be symmetric or antisymmetrical with respect to each other, giving two ≡C—H vibrations which are parallel to the axis of the molecule, $\tilde{\nu}_1$ and $\tilde{\nu}_3$, and two which are perpendicular to the axis, $\tilde{\nu}_4$ and $\tilde{\nu}_5$. Finally there is the $\tilde{\nu}_2$ vibration which is largely governed by the force constant of the —C≡C— stretching mode where the CH groups are regarded as moving as rigid units. Thus, the five vibrational modes listed may be labelled as shown. In addition $\tilde{\nu}_4$ and $\tilde{\nu}_5$ are both doubly degenerate since they can take place in a plane perpendicular to the plane in which they are represented. This gives two additional modes, and therefore the seven theoretical modes may be accounted for.

Since the $\tilde{\nu}_1$, $\tilde{\nu}_2$, and $\tilde{\nu}_4$ vibrations are completely symmetric with respect to the centre of the molecule, no change in electric dipole moment will result. $\tilde{\nu}_1$, $\tilde{\nu}_2$, and $\tilde{\nu}_4$ are therefore inactive in the infrared. The polarizability of the molecule will change during these vibrations and so these modes will be Raman active. For the $\tilde{\nu}_3$ and $\tilde{\nu}_5$ vibrations, however, there is a change of dipole moment during the vibration. There are thus two infrared active vibrations, $\tilde{\nu}_3$ and $\tilde{\nu}_5$, three Raman active vibrations, $\tilde{\nu}_1$, $\tilde{\nu}_2$, and $\tilde{\nu}_4$, and no coincidences between the two spectra.

Below 3500 cm^{-1} acetylene has three intense infrared absorption bands, at 3287, 1327, and 729 cm^{-1}. The 729 cm^{-1} band exhibits a strong Q-branch (see Fig. 4.62). Since this 729 cm^{-1} band is the only intense perpendicular band it is identified with the only perpendicular infrared active vibration $\tilde{\nu}_5$. Furthermore, a deformation vibration would be expected to be at a lower frequency than a valence vibration. The remaining infrared active fundamental $\tilde{\nu}_3$ can be identified with the 3287 cm^{-1} parallel band since it is well established that CH stretching vibrations give characteristic group frequencies in the 3000 cm^{-1} region. The remaining intense infrared 1327 cm^{-1} absorption band is therefore likely to be an overtone or a combination frequency. One possibility would be the combination band $(\tilde{\nu}_4 + \tilde{\nu}_5)$. This would make $\tilde{\nu}_4 \approx 597$ cm^{-1}, and, in fact, the value of 612 cm^{-1} is obtained as the experimental Raman value.

As regards the Raman spectrum of acetylene, there are three strong Raman frequencies observed at 612, 1974, and 3374 cm^{-1}. For the same reason as for the 3287 cm^{-1} band, the 3374 cm^{-1} frequency is identified with the CH valence vibration $\tilde{\nu}_1$, while the 1974 cm^{-1} frequency, being too large for a deformation vibration, is identified with the $\tilde{\nu}_2$ mode. The 612 cm^{-1} is the only observed Raman frequency in the characteristic region for a bending vibration, and so it is assigned to the $\tilde{\nu}_4$ mode.

The assignment of the fundamental modes of acetylene are now complete; the infrared spectrum at higher pressures shows in addition a number of weak bands which can be satisfactorily accounted for in terms of various overtone and combination modes.

Table 4.11 The fundamental frequencies (cm^{-1}) of C_2H_2, C_2HD and C_2D_2

Fundamental	C_2H_2	C_2HD	C_2D_2
$\tilde{\nu}_1$	3374	3335	2700
$\tilde{\nu}_2$	1974	1851	1762
$\tilde{\nu}_3$	3287	2584	2427
$\tilde{\nu}_4$	612	519	505
$\tilde{\nu}_5$	729	683	539

Deuteroacetylene C_2D_2 also has a centre of symmetry and can be assigned in a similar manner to C_2H_2; it should be noted, however, that the C_2HD molecule is not symmetrical (it belongs to point group $C_{\infty v}$) and therefore all fundamentals are infrared and Raman active. The fundamental frequencies for all three molecules are given in Table 4.11.

4.20.2 Vibrational analysis of carbon tetrachloride (point group T_d)

For a symmetrical tetrahedral molecule such as CCl_4, group theory predicts that the $(3N - 6) = 9$ fundamental modes of vibration belong to the symmetry classes: $a_1 + e + 2f_2$. In other words, two of the modes are triply degenerate (f_2) and a third is doubly degenerate (e). While all four modes are Raman active (and the a_1 mode will give rise to a polarized band), only the f_2 modes are infrared active.

Table 4.12 Some vibrational frequencies of liquid CCl_4

Vibrational frequencies obtained by the Raman method $\Delta\tilde{\nu}/\text{cm}^{-1}$	*Bands observed in the infrared*† $\tilde{\nu}/\text{cm}^{-1}$	*Assignment of observed frequencies*
145		$\tilde{\nu}_1 - \tilde{\nu}_4$
218, strongly depolarized		$\tilde{\nu}_2$
314, strongly depolarized	305	$\tilde{\nu}_4$
434		$2\tilde{\nu}_2$
~460, very strongly polarized		$\tilde{\nu}_1$
762, medium depolarized	768	$\tilde{\nu}_3, \tilde{\nu}_1 + \tilde{\nu}_4$
791, medium depolarized	797	
1539, very weakly polarized	1546	$2\tilde{\nu}_1 + 2\tilde{\nu}_4$

† For simplicity some of the infrared bands have been omitted.

The eight observed Raman frequencies together with the main infrared absorptions observed for liquid CCl_4 are listed in Table 4.12.

The assignment of the fundamentals can be deduced in the following way. (i) The intense Raman band at 459 cm^{-1} is almost completely polarized and has no infrared counterpart. This band can be readily assigned to the totally symmetric breathing mode $\tilde{\nu}_1(a_1)$. (ii) The strong depolarized Raman band at 218 cm^{-1} is also absent in the infrared spectrum, so this can be assigned to the doubly degenerate bending mode $\tilde{\nu}_2$. (iii) The strong, depolarized Raman bond at 314 cm^{-1} appears in the infrared spectrum and this must be the frequency of the triply degenerate bending mode $\tilde{\nu}_4$. (iv) The doublet centred on 775 cm^{-1}, which appears in both the infrared and Raman spectra, must be associated with the antisymmetric stretching mode $\tilde{\nu}_3$. The vibrational frequency of 844 cm^{-1} for the diatomic radical CCl confirms that ~ 775 cm^{-1} is an appropriate region for this type of stretching motion. The presence of a doublet, which is at first surprising, has been explained in terms of *Fermi resonance* between $\tilde{\nu}_3$ and the combination band $(\tilde{\nu}_1 + \tilde{\nu}_4)$. Fermi resonance may occur when two excited levels arising from different vibrations have almost the same energy, i.e. they are accidentally degenerate. In addition, these vibrational levels must have the same symmetry. A perturbation of the energy levels takes place and this results in a modification of both the positions and the intensities of the two observed bands. For CCl_4, $(\tilde{\nu}_1 + \tilde{\nu}_4)$ numerically has a value of 773 cm^{-1} and thus a weak band might be expected very close to the much more intense $\tilde{\nu}_3$ vibration. Both vibrational levels, however, have the same symmetry (f_2), and the perturbation of the energy levels results in two bands of medium intensity at 762 and 791 cm^{-1}.

The other weak bands listed in Table 4.12 can be interpreted on the basis of these four fundamentals. Early workers found a fine structure associated with the Raman lines $\tilde{\nu}_1$, $\tilde{\nu}_2$, and $\tilde{\nu}_4$ of CCl_4, and it was interpreted as a deviation from tetrahedral symmetry. However, the multiplets can be explained as due to the two isotopes of chlorine, ^{35}Cl and ^{37}Cl. For the totally symmetric stretching

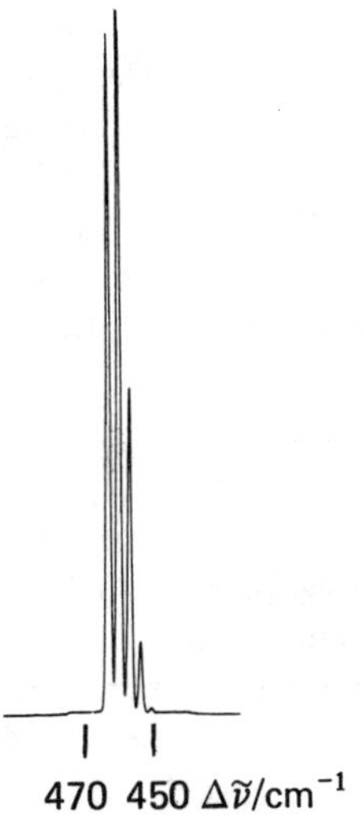

Fig. 4.63 Raman spectrum of the $\tilde{\nu}_1$ totally symmetric stretching mode for solid CCl_4 at 77 K.

Table 4.13 Selection rules for diphosphorus tetraiodide, P_2I_4

Structure	*Point group*	*Raman*	*Infrared*	*Coincidences*	*Polarized Raman bands*
trans	C_{2h}	6	6	0	2
gauche	C_2	12	12	12	3
cis	C_{2v}	12	9	9	2
Number of observed bands		6	5 or 6	0	2

mode, the isotopic modifications together with their observed frequencies are: $C^{37}Cl_4$ (0.4%) not observed, $C^{35}Cl^{37}Cl_3$ (4.7%) 452.0 cm^{-1}, $C^{35}Cl_2{}^{37}Cl_2$ (21%) 456.4 cm^{-1}, $C^{35}Cl_3{}^{37}Cl$ (42.2%) 459.4 cm^{-1}, and $C^{35}Cl_4$ (31.6%) 462.4 cm^{-1} (see Fig. 4.63).

4.20.3 Molecular structure of diphosphorus tetraiodide, P_2I_4

So far we have only considered examples where a complete vibrational analysis can be carried out. However, it is not always necessary to attempt a total assignment, and significant structural inferences can be made using group theory predictions as illustrated with P_2I_4. P_2I_4 is known to have a *trans* (C_{2h}) structure from electron diffraction studies; a vibrational study of both solid and solution show convincingly that the C_{2h} structure is also present in solution. The selection rules for the three possible conformations *trans*, *cis*, and *gauche* are summarized in Table 4.13 together with the numbers of observed bands. The small number of observed bands, the number of polarized lines, the lack of coincidences between the spectra, and the close similarity between the solid and solution phases all confirm a *trans* (C_{2h}) structure. Similar arguments have been used to conclude that P_2Cl_4 and $P_2(CF_3)_4$ have a *trans* structure in all three phases; in contrast, N_2F_4 has a *gauche* structure in the vapour and solid phases while it exists as a mixture of the *gauche* and *trans* forms in the liquid state.

4.20.4 Molecular structure of trisilylamine, $N(SiH_3)_3$

This is an example of where significant structural information can be obtained from characteristic group frequencies.

The important structural point about trisilylamine concerns the geometry of the NSi_3 skeleton. The presence of a lone pair of electrons on the nitrogen atom would normally imply a pyramidal skeleton (C_{3v}) as in ammonia, but the back donation of these electrons into empty 3d orbitals on the silicon atoms would lead to an essentially planar NSi_3 skeleton, D_{3h}.

The skeletal modes can be picked out from the SiH_3 vibrations by examining the corresponding spectrum of the deutero-compound, $N(SiD_3)_3$. The positions of the SiH_3 vibrations will be shifted to lower wavenumbers in the deutero-compound while the skeletal modes will be relatively unaffected.

For a pyramidal skeleton, the molecule would be expected to show four bands due to fundamental modes ($2a_1 + 2e$) in both the infrared and Raman. The appearance of only three skeletal fundamentals in the Raman spectrum (987, 496 pol, 204 cm^{-1}), and the absence in the infrared of the polarized Raman band at 496 cm^{-1}, provide sufficient evidence for a planar structure. This result agrees with the planar skeleton found by electron diffraction.

It is interesting to note that both $P(SiH_3)_3$ and $P(GeH_3)_3$ have a pyramidal shape.

4.20.5 Hydrogen bond studies

The hydroxyl absorption band due to the O—H stretching frequency in Z—O—H compounds is dependent on the nature of Z and on the phase in which the measurements are made. As an example of the latter, methyl alcohol may be considered. In the gaseous phase at reasonably high temperatures, or in very dilute solution, this compound has an OH band at about 3640 cm^{-1}, whereas in the pure liquid both infrared and Raman spectra indicate that it has an additional absorption band at about 3400 cm^{-1}. The 3640 cm^{-1} band is attributed to a monomeric form of CH_3OH, whereas the 3400 cm^{-1} is attributed to some form of methyl alcohol polymer.

The dependence of the OH stretching frequency on the nature of Z is given in Table 4.14.

Table 4.14 The dependence of the OH stretching frequency in some alcohols on the nature of Z [4.35]

Compound	*Molality of compound in* CCl_4	*Wavenumber of peak* $\tilde{\nu}/cm^{-1}$
Methyl alcohol	0.00324	3640
Ethyl alcohol	0.00322	3632
Isopropyl alcohol	0.00300	3620
Allyl alcohol	0.00309	3620
Cyclohexanol	0.00327	3615

In monomeric formic acid the fundamental O—H stretching frequency is found at 3570 cm^{-1}. The monomer exists only at higher temperatures, and at the lower temperature the dimer is formed, and a band is observed at 3080 cm^{-1} which has to be ascribed to the O—H stretching vibration frequency in the dimer. Thus, a substantial shift in frequency has occurred, and this is attributed to what is known as intermolecular hydrogen bonding. This type of bonding is sufficient to hold the two formic acid molecules together so that they behave as one unit. However, the strength of such intermolecular hydrogen bonds is usually of the order of 10–40 kJ mol^{-1} and this value is much less than that of a normal chemical bond. Such intermolecular hydrogen bonds exist in solutions of alcohols and phenols except the very dilute ones.

The hydrogen bond is found to occur when a hydrogen atom lies in between two highly electronegative atoms such as O and F or O and N, and would be represented as follows:

—O—H···F —O—H···N⋜

When the hydrogen bonding takes place within the molecule itself, this is known as intramolecular hydrogen bonding, and an example of this is to be found in salicylic acid:

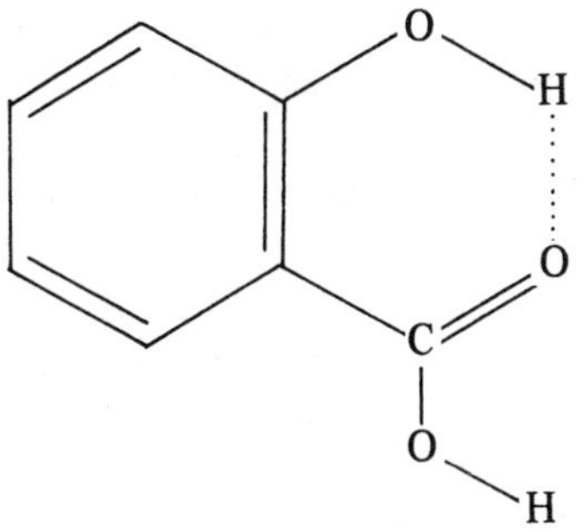

Sutherland [4.36] has listed the frequency displacements which take place on intermolecular hydrogen bonding and has given the following regions for the fundamental band of the O—H stretching frequency for the different types of hydrogen bonding:

—O—H···O	3500–3200 cm^{-1}
—O—H···O=C—	3300–2830 cm^{-1}
—N—H···O=C—	3320–3240 cm^{-1}
—N—H···N⋞	3300–3156 cm^{-1}

Hydrogen bonding may be recognized by such shifts in the fundamental frequency and by the splitting or broadening of such bands with an enhanced intensity of the vibrational transitions.

In dilute solution the free N—H fundamental bands are to be found at 3500 and 3380 cm^{-1}; these free N—H bands are usually narrow and sharp and correspond to the N—H stretching vibrations:

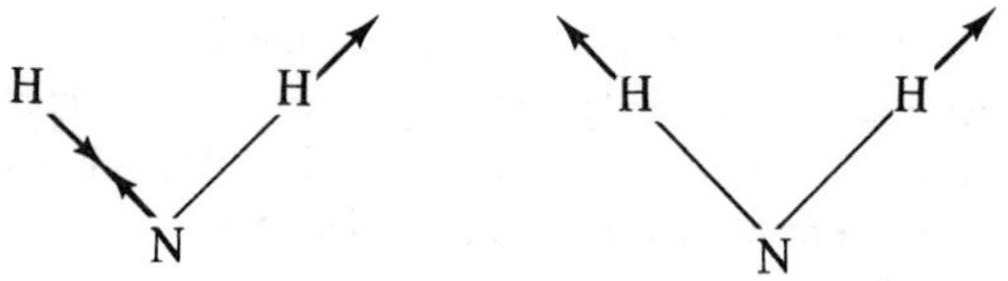

3500 cm^{-1} 3380 cm^{-1}

that is the asymmetrical and symmetrical vibrations, respectively. The N—H stretching bands have been studied by both the infrared and Raman techniques for a number of primary and secondary aliphatic amines in carbon tetrachloride solution. This approach may be employed to follow the effect of concentration on intermolecular hydrogen bonding. As the concentration is increased both the symmetrical and asymmetrical stretching frequencies shift to lower frequencies. This has been attributed to the following type of hydrogen bond formation:

—N—H···N⋞

Both the infrared and Raman methods may be employed to study the effect of hydrogen bonding on the O—H stretching frequency. In the case of alcohols the non-hydrogen bonding O—H in solution results in a sharp infrared band between about 3640 and 3610 cm^{-1}. For pure alcohols and those in sufficiently concentrated solution to be intermolecularly hydrogen-bonded a fairly broad infrared band occurs in the 3380–3300 cm^{-1} region. This behaviour of the infrared spectra of the alcohols is very similar to that of their Raman bands which are to be found between about 3630 and 3380 cm^{-1}, where the 3630 cm^{-1} region is to be attributed to the non-hydrogen-bonding hydroxyl group.

Intermolecular hydrogen bonding can be detected by examining a number of solutions of different concentrations. When the spectrum becomes independent of dilution, then it may be concluded that intermolecular hydrogen bonding is no longer detectable and its concentration is probably low or negligible. It would seem from the majority of compounds containing OH that the intermolecular hydrogen bonding is no longer present (or detectable) at concentrations less than about 0.0005 M. Broadening, strengthening, and splitting of the fundamental OH band are good indications of the presence of hydrogen bonding. Generally, as the strength of a linear hydrogen bond increases, the stretching frequency is lowered, the band is broadened in contour, and it has an enhanced integrated intensity. In addition, non-linear bonded systems are also affected in these ways, although their behaviour is less pronounced.

The intermolecular bonding can be distinguished from the intramolecular hydrogen bonding in that the intra-type is independent of dilution. Some of the strongest intramolecular hydrogen bonds occur in aromatic systems when an OH substituent is adjacent to a highly electronegative atom.

To illustrate how the method may be employed to study both inter- and intra-molecular hydrogen bonding some work by Baker and Shulgin [4.37] will now be considered. Their results for 2-allylphenol in carbon tetrachloride solutions are given in Fig. 4.64.

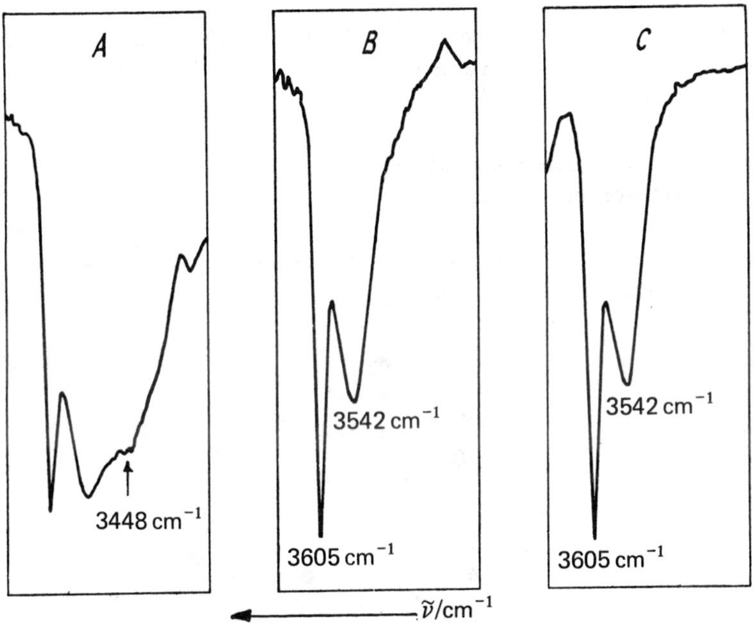

Fig. 4.64 The infrared spectrum between 4075 and 3300 cm^{-1} of 2-allylphenol in carbon tetrachloride solution: (a) 0.7 M; (b) 0.07 M; (c) 0.007 M. The ordinates are proportional to percentage transmission. (From ref. 4.37. Courtesy of the Journal of the American Chemical Society.)

In the 0.7 M solution of 2-allylphenol there is an infrared band at 3605 cm^{-1}. This band is given by each of the three solutions and is almost identical with that of the stretching frequency of the OH group in phenol. Hence, this frequency is attributed to a non-hydrogen-bonded hydroxyl group. As the concentration is diminished from 0.7 to 0.07 M the very broad band at 3448 cm^{-1} is markedly decreased, and in the 0.007 M solution has disappeared. This band is therefore attributed to intermolecular hydrogen bonding, since this would diminish with decreasing concentration. The band at 3542 cm^{-1} which is given by all three solutions and not by phenol itself is therefore attributed to the following intramolecular hydrogen bonding:

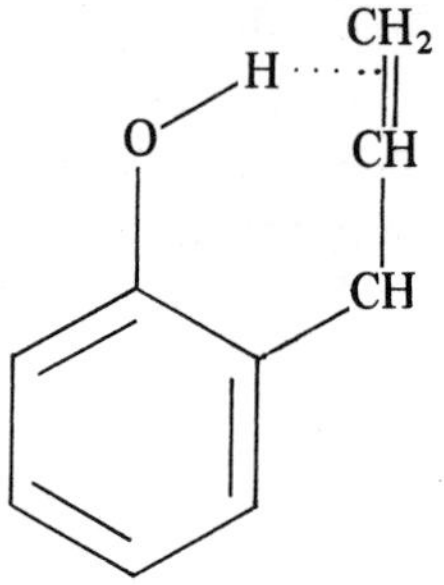

where the hydrogen in the OH group is bonded to the π-electrons of the ethylenic double bond. This results in an average shift ($\Delta\tilde{\nu}_{O-H}$) of the OH frequency of (3605 − 3542), that is 63 cm^{-1}.

4.20.6 Polarized infrared and Raman spectra of single crystals

Plane polarized light has its electric vector confined to a particular plane, which necessarily includes the direction of propagation.

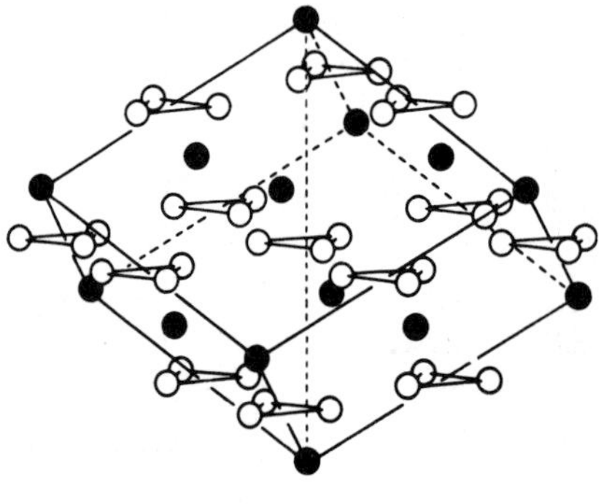

●: Ca; ○: O

Fig. 4.65 The crystal structure of calcite. The planar carbonate ions are oriented perpendicular to the crystal *c* or optic axis. (From ref. 4.38.)

Plane polarized infrared radiation is produced for a particular application using a metallic wire-grid while the Raman experiment uses a laser source which is plane polarized to start with. The windows of the laser tube are set at the

Brewster angle, and one plane of light polarization is transmitted with great efficiency.

As an example of a polarized infrared experiment suppose a single crystal of calcite (CaO_3) is examined. Figure 4.65 shows the crystal structure of calcite and the planes of the carbonate ions are stacked perpendicular to the crystal axis *c*. The three infrared active internal carbonate modes consist of an out-of-plane bend ($\tilde{\nu}_2 = 878\ cm^{-1}$) which produces a dipole moment vector parallel to the *c*-axis, an antisymmetric stretch ($\tilde{\nu}_3 = 1432\ cm^{-1}$), and an in-plane bend ($\tilde{\nu}_4 = 715\ cm^{-1}$). The last two modes give rise to an electric vector which is perpendicular to the *c*-axis. When the *c*-axis is aligned parallel to the electric vector of the polarized light, $\tilde{\nu}_2$ will appear as an intense absorption band while $\tilde{\nu}_3$ and $\tilde{\nu}_4$ will be virtually absent. If the crystal is now rotated so that the *c*-axis is perpendicular to the electric vector of the light, then $\tilde{\nu}_2$ disappears but $\tilde{\nu}_3$ and $\tilde{\nu}_4$ become intense bands. This behaviour enables the two bending modes of the carbonate anion to be assigned unambiguously to their appropriate symmetry species.

In a similar way, the plane polarized laser beam can be used to assign unambiguously the Raman active fundamentals of single crystals. As discussed earlier (see p.230), an electric dipole moment (P) is induced in a molecule when it is subjected to an electric field (E). The relationship $|P| = \alpha |E|$ involves α, the polarizability of the molecule, and for non-isotropic molecules α is a tensor which is defined by six coefficients [see Equation (4.103)]. The six components have to be determined experimentally. Once this is done, it is possible in principle to predict the observed Raman intensities, since the latter are proportional to the squares of the components of the Raman tensors.

Provided that the X-ray crystal structure of the compound is known, and assuming that single crystals of about 2 mm can be grown, polarized vibrational spectra provide a means of unambiguously assigning frequencies to particular symmetry classes. With meaningful assignments it is possible to develop adequate force fields for calculation of the spectra of related compounds.

REFERENCES

4.1 Sayce, L.A. and Jackson, A., *Molecular Spectroscopy*, Institute of Petroleum Conference, p. 54 (1955).

4.2 *Tables of Wavenumbers for the Calibration of Infrared Spectrometers*, IUPAC, Butterworths (1961).

4.3 Rao, K.N., Humphreys, C.J. and Rank, D.H., *Wavelength Standards in the Infrared*, Academic Press, New York and London (1966).

4.4 Czerny, M., *Z. Physic*, **34**, 227 (1925).

4.5 Wilkinson, G.R., Chapter 3, *Infrared Spectroscopy and Molecular Structure*, Ed. by Mansel Davies, Elsevier, Amsterdam (1963).

4.6 Palik, E.D., *J. Chem. Phys.*, **23**, 217 (1955).

4.7 Herzberg, G., *Infrared and Raman Spectra of Polyatomic Molecules*, D. Van Nostrand, Amsterdam (1945).

4.8 Hinkley, E.D. and Kelley, P.L., *Science*, **171**, 635 (1971).

4.9 Sutherland, G.B.B.M., *Infrared and Raman Spectra*, Methuen, London (1935).
4.10 Jones, W.J., Chapter 4, *Infrared Spectroscopy and Molecular Structure*, Ed. by Mansel Davies, Elsevier, Amsterdam (1963).
4.11 Gerhard, S.L. and Dennison, D.M., *Phys. Rev.*, **43**, 197 (1933).
4.12 Dennison, D.M., *Rev. Mod. Phys.*, **3**, 280 (1931); **12**, 175 (1940).
4.13 Badger, R.M. and Zumwalt, L.R., *J. Chem. Phys.*, **6**, 711 (1938).
4.14 Crawford, M.F., Welsh, H.L. and Locke, J.L., *Phys. Rev.*, **75**, 1607 (1949).
4.15 Stammreich, H., Bessi, D. and Sala, O., *Spectrochim. Acta*, **12**, 403 (1958).
4.16 Carter, R.L., *J. Chem. Educ.*, **48**, 297 (1971).
4.17 Sherwood, P.M.A., *Vibrational Spectroscopy of Solids*, Cambridge University Press, (1972).
Bhagavantam, S. and Venkatarayudu, T., *Theory of Groups and its Application to Physical Problems*, Academic Press, New York and London (1969).
4.18 Bellamy, L.J., *Infrared Spectra of Complex Molecules*, Chapman and Hall, London (1975).
Dyer, J.R., *Applications of Absorption Spectroscopy to Organic Compounds*, Prentice Hall, New Jersey (1965).
Nakamoto, K., *Infrared Spectra of Inorganic and Co-ordination Compounds*, Wiley-Interscience, New York (1970).
4.19 Ferraro, J.R., *J. Chem. Educ.*, **38**, 201 (1961).
4.20 Ebsworth, E.A.V., Chapter 9, *Infrared Spectroscopy and Molecular Structure*, Ed. by Mansel Davies, Elsevier, Amsterdam (1963).
4.21 Jones, R.N. and Sandorfy, C., *Chemical Applications of Spectroscopy*, p. 247, Interscience, New York (1956).
4.22 Wilson, E.B. and Wells, A.J., *J. Chem. Phys.* **14**, 578 (1946).
4.23 Jones, R.N., Ramsey, D.A., Keir, D.S. and Dobriner, K., *J. Amer. Chem. Soc.*, **74**, 80 (1952).
4.24 Ramsey, D.A., *J. Amer. Chem. Soc.*, **74**, 72 (1952).
4.25 Arkell, A., Reinhard, R.R. and Larson, L.P., *J. Amer. Chem. Soc.*, **87**, 1016 (1965).
4.26 (a) DeKock, R.L., *Inorg. Chem.*, **10**, 1205 (1971).
(b) Ogden, J.S. and Turner, J.J., *Chem. in Britain*, **7**, 186 (1971).
Downs, A.J. and Peake, S.C., *Molecular Spectroscopy*, Vol. 1, Chem. Soc. Spec. Periodical Reports, London (1973).
4.27 Hendra, P.J., Chapter 15, *Laboratory Methods in Infrared Spectroscopy*, Second Edition, Eds. Miller and Stace, Heyden, London (1972).
4.28 Carpenter, J., private communication.
4.29 (a) Jones, W.J., private communication.
(b) Butcher, R.J., Willets, D.V. and Jones, W.J., *Proc. Roy. Soc. A*, **324**, 231 (1971).
4.30 Stoicheff, B.P., *Advances in Spectroscopy*, Vol. 1, p. 91, Interscience (1959).
4.31 Bosworth, Y., Clark, R.J.H. and Rippon, D.M., *J. Mol. Spectrosc.*, **46**, 240 (1973).

4.32 Woodward, L.A., *Phil. Mag.*, **18**, 823 (1934).

4.33 Beattie, I.R. and Horder, J.R., *J. Chem. Soc. A*, 2433 (1970).

4.34 Barrow, G.M., *Introduction to Molecular Spectroscopy*, p. 147, McGraw-Hill, New York (1962).

4.35 Flett, M.St. C., *Spectrochim. Acta*, **10**, 21 (1957).

4.36 Sutherland, G.B.B.M., *Trans. Faraday Soc.*, **36**, 889 (1940).

4.37 Baker, A.W. and Shulgin, A.T., *J. Amer. Chem. Soc.*, **80**, 5358 (1958).

4.38 Wilkinson, G.R., Chapter 18, *Laboratory Methods in Infrared Spectroscopy*, Second Edition, Eds. Miller and Stace, Heyden, London (1972).

5 Far-infrared spectroscopy

5.1 INTRODUCTION

The International Union of Pure and Applied Chemistry defined the far-infrared region as 10–200 cm^{-1}. This definition is one of convenience, and although, on the whole, this will be adhered to in this chapter, there is obviously no point, when the vibrational motions of a molecule are being considered, to include one, say, at 190 cm^{-1} and neglect another at 210 cm^{-1}, especially if both are concerned with similar types of vibrations.

A number of studies was made of far-infrared absorption in the earlier part of this century, in particular, the work of Czerny in 1925 [5.1] on the rotational spectra of the hydrogen halides. However, it is only since about 1963, when commercial instruments became available, that there has been a rapid development in this field, and now many laboratories examine the absorption in this region whenever a detailed physical examination of a compound is being made. The subject is well catered for in the way of reviews [5.2–5.5] and books [5.6, 5.7], and this is indicative of the progress and activity in the field. In the far-infrared region intramolecular vibrations of groups the masses of which are large or/and the force constant small may be studied. Many inorganic molecules fall into this category since in many cases the atom may be heavy and the bond single as opposed to multiple. For example, the antisymmetric stretching Os–Os frequency at 121 cm^{-1} has been studied in $Os_3(CO)_{12}$.

5.2 FAR-INFRARED INTERFEROMETERS

The types of spectrometers used to study far-infrared absorption are spectrophotometers and interferometers.

Single- and double-beam spectrophotometers employing prisms and gratings have been used in the infrared region. In the present chapter the instruments employed will be of the interferometer type which are capable of achieving a greater signal/noise ratio than the single- or double-beam spectrophotometer and lead to the following advantages over the double-beam spectrophotometer: (a) higher resolution may be achieved; (b) lower frequencies can be studied, e.g., 5–40 cm^{-1} range; (c) the interferometer has a faster scan time; (d) smaller samples may be employed. Some of the disadvantages of the interferometer are: (a) it is a single-beam system; (b) to obtain a plot of intensity versus frequency the data have to be fed to a computer; (c) the spectra can be somewhat unreliable, and a noise spike which appears in the spectrum at a particular frequency can ruin the whole spectrum.

The far-infrared spectra given in Fig. 5.15 were obtained by means of a lamellar grating interferometer type LR100 supplied by the Research and Industrial Instrument Company. This instrument is useful in that it enables spectra to be recorded to wavenumbers as low as 3 cm^{-1}, thus providing an overlap with microwave apparatus. However, its upper limit is only about 70 cm^{-1}. To extend the range several other commercial interferometers or spectrophotometers may be employed, such as a Grubb Parsons NPL Cube Interferometer which overlaps the wavenumber range of the LR100 and, in fact, covers the range ~ 15–675 cm^{-1}. Both of these instruments have a high-pressure mercury lamp as source and a Golay detector, and are capable of a maximum resolution of 0.1 cm^{-1}.

The optical diagram of the LR100 lamellar grating is given in Fig. 5.1, and the main component parts are labelled. The source is a high-pressure mercury arc lamp consuming 150 W power, while the beam chopper operates at 15 Hz and supplies the reference phase signal to the phase-locked-in amplifier. The chopped radiation from the source is collimated by an aluminized off-axis paraboloid mirror and then passes to the lamellar grating which comprises two sets of parallel intermeshing plates the front surfaces of which are flat and act as two reflecting mirrors. In fact the front set of grooves is effectively equivalent to a mirror while the other set of intermeshing plates is equivalent to another mirror. Schematically the function of the lamellar grating assembly can be represented as in Fig. 5.2 where it can be seen that the reflected rays abD and dcS have different path lengths. The reflected radiation is deflected by the off-axis paraboloid reflector to a transfer mirror directly below the source spectrum, thence to two off-axis paraboloid reflectors through the cell, and then to the Golay detector. To eliminate higher frequencies, filters are employed.

Because a path difference has been introduced into the radiation by the intermeshing metal plates of the grating, the different frequencies from the source are no longer in phase and consequently some interfere constructively and others destructively with one another. When the path difference between the front set of grooves and the intermeshing plates is zero, all the frequencies are in phase, and a maximum output is recorded at the detector. However, when the two optical path lengths are no longer identical a characteristic curve is obtained on

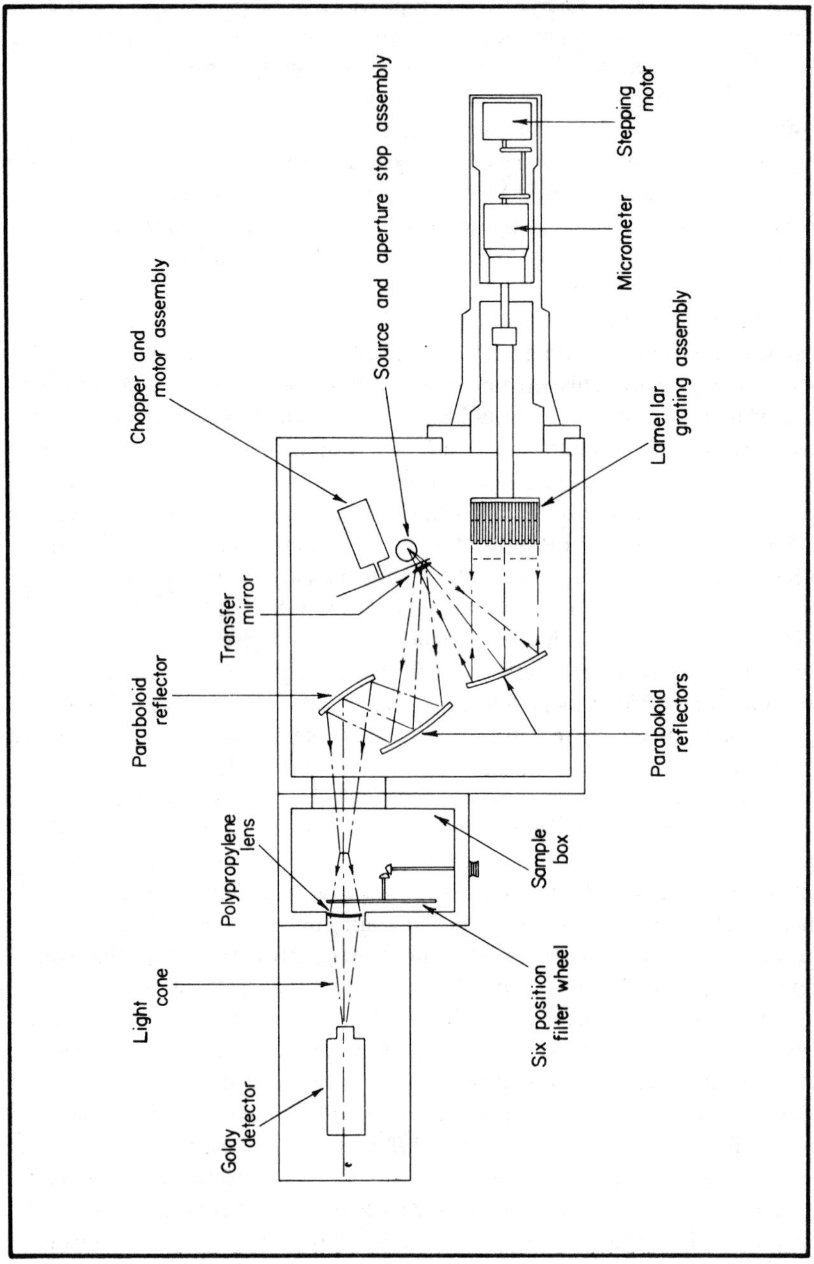

Fig. 5.1 Optical diagram of LR100 lamellar grating far-infrared spectrophotometer (Courtesy of Beckman Instrument Co.).

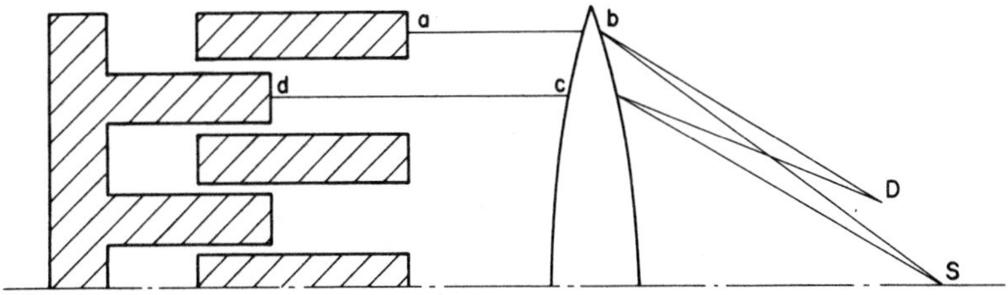

Fig. 5.2 Representation of the different path lengths in the lamellar grating spectrophotometer.

the recorder. The pattern is termed an interferogram and is really a trace of the amplified signal against path difference (see Fig. 5.3). Thus the maximum peak in Fig. 5.3 is the case where the two beams travel equal distances, while the other peaks result from cases where the two optical paths are unequal and the different frequencies are no longer in phase. This interferogram, which may be presented by a pen recorder, contains the far-infrared absorption information. However, the absorption spectrum is not apparent at this stage. The signal from the Golay detector is amplified, and punched on paper tape which is then (or later) submitted to a computer. A Fourier transform operation is carried out which yields the normal absorption spectrum (e.g., plot of absorption versus wavenumber as in Fig. 5.15).

A Michelson type of interferometer is frequently used in far-infrared absorption work, and its operation is schematically represented by the optical path in Fig. 5.4. The source of radiation is a high-pressure mercury lamp, and this radiation is divided into two beams by the beam splitter which is merely a plastic film, while the path lengths of the two beams are determined by the positions of the fixed and movable mirrors. The resulting intensity of the combined beams

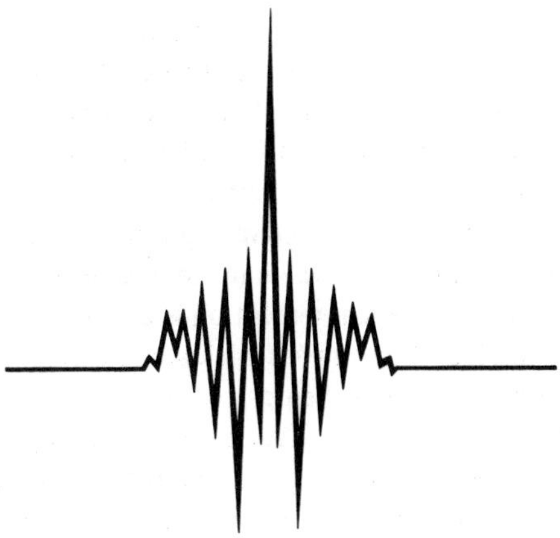

Fig. 5.3 An interferogram.

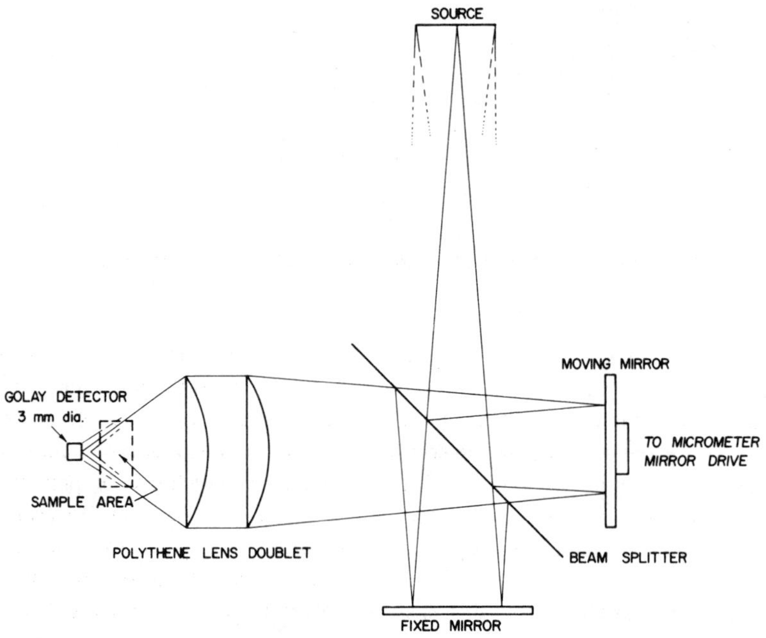

Fig. 5.4 Optical path in a Michelson interferometer.

results in an interferogram which is drawn by the attached pen recorder and is really a trace of the signal against the path difference of the beam.

The frequency range 10–500 cm^{-1} may be covered by using an R.I.I.C. spectrometer model 720 which is a Michelson type of interferometer. This instrument also uses a high-pressure mercury lamp and a Golay detector and has a resolution of 0.1 cm^{-1}. The optical diagram of this spectrometer is given in Fig. 5.4. The beam splitter is a Melinex (polyethylene terephthalate) film. By varying the thickness of this film, e.g., 12, 25, 60, or 100 μm, various parts of the region, 10–500 cm^{-1}, can be covered with higher efficiency. Part of the radiation is transmitted on to a fixed mirror, while the other part is reflected to a movable mirror to which a motor is attached which enables the mirror to travel a range of $\pm$5 cm. Thus, the movable mirror introduces a variable path difference between the two beams and an interferogram is produced. The resolution is given by 1/(2 $\times$ distance travelled by the movable mirror); thus if the mirror travels 2.5 cm then the maximum theoretical resolution is 0.2 cm^{-1}. The reflected radiation from these two mirrors which travels back along their incident paths is recombined at the beam splitter and then passed back along the optical path through the cell to the Golay detector. As with the lamellar grating the signal is then amplified and digitized on paper tape for input to a computer. More recently there is a tendency for spectrometers to have a built-in computer.

The radiation which is transmitted to the detector contains all the frequencies passed by the filter. From the interferogram $I(\alpha)$ the spectral intensity $G(\tilde{\nu})$ at each wavenumber ($\tilde{\nu}$) is obtained by means of the Fourier transformation:

$$G(\tilde{\nu}) = \int I(\alpha) \cos(2\pi\nu\chi)\, d\chi \qquad (5.1)$$

where χ is the path difference between the beams, and the computation is carried out by the use of a digital computer. The programme can be so arranged that the computer output gives a plot of percentage transmission against wavenumber on a pen recorder.

A typical sample cell would have polyethylene windows and a Teflon spacer. It has to be noted whether the liquid produces swelling of the polyethylene or bowing of the windows. If so, other material (e.g. quartz) may be used. Temperature variable cell units are commercially available. The cell may be filled from a syringe. The entire interferometer is evacuated since water vapour strongly absorbs in the far-infrared region, and as the spectrometer is a single-beam instrument, runs have to be carried out on the empty cell (yielding I_0 vs. $\tilde{\nu}$) and on the sample cell (yielding I vs. $\tilde{\nu}$). The two resultant graphs may be ratioed point by point to yield a plot of I_0/I against $\tilde{\nu}$. Finally, this latter graph may be modified to yield a plot of the absorption coefficient (α) against $\tilde{\nu}$ where:

$$\alpha = (1/l) \ln(I_0/I) \qquad (5.2)$$

5.3 INTRAMOLECULAR VIBRATIONAL MODES

A few examples of the type of study which may be carried out are now given.

5.3.1 An essential frequency in the calculation of thermodynamic functions

The low vibrational frequencies determined in the far-infrared region usually involve atoms where the masses are large or the force constant of the bond is small; thus it is very likely that the bonds tend to be single rather than multiple. It can be important to obtain such vibrational frequencies in order to complete the vibrational assignments and to calculate the thermodynamic functions such as entropy. Often these low vibrational frequencies make an appreciable contribution to the value of the thermodynamic functions (see Chapter 7).

5.3.2 Structural studies

In the determination of the structure of a simple molecule it is frequently advantageous to determine all the vibrational frequencies and to employ each of the Raman and infrared (Chapter 4) approaches. In a structural study of P_2Cl_4 the higher frequency data showed the structure to be of the type $Cl_2P—PCl_2$ as opposed to a chlorine-bridged structure [5.8]. However, this did not determine whether the $Cl_2P—P$ groups are planar or pyramidal or what the orientation is

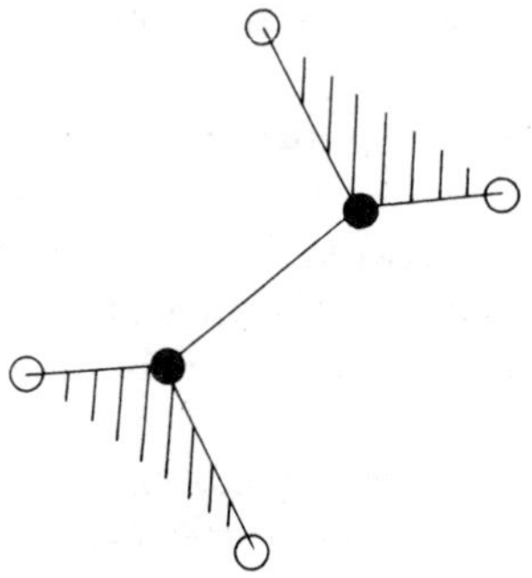

Fig. 5.5 The *trans* bent structure (symmetry type C_{2h}) of $Cl_2P—PCl_2$ (After Miller [5.8]).

around the P—P bond. A far-infrared band was obtained at 91 cm^{-1} which was ascribed to a torsional mode. From this, together with infrared and Raman data, the *trans* bent structure (see Fig. 5.5) (symmetry type C_{2h}) was judged to be the correct one.

Far-infrared studies can be of considerable help in the examination of metal–ligand systems since metal–ligand vibrations are skeletal vibrations and are likely to be sensitive to changes of mass within the complex. These metal–ligand vibration frequencies termed $\tilde{\nu}$(M–L) are sensitive to a number of factors [5.9], e.g.:

(i) They depend on the coordination number where an increase in its value leads to a decrease in $\tilde{\nu}$(M–L) as is the case in $\tilde{\nu}$(Co–Cl) where $\tilde{\nu}$ falls from 344 to 304 cm^{-1} on passing from the tetrahedral form of $CoCl_2(C_5H_5N)_2$ to the octahedral isomer. This may be accounted for by the fact that the smaller the number of bonds made by a metal to its ligand, then the more covalent the bonds are in character and the bigger the force constant and the higher the vibrational frequency.

(ii) The value of $\tilde{\nu}$(M–L) is strongly dependent on the mass of L; e.g. $\tilde{\nu}$(Ni–P) is at 262 cm^{-1} in $Ni(CO)_3PF_3$, but when the F is replaced by the appreciably heavier C_6H_5 in $Ni(CO)_3P(C_6H_5)_3$ the frequency is lowered to 192 cm^{-1}.

(iii) $\tilde{\nu}$(M–L) is also sensitive to the mass of M. Thus, for the type $[MCl_4]^{2-}$, as the atomic weight increases in the sequence Zn < Cd < Hg, the $\tilde{\nu}_3$ vibration frequency falls as follows: 271, 260, and 228 cm^{-1}

5.3.3 Study of hindered internal rotation

In Chapter 3 we have considered some aspects of hindered internal rotation and, in particular, the case of internal rotation of the CHO group in acetaldehyde about the C—C axis which involves a three-fold symmetry; the equation which adequately represents the potential function for different angles of internal rotation is:

$$V(\alpha) = (V_3/2)(1 - \cos 3\alpha) \tag{5.3}$$

The dependence of $V(\alpha)$ on α is given in Fig. 3.21. Only when $v > 2$ has the methyl group sufficient energy to rotate freely about the C—C axis. When $v = 0$, 1, or 2 the methyl group is restricted to one of the three potential valleys and can then execute only a torsional vibration about equilibrium positions. When the barrier height is low, splitting of the vibrational levels occurs, as is noted in the potential curve for acetaldehyde.

For acetaldehyde the transition from the $v = 0$ to the $v = 1$ level for the methyl torsional oscillation motion occurs in the far-infrared region at ~150 cm^{-1}. From this value and the reduced moment of inertia about the C—C axis the value of V_3 can be determined. This value is in good agreement with the value of 4889 ± 126 J mol^{-1} from microwave spectroscopy where a completely different type of transition is involved in the evaluation.

Many other sample compounds have had their energy barriers for rotation about the C—C and C—Z bond determined by this procedure and include a variety of halogenated ethanes, halogenated propanes and propenes, halogenated aldehydes, methyl silanes, and methyl alcohol.

Miller, Fateley, and Witkowski [5.10] have made studies of the torsional frequencies around the C—C single bond in a variety of aromatic compounds such as *o*-, *m*-, and *p*-F-benzaldehydes whose observed torsional bands are given in Fig. 5.6. The potential function, which is taken as a function of the angle of internal rotation, is given by:

$$V(\alpha) = \sum_{n=1}^{3} (V_n/2)(1 - \cos n\alpha) \tag{5.4}$$

They assumed that the potential function in the region of the torsional transition is harmonic, and this led to an expression for the potential being the sum of two terms in the right-hand side of the equation:

$$\tilde{\nu}^2/F = V_1 + 4V_2 \tag{5.5}$$

where $\tilde{\nu}$ is the torsional frequency (cm^{-1}) and:

$$F = h/8\pi^2 c I_r \tag{5.6}$$

and I_r is the reduced moment of inertia for the torsion.

The energy barrier V_1 has a periodicity of 2π in α and in, for example, *o*-fluorobenzaldehyde would probably be dependent on steric interaction and a small contribution from hydrogen bonding between the halogen atom and the aldehydic hydrogen atom. Thus, the V_1 term should be much more appreciable for *ortho*- than for *meta*- or *para*-derivatives, and in fact would be expected to be zero for the *para*-derivatives.

The energy barrier V_2 is a two-fold barrier and is interpreted as arising mainly from the overlap of the π-orbitals of the carbonyl bond with those of the aromatic ring. Thus, this term should be relatively important regardless of whether the substituent in benzaldehyde is of the *ortho*, *meta*, or *para* type.

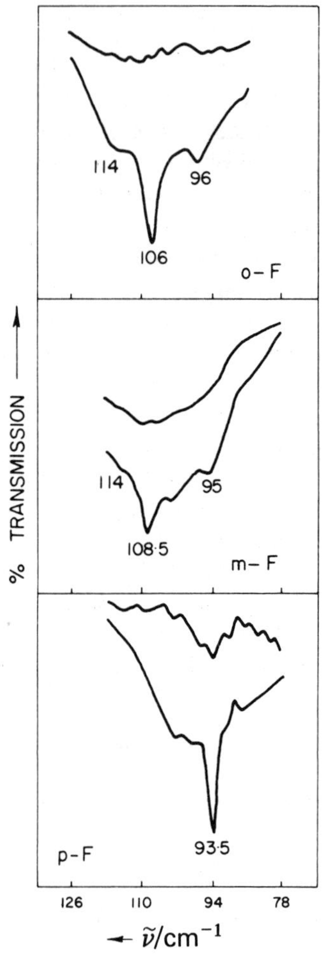

Fig. 5.6 The observed torsional bands for the *o*-, *m*-, and *p*-fluorobenzaldehydes run in a cell of 10 m path and at 2.96 K. (After Miller, Fateley, and Witkowski [5.10]).

The three-fold barrier term (V_3) was judged to be small and was taken to be zero.

For the *para*-derivatives in which the substituent is symmetrical – an atom (e.g. F) or the group CN or CH_3 – then symmetry requires that $V(\alpha) = V(\alpha \pm \pi)$ and consequently $V_1 = V_3 = 0$, and only the V_2 term is left. Consequently, for the symmetrically substituted benzaldehydes, and equally for acetophenones:

$$V_2 = \tilde{\nu}^2/4F \tag{5.7}$$

Thus, the energy barrier is determined directly from the $\tilde{\nu}$ and F values. The estimation of F requires a knowledge of bond distances and angles from other sources. Miller et al. [5.10] developed a computer programme for the evaluation of I_r which requires only the coordinate positions and masses of the atoms and which gives the centre of mass, the orientation of the three principal axes, the principal moments of inertia, and F. Since $V_2 \propto I_r$, the value of V_2 can be very sensitive to small changes in the assumed molecular geometry, and caution has to be exercised on barriers interpreted for a system where guesses have been made as to the molecular geometry.

Miller, Fateley, and Witkowski [5.10] consider that barriers obtained from condensed-state data should not be trusted since the torsional frequencies in the vapour are 20–25 per cent lower than those in the liquid state. Since the energy barrier depends on $\tilde{\nu}_2$ a very serious error may ensue in the energy barrier value if vapour-phase values of $\tilde{\nu}$ are not used; thus, for acetaldehyde, if $\tilde{\nu}$ is lowered by 1 cm^{-1}, the energy barrier is lessened by 62.8 J mol^{-1}.

For *m*-bromobenzaldehyde the dependence of the potential energy (in kJ mol^{-1}) on the internal angle of rotation about the C—C single bond is given by:

$$V(\alpha) = 6.03(1 - \cos\alpha)/2 + 16.8(1 - \cos 2\alpha)/2 \qquad (5.8)$$

and the situation is represented in Fig. 5.7 where curves A and B represent the first and second terms of this equation. Curve C, which gives the $V(\alpha)$ value, is the sum of curves A and B at each α value. It will be noted that the value of α_{max} is 95°8′ (and not 90°). The value of α_{max} is obtained from the equation:

$$\cos\alpha_{max} = -V_1/4V_2 \qquad (5.9)$$

If α_{max} is then substituted in Equation (5.8), then V_{max} is obtained where

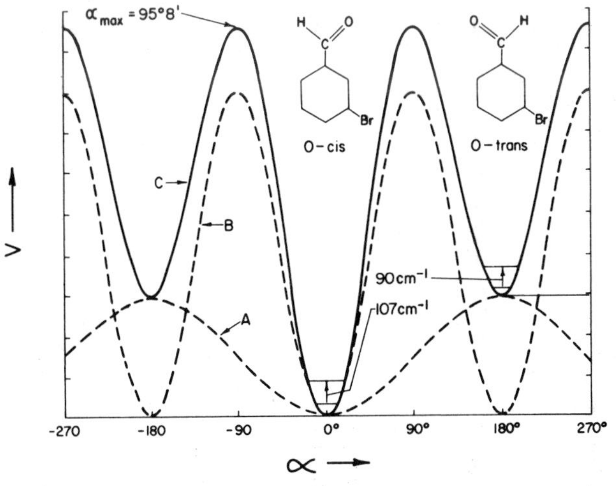

Fig. 5.7 Potential curve for the C—CHO torsion in *m*-bromobenzaldehyde where $V(\alpha)$ (kJ mol^{-1}) $= 6.03(1 - \cos\alpha)/2 + 16.8(1 - \cos 2\alpha)/2$; A and B in the diagram represent the 1st and 2nd terms in the equation. (After Miller, Fateley, and Witkowski [5.10]).

Table 5.1 Energy barrier data of some halo-benzaldehydes (After Miller, Fateley, and Witkowski [5.10])

Compound	F (cm^{-1}) O-*trans*	F (cm^{-1}) O-*cis*	*Choice*†	$\tilde{\nu}_{torsion}$	*Assigned to*†	V_1 (kJ mol^{-1})	V_2 (kJ mol^{-1})	*Stable rotamer*	*Relative population stable/unstable rotamers*	α_{max}	V_{max} (kJ mol^{-1})
m-F-benzaldehyde	1.697	1.875	A*	95	O-*trans*	5.73	17.3	O-*cis*	10	94°45′	20.3
				108.5	O-*cis*						
			B	95	O-*cis*	12.7	17.6	O-*trans*	175		
				108.5	O-*trans*						
o-F-benzaldehyde	1.888	1.602	A*	96	O-*cis*	1.2	17.5	O-*trans*	1.6	90°58′	18.1
				106	O-*trans*	12.8	17.8				
			B	106	O-*cis*	12.8	17.8	O-*cis*	180		
				96	O-*trans*						
p-F-benzaldehyde	1.743			93.5			15.0				
m-Cl-benzaldehyde	1.626	1.871	A*	94	O-*trans*	2.76	16.9	O-*cis*	2.5	92°20′	18.3
				105	O-*cis*						
			B	94	O-*cis*	9.04	18.0	O-*trans*	20		
				105	O-*trans*						
o-Cl-benzaldehyde	1.885	1.507	A*	88	O-*cis*	2.9	16.1	O-*trans*	2.6	92°36′	17.6
				103	O-*trans*						
			B	103	O-*cis*	17.5	16.7	O-*cis*	300		
				88	O-*trans*						
p-Cl-benzaldehyde	1.687			81.5			11.8				

† A and B indicate two possible assignments for the torsional oscillation frequencies; A* is the preferred one.

V_{max} is the energy required for a transition from the O-*cis* form of *m*-bromobenzaldehyde to the O-*trans*. The V_{max} value is 20 kJ mol^{-1}, and it will be noted from Fig. 5.7 and Equation (5.8) that the second term on the right-hand side of the equation is much more dominant than the first in determining the magnitude of the energy barrier, that is, the barrier is governed mostly by the overlap between the π-orbitals of the carbonyl band and those of the aromatic ring.

The O-*cis* isomer is the more stable rotamer (see Fig. 5.7), and its population relative to that of the O-*trans* for the $v = 0$ states is given by:

$$N_{\text{O-}cis}/N_{\text{O-}trans} = e^{V_1/kT} = 7/1 \tag{5.10}$$

The torsional frequencies for the O-*trans* and O-*cis* rotamers occur at 90 and 107 cm^{-1} respectively, while their F values are 1.580 and 1.867 cm^{-1}.

In Table 5.1 some of the pertinent energy barrier data are given for the halogenated benzaldehydes.

A strong incentive in this type of work would be the hope that interpretation of the relative V_2 values for such compounds would give some indication as to the nature of the C(aromatic)—C(aliphatic) bond. Certainly a comparison of the V_2 of acetaldehyde (4.939 kJ mol^{-1}) with benzaldehyde (19.5 kJ mol^{-1}) suggests that the higher energy barrier of the latter is to be related to the overlap of the π-orbitals of the ring with those of the carbonyl group.

It is interesting to correlate the V_2 values in Table 5.1 with the electron-withdrawing sequence of the halogen in different substituent positions: *para* > *ortho* > *meta*. For example, the sequence of V_2 values for the chlorobenzaldehydes is *para* < *ortho* < *meta*, which is the correct one relative to benzaldehyde.

5.3.4 Study of inversions

In some molecules inversion can lead to two distinguishable but equivalent configurations. The case of nitrogen inversion has been considered in Chapter 3.

A number of studies has been made in three-, four-, and five-membered ring compounds where ring inversion occurs. One such case is that of trimethylene oxide where seven sharp bands are observed in the vapour phase which are ascribed to a ring puckering mode [5.12]. In Fig. 5.8 a plot of the puckering potential energy against the puckering angle is made; the vibrational levels are characterized by the quantum number v, inserted for the ring puckering mode, and some of the observed transitions have been inserted. The barrier between the two minimum positions is very small, being only 15 cm^{-1}. The introduction of such a barrier proved necessary since the plot of the rotational constants against the vibrational constant of the puckering mode did not give a linear relationship.

The following potential function has been used for the ring puckering mode:

$$V = -ax^2 + bx^4 \tag{5.11}$$

where x is a ring puckering coordinate, V is thus the sum of a quadratic and a quartic term, and when it so happens that the maximum of the quadratic

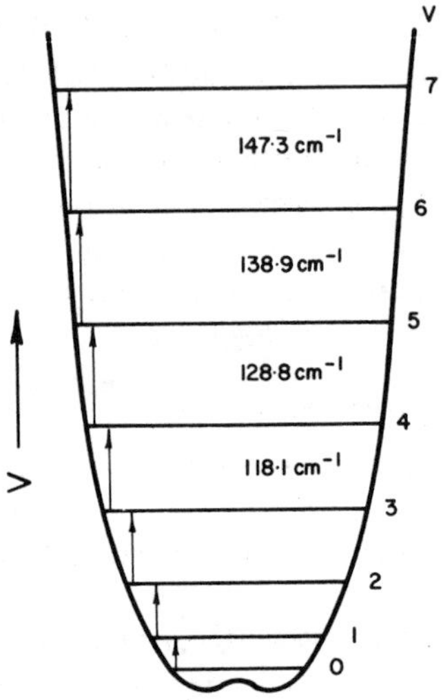

Fig. 5.8 Plot of the puckering potential against the puckering angle for trimethylene oxide. The vibrational levels for the ring puckering mode and some of the observed transitions have been inserted.

potential and the minimum of the quartic potential overlap, then a double minimum occurs, as is the case for trimethylene oxide in Fig. 5.8. When, however, the minima of the two terms overlap, the resultant potential energy curve is quadratic in form and has no double minimum.

The potential function of NH_3 is also a symmetric double minimum one [see Fig. 3.23(b)]. In this case the central maximum is above the minima of the potential energy curve by 24 kJ mol^{-1} (2070 cm^{-1}). In this case the selection rules for the vibrational transitions are $s \leftrightarrow a$ and less rigorously $\Delta v = 0, \pm 1$.

The infrared spectrum of formamide (H_2N—CHO) [5.13] yields bands at 195, 318, and 348 cm^{-1}. The likely structure of formamide is where the NCHO atoms lie in a plane with the two hydrogen atoms of the amino group slightly above or below this plane. The three bands could be attributed either to torsion around the C—N bond or inversion, or a mixture of the two types of motion.

5.4 INTERMOLECULAR VIBRATIONS AND THE STUDY OF MOLECULAR INTERACTION

From the point of view of the chemist, far-infrared studies of intermolecular modes offer information on association, in particular, on charge transfer complexes, and are a means of gaining further insight into hydrogen bonding. Far-infrared absorption may be employed to study some of these problems since, for example, in hydrogen bonding systems such as A—H· · ·Y one frequency of vibration is that of the whole unit AH and Y along the line of the hydrogen bond. The force constant of the hydrogen bond is of course weak, and the masses of AH and Y may be large; it is therefore understandable that such a stretching mode may fall in the far-infrared region.

As a consequence of the association the number of vibrational frequencies may increase. For example, when a non-linear molecule A (containing N_1 atoms) interacts with a non-linear molecule D (containing N_2 atoms) to form the associated species AD, then up to six new vibrational modes may be created, since AD now has up to $3(N_1 + N_2) - 6$ normal modes, whereas the unassociated A and D have $(3N_1 - 6)$ and $(3N_2 - 6)$ modes respectively. In practice, as a result of symmetry, the number of additional vibrational modes may be fewer than six.

The interaction of iodine with γ-picoline has been the subject of a number of interesting studies by spectroscopic and crystal structure approaches. The interaction between these two compounds is far more complicated than would be expected, and far-infrared studies assist in the elucidation of part of this complexity [5.14]. If equimolecular proportions of γ-picoline (γ-Pic) and iodine are mixed together in n-hexane:

$$\gamma\text{-Pic} + I_2 \rightleftharpoons \gamma\text{-Pic}\, I_2 \tag{5.12}$$

where the equilibrium strongly favours the charge transfer complex. However, in polar solvents partial ionization of the complex takes place:

$$2\gamma\text{-Pic}\, I_2 \rightleftharpoons (\gamma\text{-Pic}_2\, I)^+ + I_3^- \tag{5.13}$$

where the cation is the iodide-γ-picolinium ion. When I_2 is added to excess γ-picoline in an inert solvent, a solid is eventually precipitated. Before precipitation occurs, the vibrational spectra in the 500–1100 cm^{-1} region undergo a number of changes. A band at 1011 cm^{-1} due to γ-Pic I_2 disappears, as does that at 1025 cm^{-1} due to $(\gamma\text{-Pic}_2 I)^+$; a sharp band appears at 1018 cm^{-1}, and later a band at 605 cm^{-1} is also observed. The 1018 cm^{-1} band is still present in solutions which yield the 605 cm^{-1} band and is consequently not due to a species which is a forerunner of the 605 cm^{-1} product. Obviously reactions (5.12) and (5.13) are inadequate in accounting for the species present, when the proportions of γ-picoline and iodine are no longer equimolecular. The far-infrared spectrum assists in following these changes in that a strong band at 137 cm^{-1} may be ascribed to the I_3^- ion, while a band at ~ 160 cm^{-1} may be attributed to both the I—I stretching of the un-ionized complex (γ-pic I_2) and the antisymmetric (N—I—N) stretching frequency of $(\gamma\text{-Pic}\, I_2)^+$. When the iodine concentration is increased, the bands at ~ 160 cm^{-1} disappear, and the lower frequency range (100–300 cm^{-1}) becomes devoid of all bands except a weak band at ~ 210 cm^{-1}. Hence, the inadequacy of reactions (5.12) and (5.13) to account for the spectral changes is demonstrated. Thus, in the study of the interactions which take place before precipitation occurs, the far-infrared frequencies help to establish the presence of certain species and confirm some essential steps in what is an involved process.

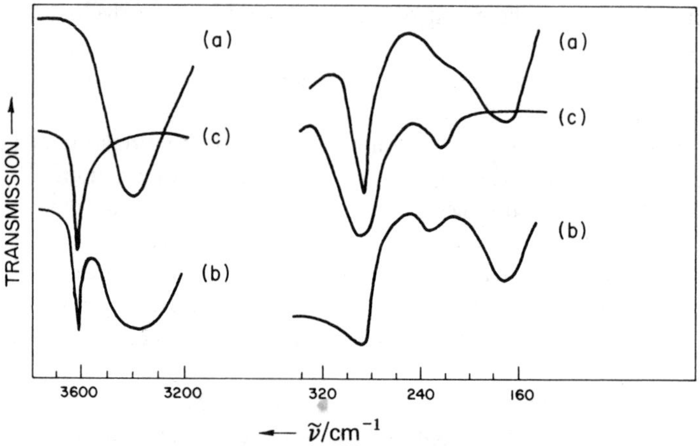

Fig. 5.9 The far-infrared spectrum of 2,6-dimethylphenol in a polyethylene matrix and the infrared spectrum in the 3500 cm^{-1} region: (a) at room temperature, (b) of a specimen first heated to 463 K then cooled to a low temperature, and (c) at 463 K (after Jakobsen, Brasch, and Mikawa [5.15]).

Many far-infrared studies have been made of the stretching mode in the hydrogen bonded species such as OH· · ·O or NH· · ·N where one molecule vibrates against the other. Often the species to be studied are dissolved in a solvent containing the solute and polyethylene. The solvent is slowly evaporated, a polyethylene film is produced, and the solute is trapped in the matrix. To illustrate the principle we shall consider a specimen of 2,6-dimethylphenol in a polyethylene matrix as reported by Jakobsen, Brasch, and Mikawa [5.15] and compare the infrared and far-infrared spectra where the work in the former region had previously clearly established when the hydrogen bonding takes place. The far-infrared spectrum is given in Fig. 5.9, together with the infrared spectrum in the O><H stretching frequency region. In Fig. 5.9 where measurements were made at room temperature the broad band at ~3400 cm^{-1} is recognized as being due to the O><H stretching frequency in the intermolecularly hydrogen bonded form O—H· · ·O, while the band at ~3600 cm^{-1} may be attributed to the O><H in the monomer itself. When the matrix is heated to 463 K, the band at ~3400 cm^{-1} disappears [Fig. 5.9(c)] and when the matrix is cooled to room temperature, only the monomer is detectable. However, on being cooled further to a low temperature the spectrum in Fig. 5.9(b) indicates the reappearance of the associated form. When the spectrum in the region 160–320 cm^{-1} is compared for each of these three stages, it becomes apparent that the band at ~160 cm^{-1} which disappears after heating the matrix to 463 K corresponds to the hydrogen bonded form and is, in fact, to be attributed to the vibration of $(ROH)_2$, i.e., of the two 2,6-dimethylphenol molecules along the hydrogen bond.

Some more detailed studies have been made of the intermolecular stretching vibration of hydrogen bonded species in solution. For example, Ginn and Wood

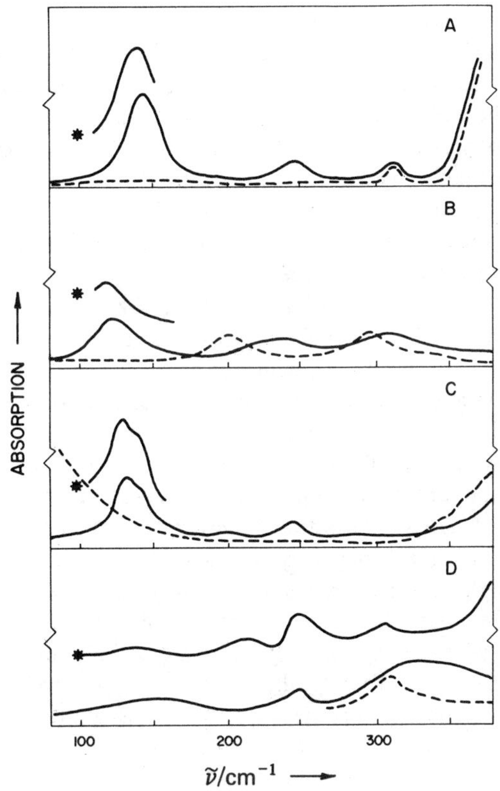

Fig. 5.10 The far-infrared spectra of mixtures of phenol with three amines. (A) ——, ~2M PhOH + ~5M Me_3N in CCl_4; *——, ~2M PhOD + ~7M Me_3N in CCl_4; – – –, ~5M Me_3N in CCl_4. (B) ——, 2.8M PhOH in Et_3N; *——, ~3M PhOD in Et_3N; – – –, liquid Et_3N. (C) ——, 2.8M PhOH in pyridine; *——, 3.5M PhOD in pyridine; – – –, liquid pyridine; (D) ——, 2.7M PhOH in CCl_4; *——, 2.8M PhOD in CCl_4; – – –, liquid CCl_4. (After Ginn and Wood [5.16]).

examined the intermolecular stretching vibrations of the type $\tilde{\nu}_{OH><N}$ and obtained bands at (a) 143 cm^{-1} for phenol–trimethylamine, (b) 123 cm^{-1} for phenol–triethylamine, and (c) 134 cm^{-1} for phenol–pyridine [5.16]. On deuteration of the O–H group the bands shifted to 141, 120, and 130 cm^{-1} respectively [5.16]; this small shift in frequency on deuteration is what would be expected in the ROH><N type of vibration since the substitution of D for H in the O–H group will not substantially alter the mass of the vibrating molecule. As was indicated earlier, the formation of the complex may lead to an addition of up to six new vibrational modes, and in these cases they can be described approximately as one stretching, four bending, and one torsional mode. The spectra obtained are represented in Fig. 5.10. It will be noted that the solutions of phenol in each of the three amines all show a medium or strong band below

150 cm^{-1}. This band is absent in the amine-plus-solvent case as is illustrated in Fig. 5.10(a) by the broken line. The band below 150 cm^{-1} was assigned to the $\tilde{\nu}_{OH><N}$, and the force constant for this vibration was evaluated for each of the cases. As would be expected (since they follow the same potential curve) the deuterated species yielded force constants with no significant variation.

The article by Jakobsen et al. [5.15] reviews "the past results and future prospects of the far-infrared studies of hydrogen bonding". They conclude: (a) There are reasonable assignments of the H-bond stretching mode and some other H-bond modes. (b) There is a reasonable picture of the modes of vibration of cyclic dimers. (c) H-bond vibrations are not localized (i.e., the whole ROH unit is involved in the vibration and not just the O—H· · ·O unit). (d) There is no direct relationship between the frequency of $\tilde{\nu}_{OH><O}$ and the $O><H$ stretching frequency. (e) A relationship between the frequency of $\tilde{\nu}_{OH><O}$ and the hydrogen bond strength has been established.

This useful summary should be consulted and serves to indicate the successes and potential of the far-infrared approach to the study of hydrogen bonding. They list the future needs in far-infrared hydrogen bond studies: (1) There is a need for corresponding low-frequency Raman data. (2) It is desirable to determine $\tilde{\nu}_{O><H}$ along with the corresponding $\tilde{\nu}_{OH><O}$ which would help in interpreting the low-frequency data as in for example the 2,6-dimethylphenol example and also when comparing $\tilde{\nu}_{OH><O}$ in a series of compounds. (3) Hydrogen bonded polymers, e.g. $(ROH)_n$ where $n > 2$, are not as well understood as H-bonded cyclic dimers. (4) There is a need for the assignment of H-bond deformations as well as H-bond stretching frequencies to obtain good thermodynamic functions which are required to characterize the hydrogen bond. To achieve this the hydrogen bond deformation frequencies have to be assigned, and the complete low-frequency spectrum of the hydrogen bond vibrations is required.

5.5 THE SPECTRA OF GASES, LIQUIDS, AND SOLIDS

5.5.1 Rotational spectra of gases

The molecules whose pure rotational spectra are studied in this region at low pressures normally have a permanent electric dipole moment (see Chapter 3).

In Chapter 3 it can be seen that the frequencies absorbed for pure rotational changes in the spectra of gaseous linear molecules are dependent on the inverse ratio of the moment of inertia which, in fact, determines whether the rotational transitions lie in the microwave region or the far-infrared, or are partly in both.

Two cases are important for such studies in the far-infrared region: (i) cases where the moment of inertia is small, such as when a light atom is chemically bound to a much heavier one (e.g., H—F and H—Cl) and many of the rotational transitions fall in the far-infrared region; (ii) transitions between higher J values for heavier molecules which may chance to fall in this region, and whose lower transitions lie in the microwave region.

A number of light diatomic and simple polyatomic molecules exhibit pure rotational changes in the 10–100 cm^{-1} region; examples are HCN, NO, CO, and H_2S_2. In many cases, though, better structural information is obtained from the microwave absorption spectra where the resolution and the accuracy of the measured frequency are greater. Once the absorption frequencies have been determined from the far-infrared spectrum, the procedure and equations employed for obtaining the moments of inertia and intermolecular distances are the same as used in the microwave region (see Chapter 3). Thus, for linear molecules the wavenumbers of the absorption lines are given by:

$$\tilde{\nu} = 2B(J+1) - 4D(J+1)^3 \tag{5.14}$$

where the rotational constant is $B = h/8\pi^2 cI$, and I, which is the moment of inertia, and the centrifugal stretching constant are evaluated for the observed rotational lines. For molecules in which B and D have been evaluated from microwave spectroscopy the accuracy is usually greater than could be achieved from the far-infrared data. Hence, for molecules where B and D are known from microwave data the formula can then be used to predict the wavenumbers of the rotational lines in the far-infrared, and these values are normally more accurate than the measured lines in the far-infrared. Hence, such (calculated) lines can be used as standards for far-infrared measurements. Suitable molecules for this purpose are CO in the 4–150 cm^{-1} region or HCN in the 4–100 cm^{-1} region.

Some symmetric-top molecules have been studied (e.g. NH_3) and the rotational lines fitted to the formula:

$$\tilde{\nu} = 2B(J+1) - D_J(J+1)^3 - 2D_{JK}(J+1)K^2 \tag{5.15}$$

where K is a second rotational quantum number whose value in $h/2\pi$ units is the component of the total angular momentum $\sqrt{[J(J+1)]}h/2\pi$ along the symmetry axis, and D_J and D_{JK} are centrifugal stretching constants. Since $K \gg J$ there are $(J+1)$ components for each value of J, and very high resolution is necessary to resolve the rotational lines. In addition, when quadrupole interaction occurs, even higher resolution is required in order to resolve them.

An asymmetric-top molecule such as water (which by definition has all its three moments of inertia different and non-zero) produces a most complex rotational spectrum which extends right across the far-infared region. The lines are of use as a secondary calibration standard.

5.5.2 Translational and rotational–translational spectra of compressed gases

Infrared and far-infrared studies of highly compressed simple gases have demonstrated that such gases have a translational spectrum. Thus, in a compressed mixture of helium and argon at room temperature a translational spectrum is obtained (see Fig. 5.11). For diatomic species such as H_2, O_2, and N_2, not only

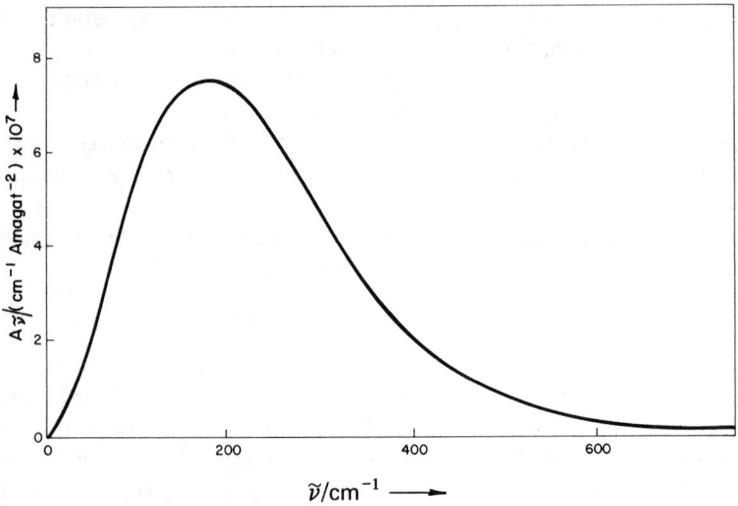

Fig. 5.11 The translational spectrum of a helium–argon mixture at room temperature at 295 K. $A_{\tilde{\nu}}$ is the absorption coefficient and is defined by Equation (5.16). (After Bosomworth and Gush [5.17]).

is there a translational contribution to the absorption in this region but a rotational translational one as well. H_2 with the smallest moment of inertia is the best case to consider for such contributions with a view to resolving these contributions into their component parts.

When two H_2 molecules collide, the induced dipole moment may be resolved into the sum of two parts: (i) the overlap dipole moment which results from short-range overlap forces; (ii) the dipole moment which results from the polarization of one molecule by the quadrupole field of the other.

Two types of absorption then become feasible; one is a pure translational change in which the rotational energy of the individual molecules remains unchanged; the other type is where rotational–translational transitions occur in which the colliding molecules change their rotational as well as their translational state, and these transitions obey the selection rule $\Delta J = 0, -2$, and $+2$, where J is the rotational quantum number corresponding to lines in the O, Q, and S branches respectively.

The absorption coefficient of a gas ($A_{\tilde{\nu}}$) is defined as:

$$A_{\tilde{\nu}} = (1/\rho^2 l)\ln(I_0/I) \qquad (5.16)$$

where I_0 is the intensity of the radiation transmitted by an empty cell at wavenumber $\tilde{\nu}$, I is the intensity of radiation transmitted by the cell filled with a gas of density ρ at the same wavenumber, and l is the path length of the cell.

The absorption coefficient of H_2 as a function of wavenumber over the 0–1400 cm^{-1} region is given in Fig. 5.12(a) [5.17]. The intensity of the estimated translational branch is readily distinguished from the rotational branch

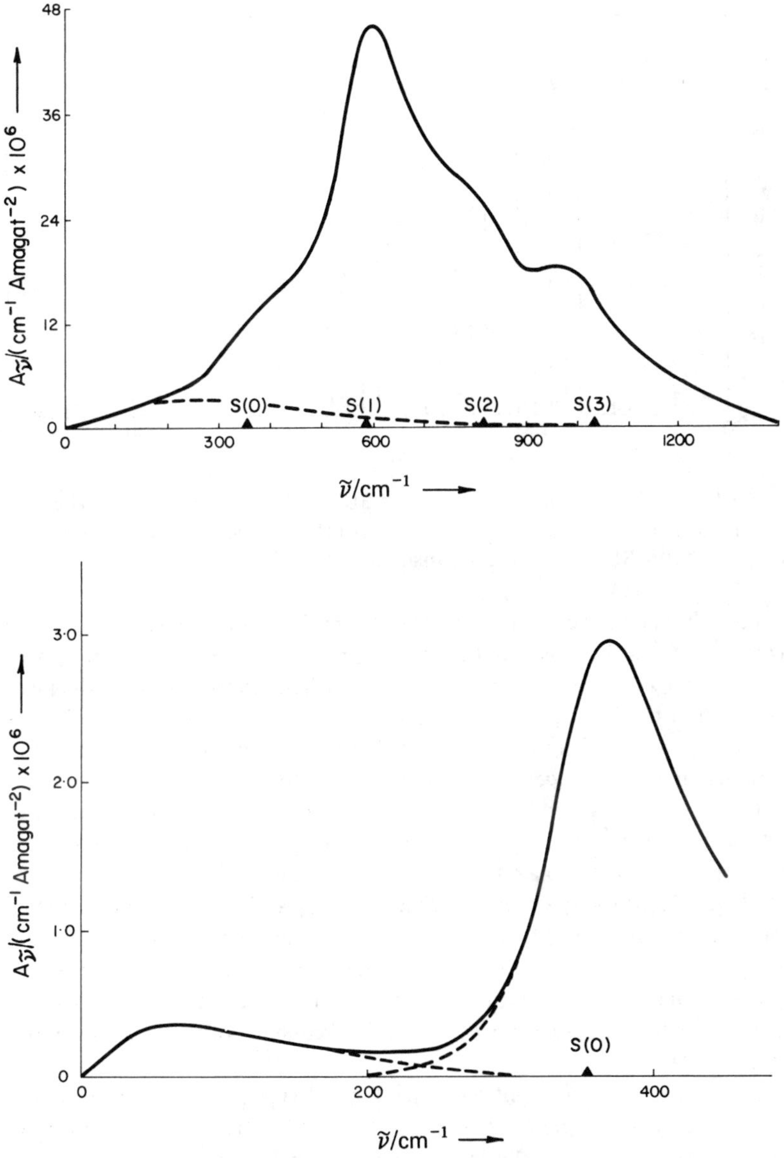

Fig. 5.12 The absorption spectrum of compressed H_2 at 300 K between 0 and 1400 cm^{-1}. The broken line indicates the separated pure translational component. (b) The absorption spectrum of compressed H_2 at 77.3 K between ~0 and 450 cm^{-1} showing the separated translational component (After Bosomworth and Gush [5.17]).

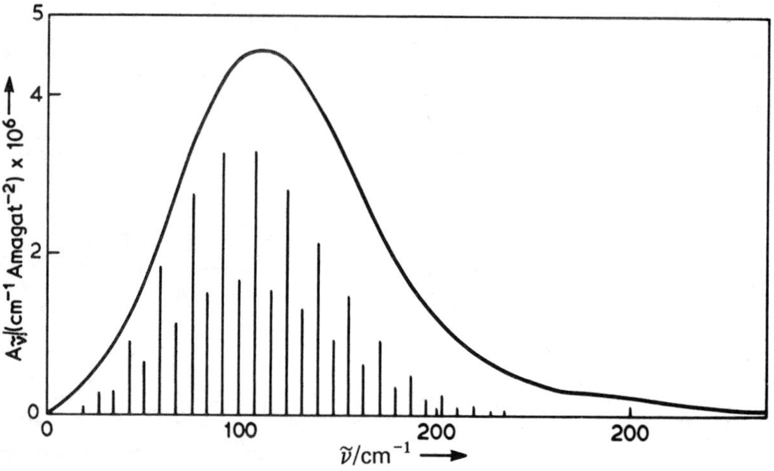

Fig. 5.13 The absorption profile of N_2 at 300 K. The vertical bars on the abscissa axis represent the calculated unbroadened rotational spectrum of the nitrogen molecule (After Bosomworth and Gush [5.17]).

because it lies at a lower frequency. The strong absorption peaks near 354 and 587 cm^{-1} are identified as the $S(0)$ and $S(1)$ rotational transitions corresponding to $J'' = 0$ and 1 in the S-branch (i.e. $\Delta J = +2$). The wavenumbers at the peak absorption of the $S(2)$ and $S(3)$ lines have also been noted. Thus, as is obvious from the figure, the absorption between 0 and 1400 cm^{-1} is predominantly rotational, and only between ~0 and 300 cm^{-1} does it appreciably influence the contour of the absorption band. Attempts were made to calculate the profile of the band. One approach is to assume that the band is the sum of a translational component and a rotational–translational component.

In Fig. 5.12(b) the spectrum of compressed H_2 at 77.3 K is given, and in contrast to the spectrum at 300 K the band is much smaller in width and there is less overlap of the rotational and the translational component. The larger peak intensity corresponds to the $S(0)$ rotational line and is located at a frequency of ~371 cm^{-1} which is 17 cm^{-1} higher than the frequency of the $S(0)$ line in the free molecule.

The far-infrared band of nitrogen (see Fig. 5.13) is quite different from the shape of the hydrogen band, since the moment of inertia of N_2 is appreciably larger than that of H_2, and this leads to a much smaller rotational frequency. Consequently, the spacing of the former lines is much narrower and the rotational lines are not resolved as they are for H_2, and the rotational–translational spectrum of N_2 has the appearance of a continuous broad absorption band (see Fig. 5.13) with a peak at ~100 cm^{-1}. From the spectrum the quadrupole moment of nitrogen can be deduced. In Fig. 5.13 the vertical bars in the abscissa axis represent the calculated unbroadened rotational spectrum of the nitrogen molecule, i.e., indicate the rotational spectrum of N_2 in the absence of broadening

The far-infrared band resulted from dipole moments which arise from quadrupole induction and overlap induction mechanisms, and the intensities of the lines – represented by the vertical bar – were calculated from intensity formulae appropriate to quadrupole induced spectra. Each rotational line is broadened by the accompanying translational transitions, and the observed spectrum follows approximately the envelope of the unbroadened spectrum. However, since the spacing of the rotational lines is small and the translational and rotational branches overlap completely, a broad band results whose half band width is of a similar order to that found in the broad band absorption exhibited by polar liquids in the far-infrared region.

5.5.3 Absorption by liquids

The rotational motion of simple gaseous molecules has been studied in detail in both the microwave and far-infrared region at low pressures and is thoroughly understood. However, the type of motion exhibited by molecules in the liquid state is less well understood except in the case of H_2 and its isotopes which are known to have quantized rotational levels in the liquid state as has H_2 in liquid argon. In the latter case the absorption between the quantized rotational levels occurs as a result of the molecular quadrupole moment of H_2 which induces a dipole in the surrounding environment. This dipole is modulated as H_2 undergoes a rotational–translational motion in the liquid state, and transitions become feasible in the far-infrared region. In the case of H_2 the rotational motion may be regarded as free rotation in the liquid state. This is exceptional though, and comes about through (i) the low moment of inertia which leads to a wide separation between the lowest rotational levels of 120 cm^{-1}, and (ii) the energy barrier to rotation for H_2 in liquid argon which is less than 25 cm^{-1} (293 kJ mol^{-1}), that is considerably less than the spacing between the rotational levels. In fact, there is so little restriction to rotation that the frequencies for the $J = 2 \leftarrow J = 0$ and $J = 3 \leftarrow J = 1$ transitions agree within the experimental error with those for the unperturbed molecule. The rotational–translational spectrum of 1 mole per cent of H_2 in liquid argon at 87 K is given in Fig. 5.14 [5.18]. The energy level diagram above the spectrum indicates the pure rotational transitions of H_2. The continuous curve is the experimental absorption, while the broken curve shows the contribution of translational fine structure to the rotational transitions. It will be observed that the rotational transitions in Fig. 5.14 are at higher frequencies than the far-infrared region as defined.

In general, though, the far-infrared absorption bands exhibited by liquids do not show any indication of a rotational profile; instead only a very broad band is obtained. In fact, it appears likely that all pure liquids exhibit just one broad absorption band in the far-infrared region unless the molecule happens to have a low-frequency intramolecular vibration or strong molecular interaction introduces associated species.

In Fig. 5.15 the far-infrared spectra of five organic compounds are given. In

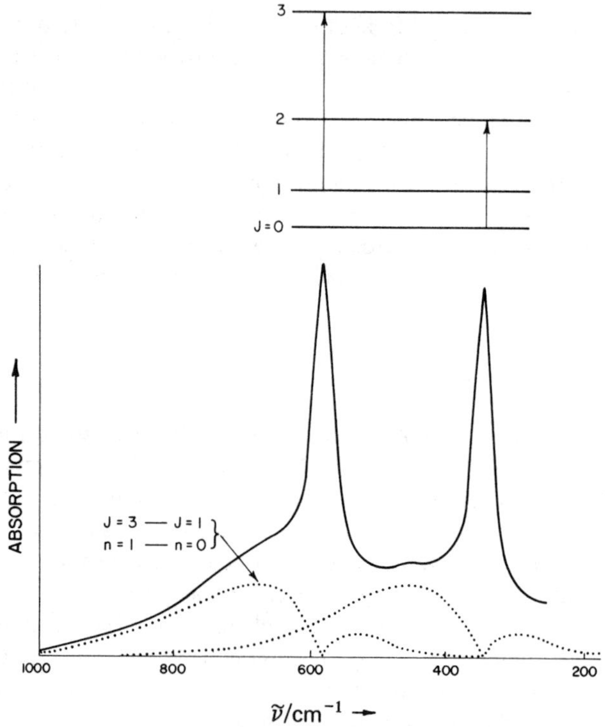

Fig. 5.14 The rotational–translational spectrum of 1 mole per cent of H_2 in liquid argon at 87 K. The continuous curve gives the measured absorption while the broken curve shows the contribution of translational fine structure to the rotational transitions. (After Holeman and Ewing [5.18]).

general each liquid whose far-infrared spectrum has been reported exhibits a broad band with its maximum somewhere in ~10–120 cm^{-1} region. This type of band has a half band width between about 45 and 90 cm^{-1}, a typical value being about 75 cm^{-1}.

In the gaseous state at low pressures the far-infrared absorption arises from rotation of the molecules, and in the non-rotator solid phase the absorption is usually ascribed to lattice vibrations. In the liquid state they have met various interpretations including (i) librational motion of the polar molecule in a cage [5.20], and (ii) rotational and translational motions of the polar molecules. Further, since the absorption maxima, both in the liquid and the solid state, have been found to occur in the same spectral region, these absorptions have also been termed 'liquid lattice' vibrations [5.21].

Which of these theories is correct has not been settled yet in that so far no theory can quantitatively account for the line shape, the half band width, the maximum value of the absorption coefficient, and the area under the broad band. A variety of studies has been made to attempt to account for the precise nature of the absorption [5.22–5.27].

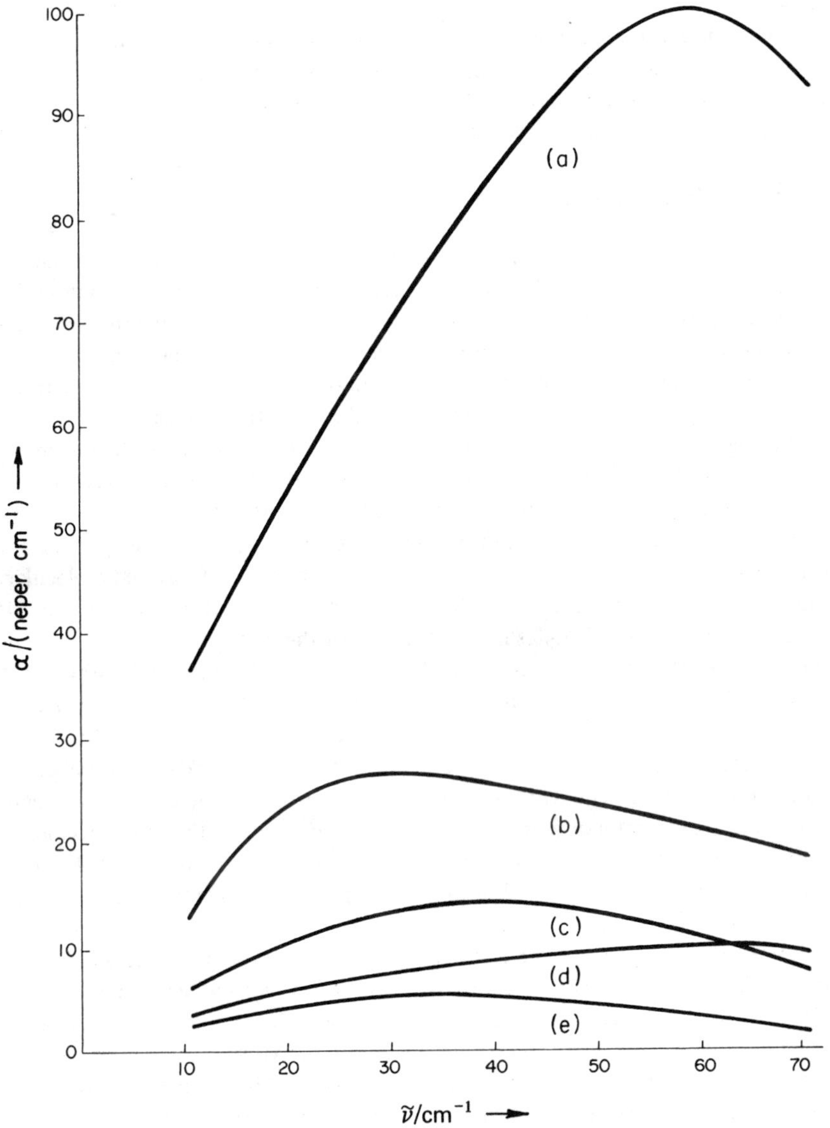

Fig. 5.15 Far-infrared spectra of some polar liquids at room temperature: (a) methyl iodide; (b) fluorobenzene; (c) iodobenzene; (d) 1-fluoronaphthalene; (e) 1-iodonaphthalene. (After Jain and Walker [5.19]).

The process responsible for this broad band absorption of non-polar liquids has been considered by many workers, and its mechanism satisfactorily explained. The absorption is attributed to the presence of induced dipole moments of the molecules which do not cancel because of long range order. In some cases

Garg et al. [5.28] have accounted for the intensities of the absorption bands in non-polar liquids on the basis of multipole-induced dipole absorption.

It may be concluded that far-infrared studies (and dielectric absorption work in the microwave region) have led to much closer insight into the motions and interactions of molecules in the liquid state.

5.5.4 Studies on (mainly) solids

If a crystal is composed only of atoms (e.g. argon), the only vibrational modes possible are lattice modes. However, for molecular crystals where groupings of molecules exist, two types of vibrations are feasible: (a) vibrational modes arising from the movement of atoms within a molecule, which are termed internal modes; (b) vibrational modes where one group of molecules moves relative to another group, which are termed external modes or lattice modes.

Normally the lattice modes are found at lower frequencies than the internal modes, since the intermolecular forces are weaker than the valence forces, and much heavier masses are involved for the molecules within their group.

For a crystal which has q atoms in each molecule and p molecules per unit cell, the degrees of freedom are as follows: (a) the number of internal molecular modes is $p(3q - 6)$; (b) rotating lattice modes $3p$; (c) translatory modes $(3p - 3)$; (d) the total number of modes is then the sum of these which is $(3pq - 3)$. Very roughly, the internal molecular modes are to be expected in the 200–4000 cm^{-1} range, while the rotary are in the 100–300 cm^{-1} range; the translatory lattice modes may lie below 100 cm^{-1}. To gain an appreciation of the number of rotating lattice and translatory modes possible it is desirable to know the number of molecules in the unit cell. It must be stressed, though, that the division between internal and external modes in a crystal is rather arbitrary, although convenient, since the associated energy levels belong to the crystal as a whole. Thus, strictly it is incorrect to speak of the vibrational energy of a molecule in a solid.

Brot et al. [5.29] have examined the far-infrared spectrum of tert-butyl chloride in the liquid state and in three crystalline states. Before this is considered further it is necessary to note the types of motions which can be exhibited by these spherical molecules in the solid state. Such molecules, which are roughly spherical in shape and on rotation about an axis sweep out a sphere, are known as globular molecules. Many such molecules have been studied by specific-heat measurements and their different phases in the solid state established. Phase I, the crystalline phase, lies below the melting point and above the first transition point, and successive solid phases which occur on a lowering of the temperature are known as phases II, III, etc. Globular molecular crystals in phase I are known as plastic crystals because of the unusual freedom of motion exhibited by the molecules, and generally they have a highly symmetric cubic packing lattice.

The results obtained by Brot et al. [5.29] for the far-infrared absorption spectrum of tert-butyl chloride are given in Fig. 5.21. In the liquid phase the

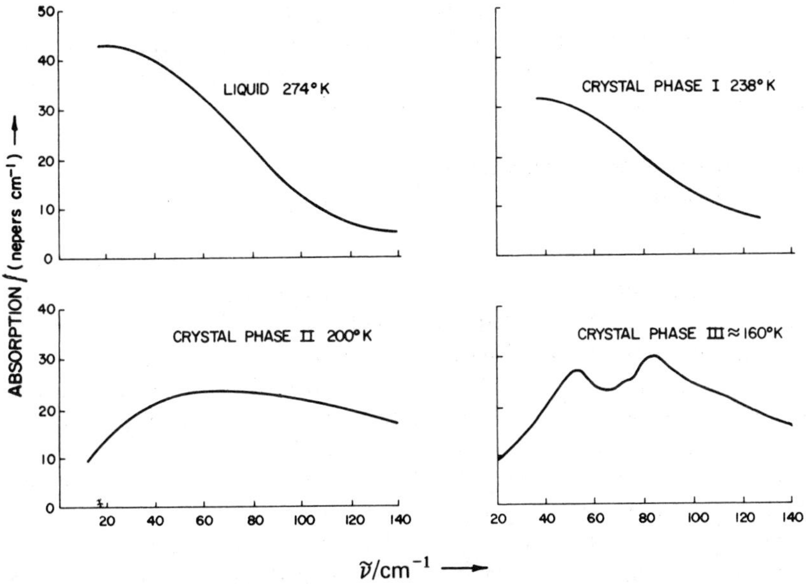

Fig. 5.21 Far-infrared spectrum of liquid and crystalline tert-butyl chloride at different temperatures. (After Brot, Lassier, Chantry and Gebbie [5.29]).

characteristic broad band is observed, although only the peak on the high-frequency side of the broad band is apparent. However, measurement at lower frequencies confirms that this is typical broad band absorption as is observed for polar liquids. In phase I at 238 K the crystals exhibit a similar shaped absorption curve to that of the liquid phase, except that the α_{max} has been reduced still further and has shifted to a higher frequency. Both phases I and II yield an absorption curve which in appearance is similar to that associated with liquids. However, at ~160 K in phase III the spectrum changes in form and yields moderately sharp bands at 52 and 83 cm^{-1}. The appearance of such bands appears to be typical of the behaviour of molecular crystals. An earlier study of tert-butyl chloride at these different temperatures had been made using a pure quadrupole magnetic resonance approach, and this had shown that in phases I and II there was some type of molecular motion. In phase I the absorption was interpreted as being translational plus rotational freedom, in phase II rotational freedom preferably about the C—Cl bond, whereas at the lowest temperature (phase III) the freedom was restricted to rotation of methyl groups within a specific molecule.

It seems that these modifications in the types of motion exhibited by tert-butyl chloride are borne out in the far-infrared spectrum. In addition, the types of motion used to explain the spectrum seem in harmony with the rotational–translational motion invoked to explain the broad band absorption within liquids.

The far-infrared absorption appears promising and important as a means of studying the molecular motions present in the solid state, and comprehensive work is required on how band shapes, band areas, and $\tilde{\nu}_{max}$ values alter with temperature variations for a wide variety of crystals.

5.5.5 Encaged molecule motion in clathrates

Davies [5.30] examined the far-infrared spectra of some different gases trapped in cages of β-quinol clathrates. The guest molecules which were trapped in the clathrate were N_2, CO_2, HCl, HBr, SO_2, and HCN, and their absorption was examined in the 10–100 cm^{-1} region. These systems may be looked upon as having a quasi-spherical orientation of solvent molecules surrounding individual solute molecules where the space left in the centre of the cage for the guest molecule is roughly a sphere of radius about 2.25 Å.

An earlier study of HCl trapped in β-quinol clathrate at 1.2 K had indicated three prominent features: (i) a β-quinol lattice vibration at 66 cm^{-1}, (ii) a translational vibration of the HCl near 52 cm^{-1}, and (iii) a rotational mode of HCl at ~20 cm^{-1}. Coupling between rotational and translational vibration was invoked as well as dipole–dipole interaction. However, Davies considered that the nature of the so-called translational motion of guest molecules in β-quinol clathrates still needed to be established.

REFERENCES

5.1 Czerny, M., *Z. Phys.*, **34**, 227 (1925); **44**, 235 (1927).

5.2 Wilkinson, G.R., Inglis, S.A., and Smart, C., in *Spectroscopy*, Ed. Wells, M.J., p. 157, Institute of Petroleum, London (1962).

5.3 Wood, J.L., *Quart. Rev. Chem. Soc.*, **17**, 362 (1963).

5.4 Brasch, J.W., Mikawa, Y., and Jakobsen, R.J., in *Applied Spectroscopy Reviews, Vol. I.*, Dekker, New York (1968).

5.5 *Molecular Spectroscopy*, Ed. Hepple, P., Institute of Petroleum, London (1968).

5.6 *Chemical Applications of Far-Infrared Spectroscopy*, Finch, A., Gates, P.N., Radcliffe, K., Dickson, F.N. and Bentley, F.F., Academic Press, New York, (1970).

5.7 *Submillimetre Spectroscopy*, Chantry, G.W., Academic Press, New York, (1971).

5.8 Miller, F.A., in *Molecular Spectroscopy*, Ed. Hepple, P., p.5, Institute of Petroleum, London (1968).

5.9 Adams, D.M., in *Molecular Spectroscopy*, Ed. Hepple, P., p.29, Institute of Petroleum, London (1968).

5.10 Miller, F.A., Fateley, W.G., and Witkowski, R., *Spectrochimica Acta*, **23A**, 891 (1967).

5.11 Packer, J. and Vaughan, J., *Modern Approach to Organic Chemistry*, Oxford University Press (1953).

5.12 Lukin, M.S., Powell, J. and Shaw, B.L., *J. Chem. Soc. A*, 1410 (1966).

5.13 Druce, P.M., Lappert, M.F., and Riley, P.N.K., *Chem Commun.*, 486 (1967).

5.14 Hague, I. and Wood, J.L., *J. Phys. Chem.*, **72**, 2438 (1968).

5.15 Jakobsen, R.J., Brasch, J.W. and Mikawa, Y., *Applied Spectroscopy*, **22**, 641 (1968).

5.16 Ginn, S.G.W. and Wood, J.L., *Spectrochimica Acta*, **23A**, 611 (1967).

5.17 Bosomworth, D.R. and Gush, H.P., *Canad. J. Phys.*, **43**, 751 (1965).

5.18 Holeman, G. and Ewing, G., *J. Chem. Phys.*, **47**, 571 (1967).

5.19 Jain, S.R. and Walker, S., *J. Phys. Chem.*, **75**, 2942 (1971).

5.20 Hill, N.E., *Proc. Phys. Soc.*, **82**, 723 (1963).

5.21 Chantry, G.W., Gebbie, H.A., Lassier, B. and Wyllie, G., *Nature*, **214**, 163 (1967).

5.22 Benson, M., Martin, G.D., Walker, S. and Warren, J., *Canad, J. Chem.*, **50**, 2610 (1972).

5.23 Kroon, G. and van der Elsken, J., *J. Chem. Phys. Lett.*, **1**, 285 (1967).

5.24 Gordon, R.G., *J. Chem. Phys.*, **38**, 1724 (1963).

5.25 Higasi, K., Minami, R., Takahasi, H., and Ohno, A., *J. Chem Soc. Faraday II*, **69**, 1579 (1973).

5.26 McLellan, C.K. and Walker, S., *J. Chem. Phys.*, **61**, 2412 (1974).

5.27 Hindle, P., Walker, S., and Warren, J., *J. Chem. Soc. Faraday II*, **71**, (1975); *J. Chem. Phys.*, **62** (1975).

5.28 Garg, S.K., Bertie, J.E., Kilp, H., and Smyth, C.R., *J. Chem. Phys.*, **49**, 2551 (1968).

5.29 Brot, C., Lassier, B., Chantry, G.W., Gebbie, H.A., *Spectrochimica Acta*, **24A**, 295 (1968).

5.30 Davies, P.R., *Trans. Faraday Soc.*, **66**, 181 (1970).

6 Force constants

6.1 INTRODUCTION

It has been shown in earlier chapters that infrared and Raman spectroscopy can give information concerning the fundamental vibrational frequencies of a molecule. These frequencies are determined by the masses of the atoms involved in the vibration and the force constants of the molecular 'springs' holding together the constituent atoms. What often particularly interests the chemist are the values of these force constants since they should yield basic information concerning the way atoms are tied together in the molecule. This chapter describes how these force constants are obtained from the experimental data. The enquiring reader who wishes to read further is referred to the excellent works by Herzberg [6.1], by Wilson, Decius, and Cross [6.2], and the very readable volume by Woodward [6.3].

6.2 THE MOLECULE AS A SET OF HARMONIC OSCILLATORS

For the diatomic molecule AB, if harmonic forces are assumed to exist between the two nuclei (i.e. constrain the molecular spring between the two atoms to obey Hooke's law), then the vibrational potential energy V may be written as:

$$2V(r) = f_{\mathrm{AB}}(r - r_{\mathrm{e}})^2 \tag{6.1}$$

where r is the AB bond length (equal to r_{e} at equilibrium) and f_{AB} is the force constant of the AB bond. By putting $(r - r_{\mathrm{e}}) = q_{\mathrm{AB}}$ then:

$$2V = f_{\mathrm{AB}}q_{\mathrm{AB}}^2 \tag{6.2}$$

The vibrational kinetic energy may similarly be written as:

$$2T = \mu_{AB}\dot{q}_{AB}^{2} \tag{6.3}$$

where μ_{AB} is the reduced mass of the AB molecule:

$$\mu_{AB} = m_A m_B/(m_A + m_B)$$

The vibrational frequency of the oscillating molecule may be simply found by recourse to Lagrange's equation of motion:

$$\frac{\partial}{\partial t}\left(\frac{\partial T}{\partial \dot{q}}\right) + \frac{\partial V}{\partial q} = 0 \tag{6.4}$$

when, after a little algebra:

$$\mu_{AB} q_{AB} + f_{AB} q_{AB} = 0 \tag{6.5}$$

If it is assumed that q_{AB} = const × $\cos(2\pi\nu t)$ (i.e. that the stretching and contracting of the AB bond is periodic with a frequency ν), then:

$$(f_{AB} - 4\pi^2\nu^2\mu_{AB})q_{AB} = 0 \tag{6.6}$$

and the vibrational frequency:

$$\nu = \frac{1}{2\pi}\left(\frac{f_{AB}}{\mu_{AB}}\right)^{1/2} \tag{6.7}$$

In this discussion of the diatomic molecule it has been convenient to describe the vibrational motion of the two atoms in terms of an *internal coordinate,* namely the change in the internuclear distance q_{AB}, rather than the considerably more complex set of Cartesian (x, y, z) coordinates of each atom. On moving to the more involved situation in polyatomic systems it is also found useful to describe the molecular vibrations, both mathematically and visually, in terms of changes in the internal coordinates of the molecule. These may be the distances (bonded or non-bonded) between two atoms in the molecule, the bond angles or dihedral angles, or, in fact, any interatomic distance or bond angle which serves to define the molecular configuration. If the molecule possesses elements of symmetry it is often useful to consider the molecular motions in terms of *symmetry coordinates.* These are simply linear combinations of internal coordinates, chosen in such a way as to reflect the symmetry of the molecule. For example, in the water molecule three valence internal coordinates may be envisaged: changes in the lengths of the two OH bonds r_1 and r_2 (OH stretching) and in the HOH angle (HOH bending). Under the C_{2v} point group these internal coordinates may be expressed in terms of three symmetry coordinates, two of species a_1 and one of species b_2:

$$a_1{:}s_1 = 2^{-1/2}(\Delta r_1 + \Delta r_2)$$

$$s_2 = \Delta\alpha$$

$$b_2{:}s_3 = 2^{-1/2}(\Delta r_1 - \Delta r_2)$$

where the vibrations are distinguished depending upon whether they are symmetric or antisymmetric with respect to the two-fold rotation axis or the plane of symmetry which bisects the HOH bond angle. In general, the symmetry coordinates either describe bond stretching or bond angle bending. Thus in the water molecule considered above no symmetry coordinate exists containing r_1, r_2, and α. This is simply a consequence of the fact that, whereas r_1 and r_2 are valence parameters related to each other by one or more symmetry elements of the molecule, no symmetry element relates the diverse parameters of bond length and bond angle. It will be shown later that, provided that bond stretching and bond bending transform as the same irreducible representation of the molecular point group, they may mix together to produce a vibrational motion of mixed stretching and bending character.

In general the potential energy may be written (assuming that the molecule is composed of a series of harmonic oscillators) as:

$$2V = \sum_{i,j} f_{ij} q_i q_j \tag{6.8}$$

where the q are some generalized coordinates which may be internal coordinates or linear combinations of these in the form of symmetry coordinates as just described. The f_{ij} $(i = j)$ are termed diagonal force constants (for reasons to become apparent later) and $\frac{1}{2} f_{jj}$ represents the change in potential energy when coordinate j is extended by one unit. The off-diagonal terms f_{ij} $(i \neq j)$ are called interaction force constants and are usually smaller than the diagonal force constants. Stretching force constants generally have the units of $\mathrm{N\,m^{-1}}$, and bending force constants the units joule radian^{-1}.†

Some of the f_{ij} are equal to zero on symmetry grounds. If the molecule contains a plane of symmetry then all the q_i may be described as either in-plane or out-of-plane coordinates depending upon whether the plane is preserved or destroyed during the motion. In general, the interaction force constant f_{ij} between an in-plane coordinate and an out-of-plane coordinate is zero, as can be seen quite simply from the following argument. Displace the in-plane coordinate q_{in} by a unit amount. The interaction energy from Equation (6.8) is then given by $f_{\mathrm{in,out}}$ q_{out} where q_{out} is the out-of-plane coordinate. If q_{out} is then displaced by $+1$ unit the energy becomes $f_{\mathrm{in,out}}$, and if q_{out} is displaced by -1 unit the energy becomes $-f_{\mathrm{in,out}}$. But since these two displacements are geometrically equivalent, being identical displacements either side of the plane of symmetry, the total energy change involved must be the same for both, i.e. $f = -f$ (the only way this condition may be fulfilled is if $f = 0$). In general, if the q_i are symmetry coordinates then $f_{ij} = 0$ unless i and j belong to the same symmetry species. This simple rule derives from the fact that the total potential energy V in (6.8) must

† The older units of m dyne Å^{-1} or dyne cm^{-1} and erg radian^{-2} are related to these by 10^{-2} m dyne Å^{-1} = 10^{3} dyne cm^{-1} = N m^{-1} and 10^{7} erg rad^{-2} = joule rad^{-2}.

be a totally symmetric function and hence must belong to the highest symmetry representation of the point group. Products of q_i and q_j which are not then totally symmetric lead to a zero value of f_{ij}.

The kinetic energy may be written in similar fashion to (6.3) for the polyatomic system as:

$$T = \frac{1}{2}\sum_{i,j} g_{ij}^{-1}\, \dot{q}_i \dot{q}_j \tag{6.9}$$

where the g_{ij}^{-1} are functions of the molecular geometry and atomic masses.

The g_{ij} can be calculated from first principles but one usually uses tabulated analytic functions from which the g_{ij} may be calculated for most types of internal coordinate [6.2]. For an internal coordinate q_i representing the stretching of an AB bond, g_{ii} is simply given by the sum of the reciprocals of the masses of atoms A and B, namely:

$$g_{ii} = 1/m_{\mathrm{A}} + 1/m_{\mathrm{B}} = \mu_{\mathrm{AB}}^{-1}$$

For bending coordinates, g_{ii} is a complex function of the masses of the atoms defining the bond angle, the bond lengths and the value of the bond angle itself. The off-diagonal terms g_{ij} are evaluated similarly. For example, the term linking two stretching coordinates (e.g. the AB bond stretching and the BC bond stretching) is given by $m_{\mathrm{B}}/\cos\phi$, where ϕ is the angle between the two bonds and m_{B} the mass of the atom B common to both coordinates. The form of all these g_{ij} functions is such that the total kinetic energy in Equation (6.9) represents only vibrational energy. Thus, during any molecular motion, the mass centre of the molecule remains in the same position in space, no translation occurs, and no rotation of the molecule around this mass centre is allowed. Application of the Lagrange recipe (6.4) to Equations (6.8) and (6.9) leads to a series of equations similar to (6.5) which determine the set of vibrational frequencies ν:

$$\sum_{i,j} (f_{ij} - \lambda_j g_{ij}^{-1})q_j = 0 \tag{6.10}$$

(where $\lambda = 4\pi^2\nu^2$), which in matrix form can be written as:

$$(\boldsymbol{F} - \lambda \boldsymbol{G}^{-1})\boldsymbol{Q} = 0 \tag{6.11}$$

The condition for compliance of the set of simultaneous equations (6.10) is:

$$|\boldsymbol{F} - \lambda \boldsymbol{G}^{-1}| = 0 \quad \text{or} \quad |\boldsymbol{GF} - \lambda \boldsymbol{E}| = 0 \tag{6.12}$$

$\boldsymbol{G}$ is called the inverse kinetic energy matrix [the reason can be seen from (6.9)], $\boldsymbol{F}$ is simply a matrix containing the force constants, and $\boldsymbol{E}$ is the diagonal unit matrix. If the force constants ($\boldsymbol{F}$) are known for the molecule, then substitution in Equation (6.12) leads to the vibrational frequencies. If the latter are known, the force constants may in principle be computed (subject to the restrictions which will be considered later). On substitution of a value of λ into (6.11) the vector $\boldsymbol{Q}$ may be obtained which describes the real molecular motion corresponding to the vibrational frequency ν. In general it is found that the *normal* mode (Q_i) is made up of contributions from the internal or symmetry modes (q_j) that were chosen in Equations (6.8) and (6.9), that is:

$$Q_i = \sum_j L_{ij} q_j \tag{6.13}$$

where the summation runs over all the internal coordinates q_j or over all the symmetry coordinates which have the same symmetry description. An example will illustrate this point. FNO is a bent triatomic molecule and the vibrational potential energy may be written as:

$$2V = f_{NO}q_{NO}^2 + f_{NF}q_{NF}^2 + f_\alpha q_\alpha^2 + f_{NO,NF}q_{NO}q_{NF} + f_{NO,\alpha}q_{NO}q_\alpha + f_{NF,\alpha}q_{NF}q_\alpha \tag{6.14}$$

where q_{NO}, q_{NF}, and q_α represent changes in the three internal coordinates (the NO and NF bond lengths and the FNO angle α). $(3N - 6) = 3$ fundamental vibrations are expected for this molecule and these have been recorded at 1877, 776, and 523 cm^{-1} in the gas-phase infrared spectrum. The high-frequency band is generally assigned as NO stretching and the low-frequency bands as NF stretching and ONF bending vibrations. On solution of Equation (6.12) the values of L_{ij} shown in Table 6.1 are found. As can be seen from the first column, which describes the molecular motion corresponding to the 1877 cm^{-1} vibration,

Table 6.1 Form of the normal modes of FNO.

$\tilde{\nu}$	1876.8 cm^{-1}	775.5 cm^{-1}	522.9 cm^{-1}
L_{NO}	0.36	−0.05	−0.02
L_{NF}	−0.02	−0.27	0.23
L_α	−0.16	−0.37	0.16

this frequency primarily arises from stretching of the NO bond but contains some admixture of the FNO bending coordinate and a small amount of stretching of the NF bond. For the other two frequencies, there is only a small contribution from NO stretching but considerably greater contributions from NF stretching and ONF angle bending. The higher frequency vibration contains more bending than stretching but the situation is reversed for the lower frequency mode. Crudely then, the 1877 cm^{-1} vibration would be described as NO stretching, the 776 cm^{-1} vibration as FNO bending, and the 523 cm^{-1} vibration as NF stretching, although it can be seen that in reality all three of these internal mode descriptions are mixed together to form the normal mode.

Further information about the normal modes may be obtained by recourse to the potential-energy distribution (PED). For a given normal vibration the potential energy may be expressed as a weighted sum of the potential energy contributions from all the internal modes comprising the normal mode. For the kth mode:

$$2V = \sum_{i,j} f_{ij} L_{ik} L_{jk} \tag{6.15}$$

Generally the terms with $i = j$ are larger than the interaction terms ($i \neq j$), and indeed in Table 6.2 for FNO the dominating terms can be seen to be associated with the diagonal force constants. The PED also leads to the same conclusion as above, namely that the two low-frequency vibrations are substantial mixtures of NF stretching and FNO bending but that the 1877 cm^{-1} vibration is almost purely due to NO stretching.

Table 6.2 Potential-energy distribution in FNO†

$\tilde{\nu}$	1876.8 cm^{-1}	775.5 cm^{-1}	522.9 cm^{-1}
V_{NO}	1.00	0.13	0.04
V_{NF}	0	0.47	0.75
V_{α}	0.02	0.69	0.30
$V_{NO,NF}$	0	−0.20	−0.17
$V_{NO,\alpha}$	−0.02	0.03	−0.01
$V_{NF,\alpha}$	0	−0.11	0.09

† The columns have been normalized such that $\Sigma V = 1.00$.

6.3 TYPES OF FORCE FIELD

In the above discussion of FNO a *general quadratic valence force field* (GQVFF) has been used where the potential energy is expressed in terms of harmonic force constants associated with the valence coordinates of the molecule (i.e. bonded distances and the angles between them). Another force field of less applicability is the *general quadratic central force field* (GQCFF) where the potential energy is expressed as:

$$2V = \sum_{i,j} f_{ij} q_{ij}^2 \tag{6.16}$$

q_{ij} being the change in the distance (bonded or non-bonded) between the atoms i and j in the molecule [6.1]. One of the obvious drawbacks of the approach is that in a linear molecule such as CO_2 no force constant is included in (6.16) to describe the OCO bending motion of the molecule. The GQVFF is considerably more popular since it is more useful in correlating force constants with molecular parameters such as bond lengths and bond dissociation energies. The *Urey–Bradley force field* (UBFF) is a hybrid of the valence and central approaches [6.4]. Bond stretching is considered from a valence basis whereas the forces operative during angle bending are considered to be determined by the van der Waals repulsions between non-bonded atoms. There are usually fewer force constants in the UBFF than in the GQVFF and it is often too simple to fit the observed frequencies accurately.

In the previous section the two frequencies of FNO which are close together

(776 and 523 cm^{-1}) were found to contain a heavy admixture of NF stretching and FNO bending, but the mode at 1877 cm^{-1} was (from the PED) almost a pure NO stretching vibration. This is quite a general observation. If two molecular vibrations are widely separated in energy, the higher frequency mode will contain only small contributions from the internal coordinates which dominate the lower frequency mode, and vice versa. As the two modes get closer in energy this becomes increasingly less true and the same internal coordinates occur in the description of both vibrations. This fact has led to some simplified force fields, so-called frequency factored force fields where only the high-frequency vibrations are considered, the presence of low-frequency modes in the molecule being ignored. In transition metal carbonyls for example, the stretching vibrations of the carbonyl group occur at around 2000 cm^{-1}, whereas the vibrations described roughly as M—C stretching, M—C—O bending, etc. occur at below 500 cm^{-1}. The PED using a GQVFF has shown that the vibrations occurring around 2000 cm^{-1} are almost 100 per cent CO stretching. The *Cotton–Kraihanzel force field* (CKFF) makes use of this fact and is concerned only with the stretching vibrations of the C—O groups in the molecule [6.5]. As it turns out, it is the force constants associated with the stretching motions of the CO groups which are of prime importance in structural and bonding considerations for these species.

6.4 DETERMINATION OF *F*

After discussion of the form of the normal vibrations and the ways of expressing the potential energy, attention can now be focussed on the actual determination of the force constants F and the problems associated with it. The elements g_{ij}^{-1} in Equations (6.10) and (6.11) can be calculated, if the molecular geometry parameters, bond lengths, angles, etc. are known, by simple substitution into formulae which are readily available [6.2]. Given a set of force constants $\boldsymbol{F}$, and the $\boldsymbol{G}$ matrix so constructed, it is a trivial matter to determine the values of λ by direct solution of (6.10) or (6.11). However, this is not usually the problem facing the spectroscopist. One generally has the reverse situation, a series of experimentally determined vibrational frequencies (corrected for Fermi resonance if it occurs), and hence values of λ, and unknown force constants $\boldsymbol{F}$. The determination of $\boldsymbol{F}$ is a much more difficult proposition than determination of the frequencies, given the force constants. What is done is to make a guess at a reasonable set of force constants f_{ij} (e.g. from diatomic molecules, and chemically related systems) and calculate a set of λ values which can be compared with the observed set. The differences between observed and calculated values of λ are then used to make adjustments in the f_{ij} in an iterative fashion and a new set of values of λ obtained from the new set of f_{ij}. The adjustment procedure continues until a least-squares fit is obtained between the observed and calculated λ values, i.e. $\Sigma(\lambda_{obs} - \lambda_{calc})^2$ is a minimum. Computer programs are available to do this [6.6]. However, for an N atom molecule there will be $(3N - 6)$ normal modes [$(3N - 5)$ if the molecule is linear], and this will require $(N/2)(N + 1)$ force constants to describe completely† the potential energy, where $\bar{N}$ is the number of normal modes with different frequencies. (This number is reduced if the molecule possesses some symmetry.) For a bent triatomic molecule (e.g. the FNO example) with $(3N-6)=3$ normal modes one needs to determine $(\bar{N}/2)(\bar{N}+1)=6$ force constants to describe the vibrational potential energy function. Unfortunately, the three observed frequencies do not provide enough information to permit determination of all six vibrational force constants. This underdetermined nature of the vibrational problem immediately makes the task more difficult. Either a source of more vibrational data has to be found if all the potential parameters are to be determined or an approximate force field has to be used by leaving some f_{ij} fixed at zero or some other value.

Using the latter approach the crudest simplification is to use a diagonal force field and neglect all the off-diagonal (interaction) force constants. For most systems this is found to be inadequate and an intermediate state is reached by which only those force constants which are most sensitive to frequency and/or are chemically most significant are included in the least-squares refinement process. These are generally all the diagonal force constants plus some off-diagonal ones. The others are either constrained to zero or held fixed at some intuitively 'sensible' value. One of the restrictions in this process is that the number of f_{ij} included must be less than or equal to the number of observed frequencies.

6.5 INFORMATION FROM ISOTOPIC MOLECULES AND OTHER SPECTROSCOPIC DATA

To a good degree of approximation the electronic structure and hence vibrational potential function of an isotopically substituted molecule should be very close to that of the parent. The force constants in the molecule should then be invariant under isotopic substitution but since the mass of one of the atoms has been changed some of the frequencies at least should be different and hence give new information. However, not all the new frequencies observed in the isotopic molecule are useful in determining the force field. For example, observation of the CO stretching vibration of $^{12}C^{16}O$ allows calculation of the one and only force constant f_{CO}, and although observation of the vibrational frequency of the isotopically substituted species $^{13}C^{16}O$ gives a check on this calculated force constant, it gives no new information about the vibrational potential energy function. Similar considerations apply to the stretching vibrations of a linear triatomic molecule. For the XYZ molecule two stretching modes may be observed which give two values of λ (λ_1, λ_2). There are, however, three force constants (f_{XY}, f_{YZ}, and $f_{XY,YZ}$) which describe the stretching vibrations of the molecule. The problem is thus underdetermined and there is not enough information to allow calculation of all three force constants. Observation of the stretching frequencies of the isotopic molecule X'YZ, however, gives two new frequencies leading to λ_1' and λ_2'. These new values of λ are, however, related to the previous ones by

† There would be $\bar{N}^2$, but $f_{ij}=f_{ji}$ and so the actual number needed is smaller.

a simple function of the atomic masses and geometric parameters. This is known as the *Teller–Redlich product rule* [6.1].

In general, for two isotopic molecules, the product of the λ/λ' values for all vibrations of a given symmetry type is independent of the potential constants and depends only on the masses of the atoms and the geometrical structure of the molecule. (Strictly speaking it is only truly applicable to harmonic vibrations.) In mathematical terms:

$$\frac{\lambda_1\lambda_2\ldots}{\lambda_1'\lambda_2'\ldots} = \left(\frac{m_1'}{m_1}\right)^{\alpha}\left(\frac{m_2'}{m_2}\right)^{\beta}\ldots\left(\frac{M}{M'}\right)^{t}\left(\frac{I_x}{I_x'}\right)^{\delta x}\left(\frac{I_y}{I_y'}\right)^{\delta y}\left(\frac{I_z}{I_z'}\right)^{\delta z}$$

where all the primed quantities refer to the isotopic molecule. The m_1, m_2, etc. are the masses of the sets of equivalent atoms in the molecule; α, β, etc. refer to the number of vibrations of this symmetry species these atoms are involved in. M is the total mass of the molecule and t is the number of translations of the molecule with this symmetry description. δx, δy, δz are 1 or 0, depending upon whether rotation about the x, y, or z axes is of this symmetry species or not. The I are the moments of inertia of the molecule. In the trivial case of CO, for example, the CO stretching vibration is of species σ^+, as is one of the translations of the molecule (in the direction of the C—O axis). There is no rotation corresponding to this symmetry type. Thus $t = 1$, all $\delta = 0$, and:

$$\frac{\lambda}{\lambda'} = \frac{m(^{13}\mathrm{C})}{m(^{12}\mathrm{C})}\cdot\frac{M(^{12}\mathrm{CO})}{M(^{13}\mathrm{CO})} = \frac{\mu(^{13}\mathrm{CO})}{\mu(^{12}\mathrm{CO})}$$

In this particularly simple case the same result could have been obtained by manipulation of Equation (6.7).

Thus not all of the isotopic frequencies may be used to give new information concerning the force field. In the above linear triatomic case the observation of the frequencies from the parent and isotopic molecule gives, in total, four pieces of data. These are interrelated via the product rule which leaves three effective pieces of information, but sufficient in principle to allow determination of all three stretching force constants. In larger molecules some vibrational frequencies are found to be insensitive to isotopic substitution simply because the molecular motion describing the mode does not involve motion of the substituted atom. This reduces further the amount of usable new information obtainable from the vibrational spectrum of the isotopic molecule.

The most popular isotopes used in such studies are those with the largest mass change, e.g. ^{1}H and ^{2}H (100%), ^{10}B and ^{11}B, ^{14}N and ^{15}N, ^{12}C and ^{13}C, ^{16}O and ^{17}O, ^{18}O etc. However, under favourable conditions the isotopic shifts of heavier atoms (e.g. first-row transition metals) may be seen in the spectrum.

Figure 6.1 shows a part of the interesting spectrum in the CO stretching region of the set of isotopically substituted molecules $\mathrm{Cr}(^{12}\mathrm{CO})_x(^{13}\mathrm{CO})_{6-x}$ isolated in solid argon at 20 K. The sharpness of the infrared absorptions means that most of the expected absorptions are in fact observed (i.e. the spectrum is not one consisting of several broad bands merged together). If a frequency factored force

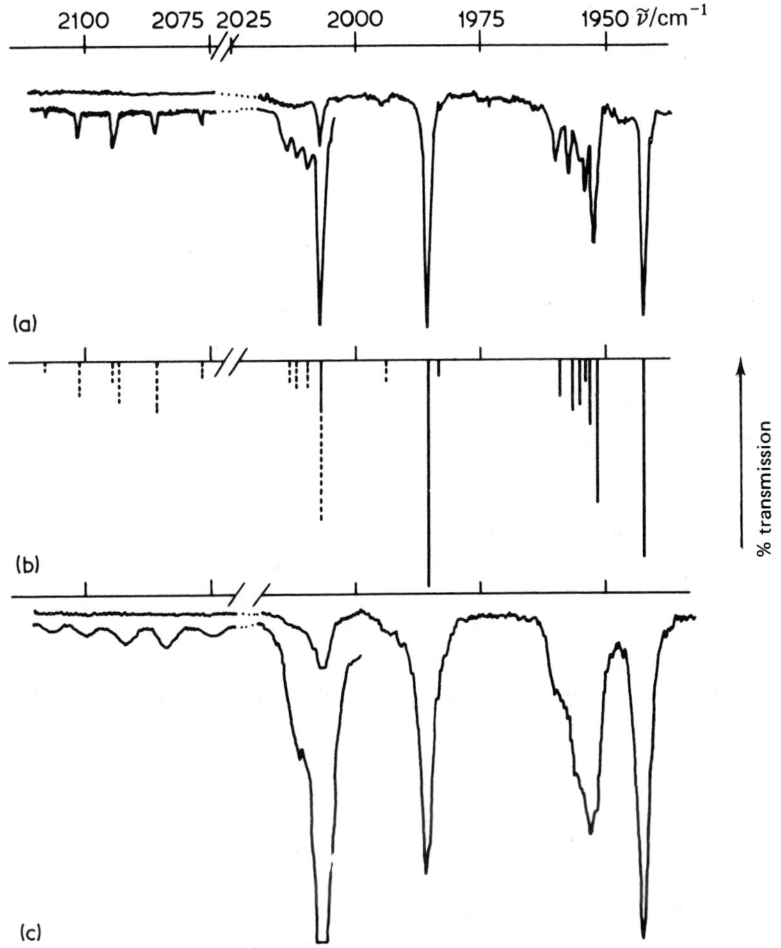

Fig. 6.1 Infrared spectra of $Cr(^{12}CO)_x(^{13}CO)_{6-x}$. (a) In an argon matrix at 20 K showing resolution of 19 fundamental absorptions; the lower trace is for a sample containing ten times as much material as the upper. (b) Calculated frequency and intensity pattern using the force constants described in the text; the broken lines correspond to the lower trace in (a), the solid lines to the upper trace in (a). (c) Spectra taken in cyclohexane solution showing the unresolved nature of the fine structure observed in (a); the lower trace is of a sample with a concentration ten times larger than the upper. (Courtesy Dr. R.N. Perutz).

field is used and only the stretching vibrations of the CO groups are considered, the vibrational potential energy of this system can be described by three force constants: f_{CO}, the stretching force constant of a CO bond (all six bonds in the molecule are chemically equivalent), and $f^{trans}_{CO,CO}$ and $f^{cis}_{CO,CO}$, the interaction force constants between two CO bonds *trans* and *cis* to one another.

The total of 19 observed bands may be used to determine these three vibrational force constants. In this particular case there is the unusual situation of a very overdetermined problem (i.e. more data than force constants). By including all 19 frequencies in a least-squares refinement process, the vibrational force constants are found to be $f_{CO} = 1644.27 \pm 0.14\,N\,m^{-1}$, $f_{CO,CO}^{cis} = 26.58 \pm 0.10\,N\,m^{-1}$, and $f_{CO,CO}^{trans} = 52.35 \pm 0.48\,N\,m^{-1}$. The standard error between observed and calculated frequencies is only $0.313\,cm^{-1}$, which is very good agreement indeed and of the order of the experimental error involved in locating the band maxima. Use of a smaller number of observed frequencies to fix the f_{ij} leads to a larger standard error between observed and calculated frequencies. This is a general observation, that the more data available the smaller the errors in the calculated force constants. If only three of these observed frequencies were used in the least-squares process, poorer agreement would be found between observed and calculated frequencies, with correspondingly larger errors in the force constants. Often the spectroscopist has to be satisfied with least-squares refinements which give standard errors between observed and calculated frequencies of the order of a few cm^{-1}. This most often happens when the problem is underdetermined and some of the f_{ij} have to be left out.

Centrifugal distortion constants, obtained from microwave spectra, are also useful in providing additional information concerning the vibrational force field. It can be readily visualized that the forces resisting centrifugal distortion as the molecule rotates will be combinations of the forces determining the frequencies of vibration and will be simply related to the f_{ij} of the vibrational potential function. Thus for small molecules with a permanent dipole moment, where centrifugal distortion constants can be experimentally determined, extra information is often available. Where the molecule has no permanent dipole moment, D values cannot be obtained from the microwave spectrum and so other methods have to be used. Coriolis coupling constants (discussed in Chapter 4) describe the coupling between two vibrations via a rotation. If a high-resolution gas-phase study of the rotational–vibrational spectrum of the molecule has been made, these constants are sometimes available and then they too may be used to help pin down the form of the vibrational potential function. Of more limited application is the use of experimentally determined electron diffraction amplitudes of vibration. The amplitude of a molecular vibration depends upon the force constants involved in the motion, but unfortunately the experimental diffraction amplitudes are rather insensitive functions of the force field. Force fields determined using such data are generally inaccurate because of this. Figure 6.2 shows an example of the determination of the force constant $f_{OCl,\alpha}$ in the bent OCl_2 molecule by a combination of vibrational data, Coriolis coupling constant, and centrifugal distortion. Here the three observed frequencies are insufficient to determine all four force constants, f_{OCl}, $f_{OCl,OCl}$, f_α, $f_{\alpha,OCl}$.

The theoretical dependence of the Coriolis coupling constant (ζ) on the value of the force constant is given by the full curve in Fig. 6.2, and the dependence of one of the centrifugal distortion constants (τ) on this force constant is given by the dashed curve. These are computed from a knowledge of the vibrational fre-

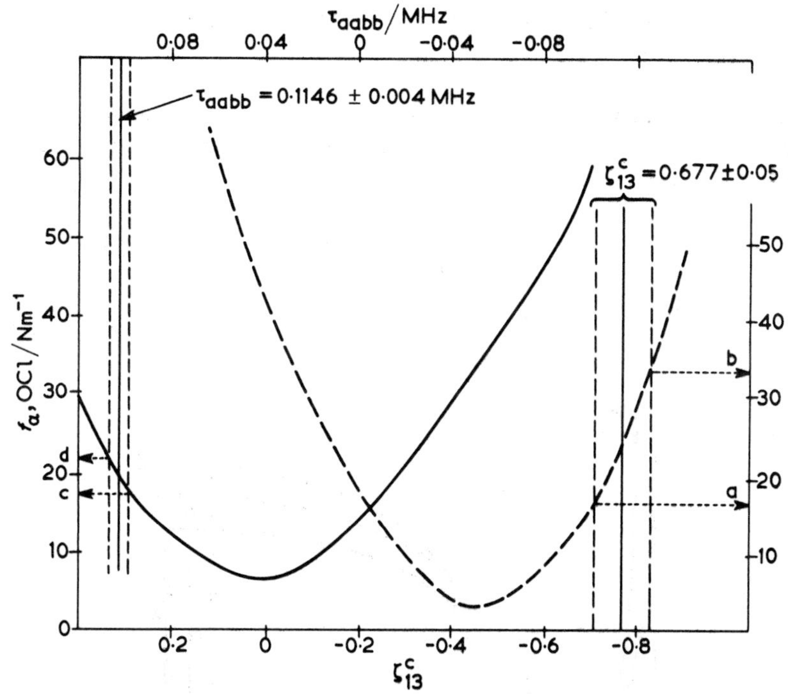

Fig. 6.2 Determination of the interaction force constant $f_{\alpha,\mathrm{OCl}}$ in OCl_2. The full curve shows the calculated dependence of the centrifugal distortion constant τ_{aabb} on $f_{\alpha,\mathrm{OCl}}$, and the dashed curve shows the calculated dependence of the Coriolis coupling constant $\zeta_{13}{}^{c}$ on this force constant. This force constant is seen to be 20 ± 2 N m^{-1} from the above plot. (Courtesy Dr. J.G. Smith).

quencies of the molecule, and its moments of inertia. The observed values of τ and of ζ are shown with their associated error limits. Using the Coriolis data it can be seen that $f_{\alpha,\mathrm{OCl}}$ must lie between points a and b, and using the centrifugal distortion data, between c and d, the latter resulting in the more accurate value of the force constant in this particular instance (20 ± 2 N m^{-1}).

This section may be concluded by saying that for many small molecules the use of isotopic, centrifugal distortion, and Coriolis data give enough information to permit determination of all the f_{ij} in the GQVFF (the most serious errors in their determination often lying in the defects of the least-squares process itself). But for larger molecules, where there are considerably more force constants and not enough data to enable all of them to be determined, an approximate force field where several of the interaction constants have been neglected is all that can be determined. Sometimes, as in the above carbonyl example, detailed information can be obtained about a part of the molecule.

6.6 THE MEANING OF VIBRATIONAL FORCE CONSTANTS

The curve labelled WXYZ in Fig. 6.3 shows a typical potential-energy curve for the diatomic molecule AB. $V(r)$ may be expressed as a general polynomial expansion about the equilibrium distance r_e:

$$V(r) = V(r_e) + \left(\frac{\partial V}{\partial r}\right)_{r=r_e}(r-r_e) + \frac{1}{2}\left(\frac{\partial^2 V}{\partial r^2}\right)_{r=r_e}(r-r_e)^2 + \frac{1}{6}\left(\frac{\partial^3 V}{\partial r^3}\right)_{r=r_e}(r-r_e)^3 + \ldots \quad (6.17)$$

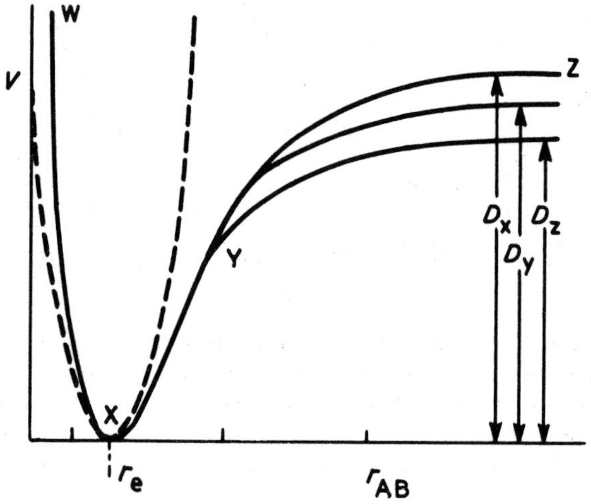

Fig. 6.3 Potential-energy curves for a system with the same value of the force constant but different values of the dissociation energy. The dashed curve represents the harmonic oscillator.

By choosing the zero of energy to occur at $r = r_e$, and noting that there is a minimum at this point, (6.17) simplifies to:

$$V(r) = \frac{1}{2}\left(\frac{\partial^2 V}{\partial r^2}\right)_{r=r_e}(r-r_e)^2 + \frac{1}{6}\left(\frac{\partial^3 V}{\partial r^3}\right)_{r=r_e}(r-r_e)^3 + \ldots \quad (6.18)$$

The approximation, made earlier, that, around the equilibrium position the forces acting within the molecule are harmonic, means that $(\partial^3 V/\partial r^3)_{r=r_e}$ and higher derivatives are zero, and comparison with Equation (6.1) leads to the relationship:

$$\left(\frac{\partial^2 V}{\partial r^2}\right)_{r=r_e} = f_{AB} \quad (6.19)$$

The force constant thus represents the curvature at the bottom of the potential well. On correlating the vibrational force constant with other molecular parameters, caution must be exercised as to which parameters are chosen. If the size of the force constant (the stiffness of the molecular spring) represents the strength of the chemical bond in some way, there should be a correlation between vibrational force constant and bond dissociation energies, bond orders (calculated by molecular orbital or other theories), and bond lengths.

Table 6.3 lists the bond dissociation energies and vibrational force constants for the halogen molecules and the hydrogen halides. It can be readily seen that, while there is a good correlation between force constant and bond dissociation energy (increasing bond dissociation energy being associated with increasing force constant) for the hydrogen halides, for the halogens themselves the low dissociation energy of F_2 is not matched by a correspondingly low vibrational force constant. This is not too surprising. It seems quite a stretch of the imagination to be able to predict the behaviour of $V(r)$ at a position a long way away from r_e (i.e. at Z) given the curvature of $V(r)$ at $r = r_e$ (at X). In Fig. 6.3 three curves are drawn, each with the same value of the force constant [i.e. the same $V(r)$ curve at the equilibrium position] but with different values of the dissociation energy (D_x, D_y, D_z). The most reliable parameters to use when correlating force constants with molecular structure are therefore those evaluated around

Table 6.3 Bond Dissociation Energies and Force Constants of the Halogen and Hydrogen Halide Molecules.

Molecule	*Bond dissociation energy/* kJ mol^{-1}	*Force constant/* N m^{-1}
F_2	153	445.4
Cl_2	242	328.7
Br_2	193	246.1
I_2	150	172.1
HF	535	965.8
HCl	404	516.3
HBr	339	411.7
HI	272	314.2

the equilibrium geometry of the molecule. If bond order (number of bonding electron pairs − number of antibonding electron pairs) is plotted against vibrational force constant for some first-row diatomic molecules, the strong correlation depicted in Fig. 6.4 is obtained. It has recently become possible to calculate vibrational force constants via molecular orbital calculations [6.7]. Similarly, the CC stretching force constants in a wide range of hydrocarbons cluster around 500 N m^{-1} (for C—C), 1000 N m^{-1} (for C=C), and 1500 N m^{-1} (for C≡C). A similar correlation exists between bond length and force constant. Several empirical relationships between the two have been derived, the best-known being that due to Badger [6.8]:

$$f_{AB} = \frac{186}{(r_e - d_{AB})^3}\ \mathrm{N\,m^{-1}}$$

where r_e is the equilibrium bond length and d_{AB} is a parameter which depends upon the position of the atoms A and B in the periodic table. Clark [6.9] and Gordy [6.10] have devised similar empirical schemes. However, the NO force constants in the NO_3^- ion obtained by the approaches of Badger, Clark, and Gordy are predicted to be significantly different: 1194, 1010, and 875 $\mathrm{N\,m^{-1}}$

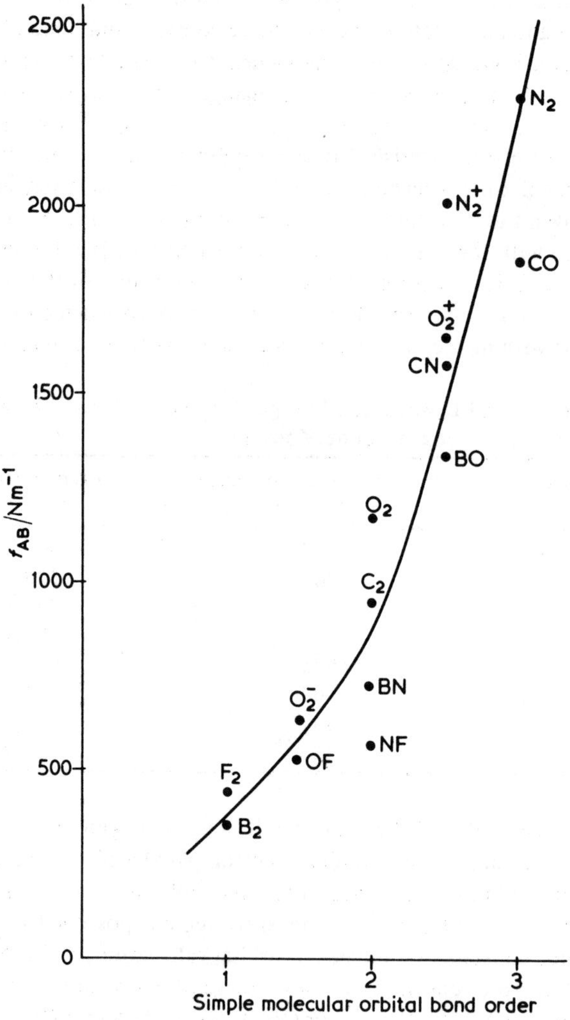

Fig. 6.4 Plot of AB stretching force constant against simple molecular orbital bond order (number of bonding electron pairs minus number of antibonding electron pairs) for some first-row diatomic molecules and ions.

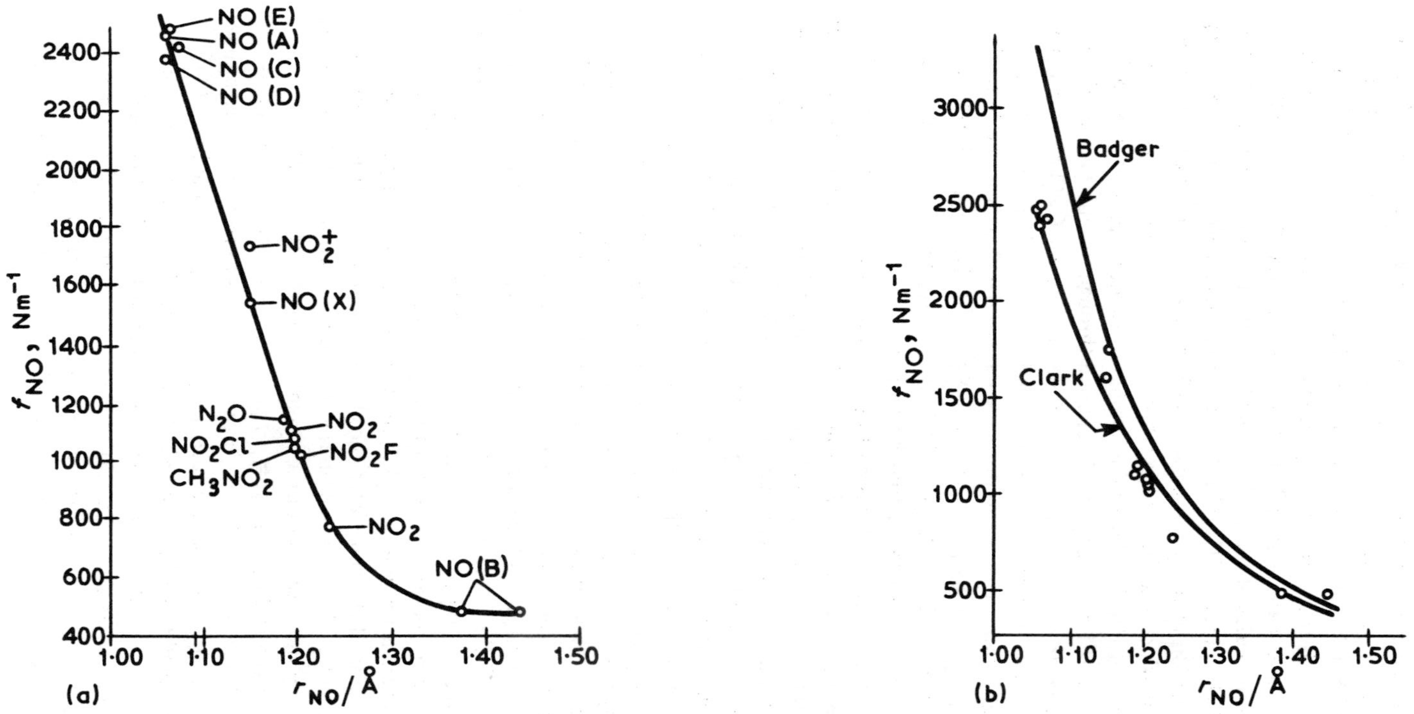

Fig. 6.5 (a) NO stretching force constants as a function of bond length in a variety of molecules containing this linkage. The solid line represents the equation $f_{NO} = 3440 r_{NO}^{-5.97}$ N m^{-1} (r_{NO} is in Å). (b) Similar plots using the force constant–bond length relationships of Badger and Clark (Adapted from J.A. Ladd and W. Orville-Thomas, *Spectrochim. Acta* (1966), 22, 919). The ground electronic state of NO is represented as NO (X), higher energy states as NO (A), NO (B), etc.

respectively. In spite of this quantitative drawback, the general rule that shorter bonds give rise to higher force constants is well proven. Figure 6.5 shows a typical plot of NO force constant against bond length for a series of NO containing species. It is found that the equation $f_{NO} = 3440r^{-5.97}$ N m^{-1} fits the data well. Similar equations specific to a given AB linkage also work well.

An interesting series of species from the vibrational point of view are the substituted methyl radicals $CClH_2$, $CHCl_2$, and CCl_3 which have been studied using the matrix isolation technique discussed in Chapter 4. They each contain an unpaired electron which may be used in π-bonding between the central carbon atom and the Cl atoms. The larger the number of Cl atoms present, the smaller the amount of C—Cl π-bonding per C—Cl bond through this mechanism. (The one unpaired electron can be exclusively owned by the one C—Cl bond in CH_2Cl but has to be shared between three C—Cl bonds in CCl_3.) The C—Cl stretching force constant is thus expected to drop along the series. This is found to be the case [6.11] : $f_{C\,Cl} = 409$ N m^{-1} (CH_2Cl), 398 N m^{-1} ($CHCl_2$), and 360 N m^{-1} (CCl_3) to be compared with $f_{C\,Cl} = 308$ N m^{-1} (CCl_4) where the unpaired electron is tied up in bonding to the fourth Cl atom.

In the infrared absorption spectra of hydrocarbons the C—H stretching vibrations of alkanes are generally found around 2900 cm^{-1}, those of alkenes between 3000 and 3100 cm^{-1}, and those of alkynes at around 3300 cm^{-1}. These correspond to CH stretching force constants of about 480, 510, and 590 N m^{-1} respectively, and indicate the increasing strength of the CH bond in the order $sp^3 < sp^2 < sp$, a trend which can be verified from bond length measurements as well. In general bond strengths increase with increasing s-character.

Valuable information concerning the forces holding atoms together in a molecule may also be derived from interaction force constants. Considerable care, however, must be exercised with the interaction force constants of polyatomic molecules, where, since approximate force fields are often used, the uncertainties in magnitude and sign are often considerable. For small molecules and systems where all the f_{ij} are well defined, however, this objection is less important. If the molecule experiences a unit positive displacement in the coordinate q_k, $q_k = 1$ then the potential energy of the system has risen from zero to $V = f_{kk}/2$. The system is not now in equilibrium and strains are set up all over the molecule, because of the new electronic charge distribution demanded by the distorted geometry. The force within the coordinate q_j arising from this displacement is given by the derivative of the potential energy with respect to q_j, i.e., force $= -\partial V/\partial q_j = \partial(f_{jk}q_kq_j)/\partial q_j = -f_{jk}q_k$. If f_{jk} is positive, it means that a unit positive change in q_k leads to a force which tends to reduce the magnitude of the coordinate q_j. For example [6.12] , f_{CO} and $f_{CO,CO}$ in the linear CO_2 molecule are found to be 1603 and 126 N m^{-1} respectively. Stretching one CO bond by a small amount then requires contraction of the other CO bond ($f_{CO,CO} > 0$) by an amount $126/1603 \approx 8$ per cent of the initial extension of the CO bond, to remove this strain in the system. This may be interpreted in the following way. As one CO bond is stretched (and thus weakened) the π-electron

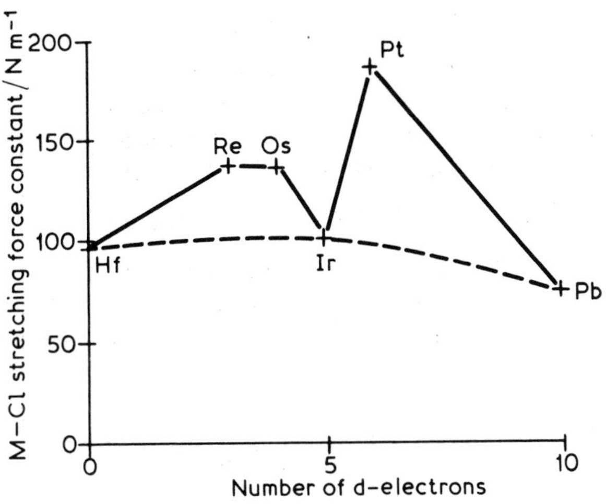

Fig. 6.6 Plot of M—Cl stretching force constant as a function of the number of d-electrons (adapted from Ref. 6.13).

density released may be used to strengthen the other CO linkage, which now requires a shorter CO bond. In other words, the total 'valency' of the carbon atom remains unchanged during the molecular vibration.

The molecular vibrations of octahedral halo-complexes $MX_6{}^{n-}$ have been studied over the years. In Fig. 6.6 the M—X bond stretching force constant is plotted against the number of d-electrons for $MCl_6{}^{-2}$ compounds. The curve that is produced [6.13] is very similar to that found for the ligand field stabilization energy as a function of number of d-electrons [6.14]. In the latter, a maximum in energy is found at $n = 3,8$ and a minimum at $n = 5$ for octahedral complexes. It is also found that the M—X force constant decreases in the halogen order $X = F > Cl > Br$, and also decreases with the oxidation number of the metal in the complex. These trends are useful in understanding the nature of the M—X bond in these molecules.

6.7 ANHARMONIC VIBRATIONS

The discussion so far has centred around the assumption that the forces between the nuclei are harmonic, namely that the shape of the curve of Fig. 6.3 around the minimum ($r = r_e$) is a good approximation to a parabola. However, as the molecule is displaced (even slightly) from its equilibrium position, the electronic charge distribution in the molecule will rearrange to accommodate the new configuration. Thus the force constant f_{AB} of Equation (6.2) is itself dependent in some way on the displacement q_{AB}. The extent of the dependence will be set by how much the electronic distribution 'relaxes' (or readjusts) as the molecule departs from its equilibrium position. Thus:

$$f_{AB} = f_{AB}^{e} + g(q_{AB}) \tag{6.20}$$

where f_{AB}^{e} is the force constant at the very bottom of the potential well (at $r = r_e$). f_{AB} of course is the force constant calculated via the observed vibrational transition $v = 0 \rightarrow v = 1$, and is thus a force constant measured higher up the potential well. The potential energy then becomes:

$$V(q_{AB}) = \tfrac{1}{2} f_{AB}^{e} q_{AB}^{2} + q_{AB}^{2} \cdot h(q_{AB}) \tag{6.21}$$

where g and h in (6.20) and (6.21) are two functions which tend to zero as q_{AB} tends to zero. The force constant really needed is of course f_{AB}^{e}. For a polyatomic molecule a more complex form of Equation (6.8) may be written which includes these ideas of anharmonicity:

$$V = \sum_{i,j} f_{ij} q_i q_j + \sum_{i,j,k} f_{ijk} q_i q_j q_k + \sum_{i,j,k,m} f_{ijkm} q_i q_j q_k q_m + \ldots \tag{6.22}$$

where the f_{ijk}, f_{ijkm}, and higher terms are the anharmonic constants. The general success of the force constant treatment described in previous sections, however, implies that the terms f_{ij} are large compared with the anharmonic terms. As an example, consider the simple case of a diatomic molecule. In Chapter 4 the energy levels of the anharmonic oscillator were written:

$$G(v) = \omega_e (v + 1/2) - \omega_e x_e (v + 1/2)^2 + \omega_e y_e (v + 1/2)^3$$
$$- \omega_e z_e (v + 1/2)^4 \quad \text{etc.} \tag{6.23}$$

For this system (6.23) may be rewritten after the style of Equation (6.22) as:

$$2V = f_{ii} q_i^{2} + f_{iii} q_i^{3} + f_{iiii} q_i^{4} + \ldots \tag{6.24}$$

By means of some rather complicated mathematics, the terms of Equation (6.23) may be related to the general anharmonic potential constants [6.1]. The f_{iii}, f_{iiii}, etc. of (6.24) may also be identified with the derivatives occurring in Equation (6.17) and (6.18).

By the analysis of the vibrational transitions of $H^{35}Cl$ up to $v = 5$ the constants of Table 6.4 can be determined. ω_e is called the fundamental harmonic

Table 6.4 Vibrational Potential Constants of $H^{35}Cl$

ω_e	$= 2991.03\ cm^{-1}$
$\omega_e x_e$	$= 52.89\ cm^{-1}$
$\omega_e y_e$	$= 0.247\ cm^{-1}$
$\omega_e z_e$	$= 0.0146\ cm^{-1}$

frequency of the anharmonic oscillator. $\tilde{\nu}$, the observed transition $v = 0 \rightarrow v = 1$, occurs at $2885.9\ cm^{-1}$.

There are thus two main points to note. Firstly, from Table 6.4 the size of the

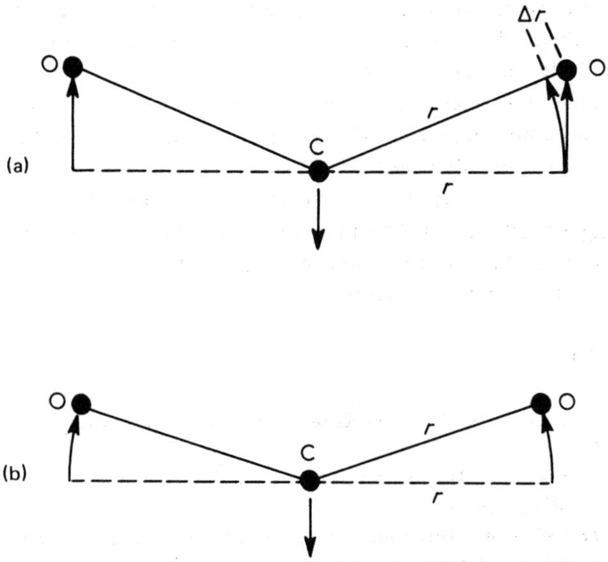

Fig. 6.7 Rectilinear and curvilinear bending coordinates for CO_2.

anharmonic constants decreases with their order. (The potential constants of Equation (6.24) decrease in this manner as well.) Secondly, the observed vibrational fundamental ($\tilde{\nu}$) at 2885.9 cm^{-1} is over 100 cm^{-1} lower than the true harmonic frequency (ω_e) of 2991.03 cm^{-1}. Thus the neglect of anharmonicity in the HCl case leads to an error of about 3.5 per cent in the harmonic frequency and about 7 per cent in the harmonic force constant, which could be a serious discrepancy if detailed correlations between the force constant and other molecular parameters are required. In general, however, non-hydrogen-containing molecules have smaller anharmonicities. For the CO molecule $\omega_e x_e = 12\ cm^{-1}$ and $\omega_e y_e = 0.01\ cm^{-1}$ for example.

For diatomic molecules, as Table 6.4 shows, the introduction of anharmonicity leads to a proliferation of potential constants even for this simple system. As the molecule gets more complicated it was shown previously that more force constants f_{ij} are introduced at an ever-increasing rate to describe the vibrational potential function such that it was often impossible to determine all of them from the observed data. If the anharmonic terms f_{ijk} are included there are even more force constants to be determined and the situation presents an impossible task for all but the simplest molecules. The situation at present is that force fields containing some anharmonic constants are available for very small systems where there are sufficient data to fix all the f_{ij} and some of the higher terms. As the molecule gets bigger the first approximation made is to neglect the f_{ijk} and higher terms and to assume harmonic forces. If there is not enough data to fix all the force constants, steps are taken to prune the number of f_{ij} down to a manageable size. Sometimes, however, it is found that sufficient data exist to fix the vibrational potential function, associated with one particular part of the molecule, or perhaps of one set of vibrations of given symmetry up to an including some of the third-order constants (f_{ijk}).

For polyatomic molecules large anharmonic terms are often due to inept choice of internal coordinate. For CO_2, for example, the rectilinear bending coordinate of Fig. 6.7(a) requires the presence of a small amount of C—O stretching coordinate. For the more true-to-life curvilinear coordinate of Fig. 6.7(b), no C—O stretching is needed at all. The anharmonic terms of Equation (6.22) are considerably smaller (by an order of magnitude) for CO_2 if the curvilinear rather than rectilinear coordinate is chosen [6.15].

REFERENCES

6.1 Herzberg, G., *Infra-Red and Raman Spectra,* Van Nostrand, London (1945).

6.2 Wilson, E.B. Jnr., Decius, J.C. and Cross, P.C., *Molecular Vibrations,* McGraw-Hill, (1955).

6.3 Woodwood, L.A., 'Introduction to the Theory of Molecular Vibrations and Vibrational Spectroscopy', Oxford University Press, (1972).

6.4 Nakamoto, K. 'Infra-Red Spectra of Inorganic and Coordination Compounds', Wiley (1970).

6.5 Cotton, F.A., Kraihanzel, C.S., *J. Amer. Chem. Soc.,* **84**, 4432, (1962).

6.6 Schachtschneider, J.H. and Snyder, R.G., *Spectrochim. Acta* **19**, 117, (1963).

6.7 Schutte. C.J.H., *Structure and Bonding,* **9**, 213 (1971).

6.8 Badger, R.M., *J. Chem. Phys.,* 2, 128 (1934); **3**, 710 (1935).

6.9 Clark, C.H.D., *Phil. Mag.,* **19**, 476 (1935); **22**, 1137 (1938).

6.10 Gordy, W., *J. Chem. Phys.,* **14**, 305 (1946).

6.11 Andrews, L. and Smith, D.W., *J. Chem. Phys.,* **53**, 2956 (1970).

6.12 Jones, L.H., *Inorganic Vibrational Spectroscopy,* Vol. 1, Dekker, New York (1971).

6.13 Labonville, P., Ferraro, J.R., Wall, M.C. and Basile, L.J., *Coord. Chem. Rev.,* **7**, 257 (1972).

6.14 Kettle, S.F.A., *Coordination Compounds*, Nelson, London (1970).

6.15 Pariseau, M.A., Suzuki, I. and Overend, J., *J. Chem. Phys.,* **42**, 2335 (1965)

7 Thermodynamic functions from spectroscopic data

7.1 INTRODUCTION

A knowledge of the energy levels of molecular systems allows the calculation of a large number of bulk thermodynamic properties of interest to chemists as a function of temperature (T). These are $[U^\circ(T) - U^\circ(0)]$, the standard molar internal energy relative to the absolute zero energy, the function $[H^\circ(T) - H^\circ(0)]/T$ where H is the molar enthalpy, the function $[G^\circ(T) - H^\circ(0)]/T$ where G is the Gibbs free energy, the entropy $S^\circ(T)$, and the heat capacity at constant pressure $C_p{}^\circ(T)$. All these values relate to the ideal gas state at one atmosphere pressure (101 325 Pa). By the knowledge of the heat of reaction at one particular temperature the functions $\Delta G_f{}^\circ$, $\Delta H_f{}^\circ$, and K° the equilibrium constant may also be calculated for a chemical process. By knowing the observed thermodynamic properties of simple organic molecules it is possible by statistical thermodynamical methods to calculate the corresponding functions for whole homologous series for example.

The statistical methods by which such thermodynamic properties may be calculated from spectroscopic data were developed in the years prior to 1950. Since then the greater sophistication of spectroscopic techniques has led to an increased knowledge of molecular energy levels and facilitated the treatment of larger and more complex molecules. For a modern review of the subject the reader is referred to the excellent discourse by Frankiss and Green [7.1].

7.2 THE PARTITION FUNCTION AND THERMODYNAMIC FUNCTIONS

The basis of the use of spectroscopic data to determine molecular thermodynamic properties such as specific heats, entropies, free energies, etc. is the ability to determine the total partition function of the system (Z). According to the Maxwell–Boltzmann distribution the number of molecules N_i populating a level of energy ϵ_i which has a degeneracy (or statistical weight) g_i is given by:

$$N_i = N_0 g_i \exp(-\epsilon_i/kT) \tag{7.1}$$

where k is Boltzmann's constant, N_0 is the number populating some reference level, and T is the absolute temperature. The total number of molecules in a given volume is then:

$$N = \sum_i N_i = N_0 \sum_i g_i \exp(-\epsilon_i/kT) = N_0 Z \tag{7.2}$$

and

$$N_i = (N/Z) g_i \exp(-\epsilon_i/kT) \tag{7.3}$$

Here, Z is the *molecular partition function* and all thermodynamic properties may be calculated from it [7.2, 7.3]. For example:

$$\frac{\mathrm{d}Z}{\mathrm{d}T} = \sum_i g_i \frac{\mathrm{d}}{\mathrm{d}T} \exp(-\epsilon_i/kT)$$

$$= \frac{1}{kT^2} \sum_i g_i \epsilon_i \exp(-\epsilon_i/kT) \tag{7.4}$$

Since the total internal energy of the system U is equal to $\sum_i N_i \epsilon_i$ this can simply be written in terms of Z:

$$U = NkT^2 \frac{\mathrm{d}(\ln Z)}{\mathrm{d}T} \tag{7.5}$$

The enthalpy H is related to U for the ideal gas state by the relationship:

$$H = U + PV \quad (H = U + RT)$$

and thus the molar heat capacities at constant pressure $[(\partial H/\partial T)_p]$ and at constant volume $[(\partial U/\partial T)_v]$ are given by:

$$C_p^\circ = R + R\frac{\mathrm{d}}{\mathrm{d}T}\left[T^2 \frac{\mathrm{d}(\ln Z)}{\mathrm{d}T}\right]$$

and

$$C_v^\circ = R\frac{\mathrm{d}}{\mathrm{d}T}\left[T^2 \frac{\mathrm{d}(\ln Z)}{\mathrm{d}T}\right] \tag{7.6}$$

Similarly the entropy is given by:

$$S^\circ = R\left[1 + \ln(Z/L) + T\left(\frac{\partial \ln Z}{\partial T}\right)_v\right] \tag{7.7}$$

where L is Avogadro's number. Determination of Z for the system therefore allows ready computation of all the thermodynamic properties of chemical interest. Often for small molecules data calculated this way are more accurate than those determined from thermal measurements. Sometimes the latter are either difficult or impossible to carry out (for unstable species for example) and the spectroscopic route is the only one available.

7.3 DETERMINATION OF Z

For a perfect gas, since the translational and internal energies are independent of each other:

$$E_{\text{total}} = E_{\text{trans}} + E_{\text{int}} \tag{7.8}$$

Such an expression means that the partition function can be similarly separated into a product of two parts:

$$Z_{\text{total}} = Z_{\text{trans}} Z_{\text{int}} \tag{7.9}$$

where the summation in Z_{trans} runs over all translational energy levels and the summation Z_{int} runs over all the internal (rotational, vibrational, nuclear, and electronic) energy levels of the molecule. The observation of transitions between these energy levels has been the subject of this book. Since nuclear spins are conserved in virtually all systems (exceptions include reactions involving hydrogen and its isotopes, and transmutation) the nuclear spin contribution to Z_{total} is generally ignored.

For a molecule of mass m in a one-dimensional box of length l_x the translational energy levels are simply given by:

$$\epsilon_i = \frac{p_i^2 h^2}{8ml_x^2} \tag{7.10}$$

Using this equation the translational partition function of an ideal gas occupying a volume V can be shown to be given by:

$$Z_{\text{trans}} = V\left(\frac{2m\pi kT}{h^2}\right)^{3/2} \tag{7.11}$$

The internal partition function is similarly the product of three contributions, arising from population of the electronic, vibrational, and rotational energy levels. With some exceptions the Boltzmann factors of excited electronic states ($\epsilon_i > 10^4\ \text{cm}^{-1}$) are entirely negligible compared with those of the ground state. The relative population of an excited state separated from the ground state by $10^4\ \text{cm}^{-1}$ is $\sim 10^{-21}$ at room temperature. Only contributions from the thermally more accessible vibrational (10^2–$10^3\ \text{cm}^{-1}$) and rotational ($< 10\ \text{cm}^{-1}$) energy levels will be considered in detail, where relative excited state populations are 10^{-3}–10^{-1} at room temperature and thus important in determining Z.

The internal energy is thus the sum of the vibrational energy G (dependent upon the vibrational quantum numbers v_1, v_2, etc. of all the vibrational modes) and the rotational energy F_v (dependent upon the rotational quantum numbers J, K, etc.) and the vibrational level it is associated with.

$$E_{\text{int}} = G(v_1, v_2, \ldots) + F_v(J, K, \ldots) \tag{7.12}$$

Then:

$$Z_{\text{int}} = \sum_v g_v \exp[-G(v_1, v_2, \ldots)/kT] \sum_r g_r \exp[-F_v(J, K, \ldots)/kT] \tag{7.13}$$

where for each vibrational level v the summation runs over all rotational sublevels r. This is obviously quite a complicated function because of the dependence of F_v on v. If this interaction between rotation and vibration is neglected, however, a considerable simplification is gained since now $Z_{\text{int}} = Z_{\text{vib}} Z_{\text{rot}}$ or more fully:

$$Z_{\text{int}} = \sum_v g_v \exp[-G(v_1, v_2, \ldots)/kT] \sum_r g_r \exp[-F(J, K, \ldots)/kT] \quad (7.14)$$

where each term can now be calculated separately.

If it is assumed that the molecule consists of a series of harmonic oscillators, the vibrational energy referred to the lowest level† is simply:

$$G(v_1, v_2, \ldots) = \sum_i (v_i) h\nu \quad (7.15)$$

and the vibrational partition function becomes:

$$Z_{\text{vib}} = g_1 \sum_{v_1} \exp(-v_i h\nu_1/kT) \cdot g_2 \sum_{v_2} \exp(-v_2 h\nu_2/kT) \ldots \quad (7.16)$$

where each summation is over all the vibrational levels of the vibrational mode. [There will be (3N − 6) or (3N − 5) such terms depending upon whether the molecule is linear or not; N is the number of atoms in the molecule.]

After using the identity for the geometric progression:

$$\sum_{n=0}^{\infty} e^{-an} = (1 - e^{-a})^{-1}$$

it follows that:

$$Z_{\text{vib}} = [1 - \exp(-h\nu_1/kT)]^{-g_1} \cdot [1 - \exp(-h\nu_2/kT)]^{-g_2} \ldots \quad (7.17)$$

Thus all that is needed to calculate the vibrational partition function is the set of ν_i taken from the infrared and Raman spectra, and possibly from the far-infrared for torsional and bending vibrations, and the degeneracy g_i of each of the vibrations. Nothing need be known about their origin, whether a particular ν_i corresponds to a predominantly stretching or bending motion for example, although an assignment is obviously desirable. Table 7.1 shows the calculated values of the vibrational partition function for the three fundamentals of CO_2 where $\tilde{\nu}_1$ (symmetric CO stretching) = 1388.3 cm^{-1}, $\tilde{\nu}_2$ (doubly degenerate OCO bending) = 667.3 cm^{-1}, and $\tilde{\nu}_3$ (antisymmetric CO stretching) = 2349 cm^{-1} in the gaseous phase. As can be seen, the higher the vibrational frequency, the closer the contribution $[1 - \exp(-h\nu_i/kT)]^{-g_i}$ to Z_{vib} approaches unity. The largest contribution to Z_{vib} is thus the low-frequency bending mode in this case but the higher frequency vibrations become more important at higher temperatures. The main error in the vibrational partition function calculated by this route seems to be the neglect of anharmonicity. In favourable circumstances this problem may

† It is conventionally accepted in statistical mechanics to refer all energies ϵ_i to the lowest occupied level in the set of ϵ_i.

Table 7.1 Calculated vibrational partition function for the fundamental vibrations of CO_2.

Frequency	g_i	*500 K*	*1000 K*	*1500 K*	*2000 K*
ν_1	1	1.01874	1.15692	1.35868	1.58301
ν_2	2	1.37287	2.62511	4.47360	6.87834
ν_3	1	1.00116	1.03522	1.11732	1.22618

be overcome by direct summation of the vibrational part of Equation (7.14) over all the vibrational levels where $\exp[-G(v_1, v_2, \ldots)/kT]$ is not negligibly small. Thus for the bending mode of CO_2, experimentally it is found that the $v = 0 \rightarrow 1$ transition occurs at 667.3 cm^{-1}, $v = 0 \rightarrow 2$ occurs at 1285.05 cm^{-1}, and $v = 0 \rightarrow 3$ occurs at 1932.5 cm^{-1}. At $T = 500$ K:

$$\sum_{v=0}^{3} g_i \exp(-h\nu_2 v/kT)$$

is then calculated to be 1.3502. Inclusion of the $v = 0 \rightarrow 4$ transition only alters this figure by about 0.0015 and the $v = 0 \rightarrow 5$ and the $v = 0 \rightarrow 6$ transitions affect only the fourth and fifth decimal places. The vibrational partition function for the bending mode of CO_2 is thus affected by anharmonicity to an extent of about 2 per cent. This direct summation, however, can only be performed where either all the anharmonic force constants are known (a rare situation as seen in Chapter 6) or where all the overtone levels $v_i \nu_i$ have been observed, up to a level such that further levels make a negligible contribution to the partition function. Thus a level 3800 cm^{-1} above the energy zero ($v = 0 \rightarrow 6$ for the CO_2 case above) makes a negligible contribution to the partition function at 500 K of 1.8×10^{-5} but at 3000 K has a non-negligible effect with a contribution of 0.16. If the thermodynamic properties only at 500 K are of interest, obviously the contributions from such high-lying levels need not be included. If the temperature of interest is 3000 K, neglect of such levels will introduce errors.

The rotational partition function will be different in form depending upon whether the molecule is linear or is a symmetric, asymmetric, or spherical top.

For a rigid linear rotor the rotational energy levels are given in terms of the rotational quantum number J and rotational constant B as:

$$F(J) = BJ(J+1)h \tag{7.18}$$

where any vibrational dependence of B has been neglected. The degeneracy of each J level is $(2J + 1)$ and so:

$$Z_{\text{rot}} = \sum_J (2J+1) \exp[-BhJ(J+1)/kT] \tag{7.19}$$

If the temperature is low and B is large, then (7.19) is best evaluated by direct summation since under these conditions the population of all but the first few rotational levels is negligible. For ordinary temperatures, however, the summation in (7.19) may be replaced by an integral. Putting $x = J(J + 1)$:

$$Z_{rot} = \int_0^\infty \exp(-xBh/kT)\,dx = kT/Bh \tag{7.20}$$

This is quite a good approximation; for HCl ($B \sim 10\ cm^{-1}$) at 50 K the rotational partition function evaluated from (7.20) is 98 per cent of that calculated by direct summation. Usually thermodynamic data are required for larger molecules at higher temperatures where the integrated approximation is very good indeed. For symmetric-top molecules a similar integration produces:

$$Z_{rot} = (\pi/B^2A)^{1/2}(kT/h)^{3/2} \tag{7.21}$$

For asymmetric-top molecules no analytic formula exists for the rotational energy levels but at high temperatures, or where B/T is small:

$$Z_{rot} = (\pi/ABC)^{1/2}(kT/h)^{3/2} \tag{7.22}$$

This is generally good to an accuracy of a few tenths of a per cent even at 100 K. For spherical-top molecules the rotational energy levels are given by a formula similar to that for linear molecules (see Chapter 4) and thus Z_{rot} is given by a function similar to (7.20).

The rotational partition function may then be quite accurately calculated if the moments of inertia of the molecule under consideration are known. If the centrifugal distortion constants are available, corrections may be applied to the rigid rotor model and small corrections (generally of a few tenths of a per cent) applied to Z_{rot}.

One other factor needs to be included in Z_{rot}. For linear molecules with ${}^1\Sigma_g{}^+$ electronic ground states (i.e. with two or more symmetry-related identical nuclei, e.g. CO_2 and C_2H_2) it was shown in Chapter 4 that the even rotational levels were symmetric and the odd levels antisymmetric. If the spin of these identical nuclei is zero (e.g. ${}^{16}O$) then only one set of levels is occupied and the evaluation of the rotational partition function by the methods described in this chapter will be in error. The same arguments hold for the rotational partition functions of non-linear polyatomic molecules containing symmetry-related nuclei. The overall result is to divide the calculated Z_{rot} by σ, which is called the symmetry number [7.2] and is equal to the number of indistinguishable positions into which the molecule can be turned by simple rotation. Thus $\sigma = 1$ for COS (i.e. no equivalent nuclei related by a rotation of the molecule), $\sigma = 2$ for CO_2 and $\sigma = 12$ for benzene (in plane rotations of $\pi/3$, $2\pi/3$, $4\pi/3$, $5\pi/3$, 2π and the six rotations of the plane about the pairs of diagonally opposed carbon atoms).

7.4 INTERNAL ROTATIONS

Many molecules contain bonds about which one part of the molecule may rotate, either completely freely or in a partially or totally restricted fashion. For example, each of the CH_3 groups of ethane may rotate about the C—C bond with respect to the other, in a hindered manner since the CH_3 groups are close to each other. On the other hand the CH_3 groups in dimethylacetylene are sufficiently well separated in space to experience a much smaller hindering potential to their relative rotation. If the rotation is highly hindered (i.e. there is a high potential hill preventing free rotation), the vibrational levels of this torsional oscillation (as it has now become) may be included in the vibrational partition function in the normal way. If, however, there is completely free rotation about the $C—CH_3$ bond, as in dimethylacetylene, this torsional degree of freedom has to be removed from the vibrational partition function and an extra term Z_{tors} introduced by summation over the rotational energy levels of this internal rotor. Z_{tors} is given by $(8\pi^3 I_{red} kT)^{1/2}/h\sigma$ where I_{red} is the reduced moment of inertia of that part of the molecule which is rotating, and σ is the symmetry number, the number of identical configurations into which the molecule can be transformed by internal rotation. For such a molecule:

$$Z_{int} = Z_{vib} Z_{rot} Z_{tors} \tag{7.23}$$

However, the situation most often encountered lies somewhere between these two extremes, i.e. where the situation is intermediate between a vibrational partition function $[1 - \exp(-h\nu_{tors}/kT)]^{-1}$ and a rotational contribution containing values of the moments of inertia. It is found that for barriers of less than about 200 cm^{-1} (600 cal mol^{-1} or 2.5 kJ mol^{-1}) the partition function is close to the free-rotation value but that barriers higher than this give rise to partition functions approaching those of torsional vibrations. Tables are available to calculate the relevant thermodynamic properties given the moments of inertia of the internal rotor and the barrier height to rotation [7.4].

7.5 ELECTRONIC CONTRIBUTION TO THE THERMODYNAMIC PROPERTIES.

The electronic partition function for a molecule with a state of degeneracy g_1 and energy E_1 above the electronic ground state of degeneracy g_0 is:

$$Z_{el} = g_0 + g_1 \exp(-E_1/kT) \tag{7.24}$$

The majority of polyatomic systems whose thermodynamic properties are of interest have a non-degenerate ground state $g_0 = 1$ and contain higher electronic states at sufficiently large energies so that their contribution to Z_{el} is negligible, as mentioned in Section 7.2.

Diatomic molecules however often have a degenerate electronic ground state. Here the total degeneracy = spin multiplicity $(2S + 1) \times$ orbital degeneracy. Thus a $^3\Pi$ state has a degeneracy of $g = 3 \times 2 = 6$. The degeneracy of the electronic states is important in determining amongst other things the entropy. For a system described by (7.24) the electronic contribution to the entropy will be $R \ln g_0$ at low temperatures and $R \ln(g_0 + g_1)$ at high temperatures. Thus at low temperatures the entropy contribution from non-degenerate states is zero. For polyatomic molecules containing unpaired electrons however spin degeneracy occurs. Thus NO_2 has to have a term $R \ln 2$ added to $S°(T)$ to take account of this.

7.6 ENTHALPY AND SPECIFIC HEATS FROM SPECTROSCOPIC DATA

It was shown in Section 7.2 how various thermodynamic properties were related to the total partition function Z. As an example, consider the evaluation of the heat capacity or specific heat at constant pressure from observed spectroscopic data. For the ideal gas state, $C_p°$ is given by the expression [see (7.6)]:

$$C_p° = R + R\frac{\mathrm{d}}{\mathrm{d}T}\left[T^2\frac{\mathrm{d}(\ln Z)}{\mathrm{d}T}\right] \tag{7.25}$$

But Z may be written as:

$$Z = Z_{\text{trans}}Z_{\text{rot}}Z_{\text{vib}} \tag{7.26}$$

and thus:

$$C_p° = R + R\frac{\mathrm{d}}{\mathrm{d}T}\left[T^2\frac{\mathrm{d}}{\mathrm{d}T}(\ln Z_{\text{trans}} + \ln Z_{\text{rot}} + \ln Z_{\text{vib}})\right] \tag{7.27}$$

or

$$C_p° = R + C_p(\text{trans}) + C_p(\text{rot}) + C_p(\text{vib}) \tag{7.28}$$

Using the expression for the translational partition function:

$$C_p°(\text{trans}) = R\frac{\mathrm{d}}{\mathrm{d}T}\left[T^2\frac{\mathrm{d}}{\mathrm{d}T}(3/2 \ln T)\right] \tag{7.29}$$

$$= 3/2\, R$$

This is the classical result that the specific heat $C_p°$ of a gas with no rotational or vibrational degrees of freedom (i.e. a monatomic system) is $5/2\, R$, and is the result obtained by applying the Boltzmann equipartition law. $U°$ then becomes $3/2\, RT$ (and $C_v° = 3/2\, R$) and $H° = 5/2\, RT$. For high temperatures the rotational partition function for a rigid molecule is given by (3.13), (3.14), or (3.15) depending upon whether the molecule is a linear or spherical top molecule, a symmetric top, or an asymmetric top respectively. Since Z_{rot} occurs in H, U, C_p, and C_v in the form $\mathrm{d}(\ln Z_{\text{rot}})/\mathrm{d}T$, all the temperature independent factors drop out on differentiation. (Free energy functions and the entropy, by way of contrast, do contain rotational contributions containing the moments of inertia of the molecule.)

$$U°_{\text{rot}} = H°_{\text{rot}} = RT \text{ (for a linear molecule)}$$

$$= 3/2\, RT \text{ (for other molecules)} \tag{7.30}$$

and

$$C_p°{}_{\text{rot}} = C_v°{}_{\text{rot}} = R \text{ (for linear molecules)}$$

$$= 3/2\, R \text{ (for other molecules)} \tag{7.31}$$

For lower temperatures, or if the effects of centrifugal distortion are to be

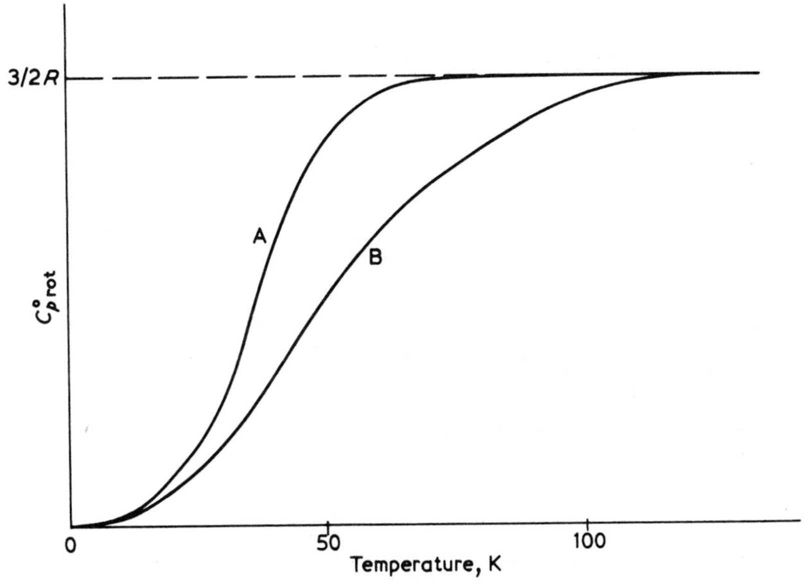

Fig. 7.1 Calculated $C_p{}^\circ{}_{\mathrm{rot}}$ curves for two triatomic molecules showing the high temperature limit of 3/2 R. This is reached at a lower temperature for A than for B (moment of inertia of A is greater than moment of inertia of B).

included, the summation (7.19) has to be used. The values given in (7.30) and (7.31) are thus the asymptotic high temperature limits.

For most gases the maximum rotational contribution to these quantities has been reached by room temperature. Figure 7.1 shows the calculated $C_p{}^\circ{}_{\mathrm{rot}}$ curve for two typical small triatomic molecules. The moments of inertia of A are larger than those of B, and it can be seen that for the former the asymptotic value is reached at a lower temperature than for the latter. Also note that the value of $C_p{}^\circ = 3/2\,R$ is independent of the moments of inertia of the molecule.

The vibrational contribution is found by substitution of (7.17) into (7.5) and (7.6), and for the harmonic case:

$$U^\circ{}_{\mathrm{vib}} = H^\circ{}_{\mathrm{vib}} = \frac{Rh}{k}\sum_i \frac{g_i\nu_i \exp(-h\nu_i/kT)}{1-\exp(-h\nu_i/kT)} \tag{7.32}$$

and

$$C_v{}^\circ{}_{\mathrm{vib}} = C_p{}^\circ{}_{\mathrm{vib}} = R\left(\frac{h}{kT}\right)^2 \sum_i \frac{g_i\nu_i{}^2 \exp(-h\nu_i/kT)}{[1-\exp(-h\nu_i/kT)]^2} \tag{7.33}$$

Hence, if the fundamental frequencies ν_i and their degeneracies are known, these thermodynamic quantities can be simply calculated. Tables of the contributions from each fundamental to $U^\circ{}_{\mathrm{vib}}(H^\circ{}_{\mathrm{vib}})$ and $C_v{}^\circ{}_{\mathrm{vib}}(C_p{}^\circ{}_{\mathrm{vib}})$ are available [7.5] and give the contributions in terms of the ratio ν_i/T. If anharmonic vibrations are considered, the situation becomes more complex but as noted previously the

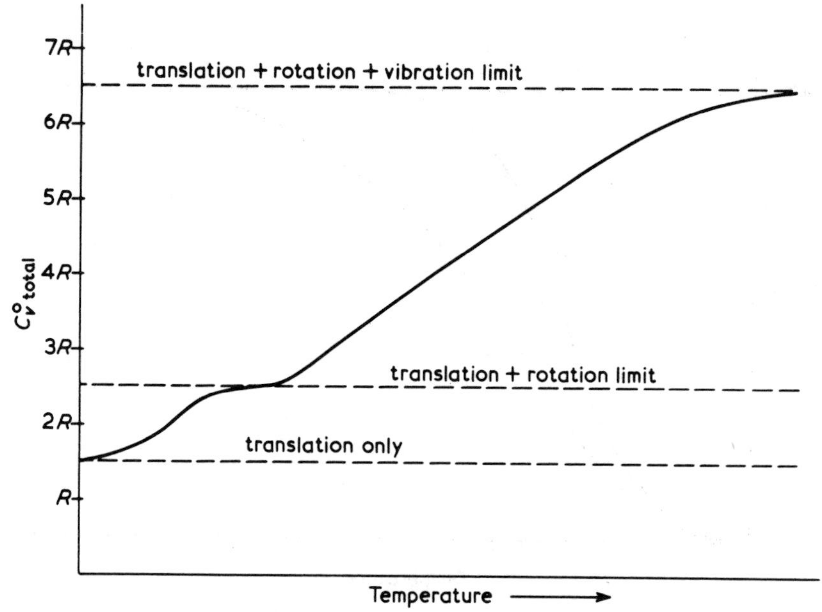

Fig. 7.2 Typical $C_v^\circ{}_{\text{total}}$ curve for a linear triatomic molecule showing how the rotational and then the vibrational high temperature limits are reached as the temperature is raised.

harmonic oscillator (for Z_{vib}) and rigid rotor (for Z_{rot}) models are generally quite satisfactory for most temperatures. The correction to be added if an anharmonic oscillator is considered is only of the order of a few tenths of a per cent in S°, C_p°, etc. Table 7.2 shows the effect of such modifications for the N_2 molecule. The high temperature limits of Equations (7.32) and (7.33) are:

$$\begin{aligned} U^\circ{}_{\text{vib}}(H^\circ{}_{\text{vib}}) &\to RT \sum_i g_i, \\ C_v^\circ{}_{\text{vib}}(C_p^\circ{}_{\text{vib}}) &\to R \sum_i g_i \end{aligned} \tag{7.34}$$

Each non-degenerate vibration therefore contributes R towards the specific heat of the molecule as reckoned using the Boltzmann equipartition law. However, at normal temperatures for most molecules the calculated values fall short of these limits. Figure 7.2 shows a typical $C_v^\circ{}_{\text{total}}$ curve for a linear triatomic molecule. At low temperatures the specific heat is close to the translational only contribution of $3/2\,R$. As the temperature increases, the rotational contribution increases in importance and a value of $C_v^\circ{}_{\text{total}} = C_v^\circ{}_{\text{trans}} + C_v^\circ{}_{\text{rot}}$ (linear molecule) $= 3/2\,R + 2/2\,R = 5/2\,R$ is found at temperatures around 100 K where the rotational high temperature limit is reached. As the temperature is raised the various vibrational modes begin to contribute to the specific heat, the low frequency modes having the larger effect at low temperature (as can be seen from Table 7.1). Eventually at temperatures of around 1000 K or higher (depending upon ν_i) the high temperature vibrational limit is reached. For the triatomic

Table 7.2 Contribution to the thermodynamic properties of N_2 at $T = 2000$ K [7.1]

	$-[G^\circ(T) - H^\circ(0)]/T/$ J K^{-1} mol^{-1}	$C_p^\circ(T)/$ J K^{-1} mol^{-1}
Translation	169.088	20.786
Rigid rotation	48.647	8.314
Harmonic oscillator	1.720	6.611
Vibrational anharmonicity	0.008	0.096
Rotation–vibration	0.017	0.088
Centrifugal distortion	0.033	0.067
Total	219.513	35.962

molecule under consideration $C_v{}^\circ{}_{\text{vib}}(C_p{}^\circ{}_{\text{vib}}) \to 4R$ and the total specific heat tends to $6.5R$.

For molecules with internal rotations, once the size of the barrier to rotation has been decided (conversely this barrier height may be calculated from observed heat capacity data) the appropriate partition function may be calculated and included in place of the relevant vibrational degree of freedom. Figure 7.3 shows the calculated contributions of $C_p{}^\circ$ for (a) completely free rotation, (b) a barrier height of about 500 cm^{-1}, both with high temperature limit of $1.2R$, and (c) a very large barrier height (i.e. a torsional vibrational motion with a high temperature limit of R). As can be seen the intermediate case behaves like a vibration at low temperatures (where there is not enough energy to surmount the potential hill) and like a rotation at higher temperatures.

In a fashion similar to the methods described above, the molecular entropies and free energy functions can be calculated. In general the functions $(H_T{}^\circ - H_0{}^\circ)$ or $(G_T{}^\circ - G_0{}^\circ)$ are tabulated rather than the functions $H_T{}^\circ$ or $G_T{}^\circ$. In calculating the partition functions leading to evaluation of the relevant thermodynamic properties, the energy zero was assigned to the lowest energy level in the set under consideration. In order to remove this arbitrariness each of the thermodynamic functions is referred to its value calculated at absolute zero.

7.7 THERMODYNAMIC PROPERTIES OF TOLUENE

As a specific example the determination of the thermodynamic properties of toluene by Scott et al. [7.6] will be examined in a little detail. This molecule has 15 atoms (C_7H_8) and its 45 degrees of freedom may be broken down in the following way: 3 translations, 3 overall rotations, 1 internal rotation (of the CH_3 group relative to the benzene ring), and 38 vibrations. The translational and

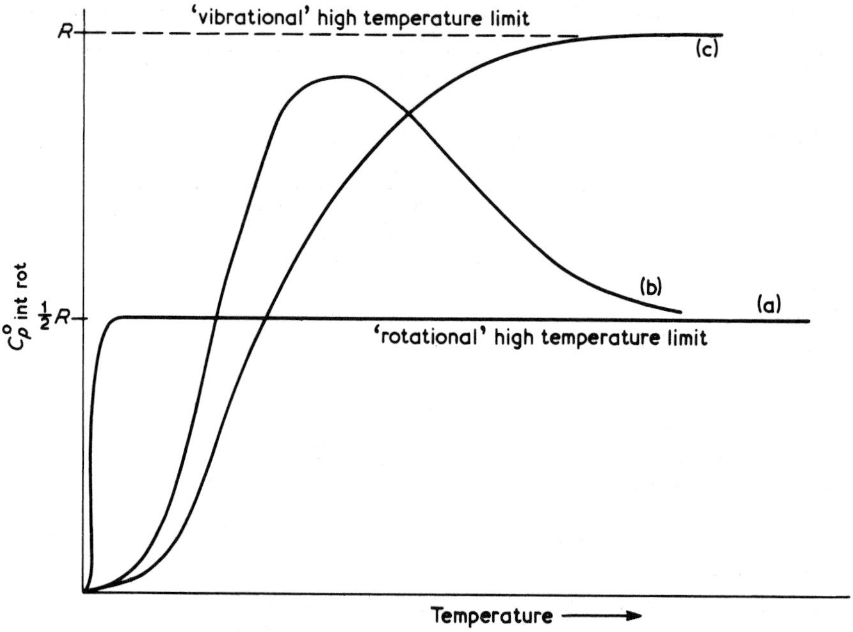

Fig. 7.3 Dependence of $C_p^{\circ}{}_{\text{int rot}}$ on temperature. (a) Zero barrier to rotation; high (rotational) temperature limit of 1/2 R reached quickly. (b) Barrier height of 500 cm^{-1}; at low temperatures behaves as a pure vibration until the rotor possesses enough energy to surmount the barrier. $C_p^{\circ}{}_{\text{int rot}}$ then decreases to free rotation value. (c) Large barrier height; internal rotor behaves as a pure vibration; even at high temperatures there is not enough energy to surmount the potential barrier to rotation.

overall rotational contributions to the thermodynamic functions were readily calculated from a knowledge of the overall mass of the molecule and its molecular dimensions from the formulae presented in earlier sections. In order to calculate the vibrational contribution a list of the fundamental vibrational frequencies is needed. Although it was noted earlier that the assignment of these fundamentals to particular molecular motions was not necessary in determining the thermodynamic functions, a vibrational assignment is usually made so as to make sure all the fundamentals are included and no combination or overtone frequencies. For toluene the fundamental vibrations were assigned and their values measured in the gaseous phase. Liquid phase figures are often slightly different from their gaseous phase counterparts and for the low frequency modes may exert a significant effect on the calculated thermodynamic properties. [The lower the vibrational frequency, the larger the effect on these functions via the exponential term appearing in the expressions of Equation (7.17)]. Thus all low frequency modes at least should be measured under the same conditions. For the thermodynamically unimportant high frequency CH stretching modes for the toluene molecule uncertainty in their actual position exists due to poor resolution and Fermi resonance problems. Once the vibrational assignment was made calculation of the vibrational contribution to the thermodynamic properties was straightforward.

Internal rotation in toluene has six-fold symmetry and so $\sigma = 6$. For this molecule there is evidence that the barrier to internal rotation should be low. In fact, comparison between experimental data and calculated thermodynamic properties based on this assumption shows the barrier height to be so low as to be thermodynamically unimportant. The internal rotation contribution is then readily calculated using Z_{tors} given in Section 7.3.

For molecules of this size empirical anharmonicity corrections are generally made since as seen in Chapter 6 actual anharmonicity constants are generally unavailable. The corrections included led to small differences at 300 K but were larger as the temperature increased (Table 7.3). The agreement between the observed and calculated values of the entropy and specific heat shows that the internal rotation is essentially free.

Table 7.3 Calculated and observed thermodynamic properties of toluene (from refs. 7.1 and 7.6); units J K^{-1} mol^{-1}

	$S°$ (298.15 K)	$S°$ (1000 K)	$C_p°$ (427.20 K)	$C_p°$ (1000 K)
Translational	165.16	190.31	20.79	20.79
Overall rotation	107.08	122.18	12.47	12.47
Free internal rotation	14.80	19.83	4.16	4.16
Vibrations	33.62	209.04	111.15	223.50
Anharmonicity	0.01	1.89	0.60	4.04
Total calculated	320.67	543.25	149.17	264.96
Observed	321.21		149.16	

7.8 EQUILIBRIUM CONSTANTS

The importance of the calculated free energy values lies in the theoretical determination of the positions of equilibria in chemical processes. Quite often, although spectroscopic data are available for some of the reactants and products, experimentally determined thermodynamic data are not. Hence, the only route to values of the thermodynamic quantities is in the theory outlined in this chapter. As an example consider the formation of ethanol from its elements [7.7]:

$$2C(\text{graphite}) + 3H_2(g) + \tfrac{1}{2}O_2(g) \rightleftharpoons C_2H_6O(g)$$

The values for the thermodynamic functions of the elements at the temperature T can be found from tables, and in principle from spectroscopic data, values of the free energy, specific heats, etc. referred to the lowest energy level of ethanol can be calculated. In order to be able directly to compare the thermodynamic properties of the left-hand side with those of the right-hand side, one linking piece of data is needed. This is the standard heat of formation of ethanol from its constituent elements at any temperature. Once this is known, all the other thermodynamic quantities ΔG_f°, K_f° etc. can be readily calculated at any desired temperature.

REFERENCES

7.1 Frankiss, S.G. and Green, J.H.S., Chemical Society Specialist Periodical Reports, Chemical Thermodynamics Vol. I, p.268, (1973).

7.2 Fowler, R.H. and Guggenheim, E.A., 'Statistical Thermodynamics', C.U.P. Cambridge (1939).

7.3 Aston, J.G. and Fritz, J.J., 'Thermodynamics and Statistical Thermodynamics', Wiley, New York (1959).

7.4 Lewis, G.N. and Randall, M., 'Thermodynamics' McGraw-Hill (1961); Pitzer, K.S. and Gwinn, W.D., *J. Chem. Phys.* **10**, 248 (1942).

7.5 Johnston, H.L., Savedorff, L. and Beltzer, J., Contributions to the Thermodynamic Functions by a Planck–Einstein Oscillator in One Degree of Freedom. Office of Naval Research, Washington D.C., (1949).

7.6 Scott, D.W., Guthrie, G.B., Messerly, J.F., Todd, S.S., Berg, W.T., Hossenlopp, I.A. and McCullough, J.P., *J. Phys. Chem.* **66**, 911 (1962).

7.7 Green, J.H.S., *Trans. Faraday Soc.* **57**, 2132 (1961).

A Appendix

The object of this chapter is to indicate how vibrational and rotational wave-functions and energy levels and the various spectroscopic selection rules may be derived in terms more mathematical than in the main body of the text. However, this chapter still only indicates the route along which this information emerges; for a fuller treatment the reader is referred to references [1–6] at the end of the chapter.

A.1 ABSORPTION AND EMISSION OF RADIATION

By using time-dependent perturbation theory, the rate of absorption of radiation via an electric dipole process from a lower state l to an upper state u can be readily calculated. For a molecular system bathed in radiation of energy density ρ of frequency ν_{lu} corresponding to the transition $l \rightarrow u$, this rate is given by $n_l B_{ul} \rho$. B is called *Einstein's coefficient of induced absorption* and is equal to:

$$B_{ul} = \frac{8\pi^2}{3h^2} \left| \int \psi_l^* \mu \psi_u \, \mathrm{d}\tau \right|^2 = \frac{8\pi^2}{3h^2} R_{ul}^2 \qquad \text{(A.1)}$$

where the integral R_{ul} is called the *transition moment*. n_l is the fraction of molecules in the lower state l and conventionally the first subscript of B_{ul} corresponds to the final state and the second to the initial state. μ is the dipole moment. If n is the number of molecules per litre of sample irradiated, then the change in intensity of the light beam after it has passed through a length δx is given by:

$$-\delta I = \frac{8\pi^2}{3h^2} R_{ul}^2 \rho h \nu_{lu} n \delta x \qquad \text{(A.2)}$$

The energy density is simply related to the intensity of the radiation by the velocity of light, $I = c\rho$, and thus:

$$-\delta I = \frac{8\pi^2}{3h^2} R_{ul}^2 \frac{I}{c} h\nu_{lu} n \delta x \tag{A.3}$$

If C is the concentration in moles dm^{-3}, then:

$$n = CN \tag{A.4}$$

where N is Avogadro's number, and:

$$\delta I = -\frac{8\pi^3}{3h^2} R_{ul}^2 \frac{I}{c} h\nu_{lu} NC\delta x \tag{A.5}$$

This represents the amount of light absorption over the entire absorption band due to the transition $l \rightarrow u$. Beer's law $I(x)/I(0) = \exp[-\epsilon(\nu)\cdot C\cdot x]$ may be written in differential form as:

$$\delta I = -\epsilon(\nu) IC\delta x \tag{A.6}$$

and the terms compared between Equations (8.5) and (8.6). The integrated absorption intensity $A = \int \epsilon(\nu)\,\mathrm{d}\nu$ is simply obtained as:

$$A = \frac{8\pi^3 N}{3hc} R_{ul}^2 \nu_{lu} \tag{A.7}$$

For electronic transitions, R_{ul} represents the transition dipole moment between the two electronic states; for rotational and rotational-vibrational transitions the dipole moment needs to be expanded; for the simple case of a diatomic system as:

$$\mu = \mu_0 + \left(\frac{\partial \mu}{\partial x}\right)_{x=0} x + \frac{1}{2}\left(\frac{\partial^2 \mu}{\partial x^2}\right)_{x=0} x^2 + \ldots \tag{A.8}$$

where x is the displacement of the system from equilibrium $(r - r_e)$. (At the equilibrium position $\mu = \mu_0$, the permanent dipole moment.) The application of these equations to electronic and vibration transitions will be considered below.

Decay from an excited state to a state of lower energy may occur by two radiative processes, *spontaneous* and *stimulated* emission. Stimulated emission is entirely analogous to the induced emission process introduced above, in that the rate of emission of quanta is proportional to the radiation density of the transition frequency ν_{lu} and to the fraction of molecules in the excited state. The rate of emission by this route is thus equal to $n_u B_{lu} \rho(\nu_{lm})$ where B_{lu} is *Einstein's coefficient of stimulated emission.* Spontaneous emission occurs via a pathway which is independent of the power density of the radiation and occurs at a rate $n_u A_{lu}$ where A_{lu} is *Einstein's coefficient of spontaneous emission*. This is equal to:

$$A_{lu} = \frac{64\pi^4 \nu_{ul}^3}{3h} \left| \int \psi_u^* \mu \psi_l \, d\tau \right|^2 \tag{A.9}$$

At any time for a system in equilibrium the number of molecules in the states l and u must be constant, i.e. the rate at which molecules enter the state u must equal the number leaving it. Thus:

$$n_u A_{lu} + n_u B_{lu} \rho(\nu_{lu}) = n_l B_{ul} \rho(\nu_{lu}) \tag{A.10}$$

For non-degenerate states l and u the Boltzmann distribution gives a relationship between n_l and n_u:

$$n_u = n_l \exp\left(-\frac{h\nu_{lu}}{kT}\right)$$

and for a system in thermal equilibrium with the radiation, it can be shown that for non-degenerate levels:

$$B_{lu} = B_{ul} \tag{A.11}$$

and

$$\frac{A_{lu}}{B_{lu}} = \frac{8\pi h}{c^3} \nu_{lu}^3 \tag{A.12}$$

from which it can be seen that the probability of spontaneous emission depends upon ν^3. This spontaneous emission is likely for electronically excited states, but for vibrational transitions, or rotational transitions, where ν^3 becomes very small, spontaneous emission processes have very low probability.

The relative rate of decay from u to l is given by:

$$\frac{A_{lu}}{B_{lu}\rho(\nu_{lu})} = \exp\left(\frac{h\nu_{lu}}{kT}\right) - 1 \tag{A.13}$$

a relationship of great importance in understanding the operation of a laser or maser. For a system in thermal equilibrium with its room temperature surroundings, the two processes (spontaneous and stimulated) proceed at comparable rates for frequencies (ν_{lu}) in the microwave region. Thus the presence of a large radiation density at long wavelengths is often sufficient to produce a large proportion of stimulated compared with spontaneous emission, and maser (microwave) action may be produced. For systems where the transition frequency ν_{lm} lies to higher energy, the stimulated route is of negligible importance and a *population inversion* ($n_u > n_l$) has to be artificially forced upon the system in order to promote laser action.

It can be seen from Equation (A.10) that the application of a high radiation density ρ can only lead to the populations of upper and lower levels n_u and n_l becoming equal in the two-level situation, in which case the transition is saturated. It is only possible therefore to observe net stimulated emission from such a system if a pumping mechanism other than optical excitation is used to create a population inversion with respect to the lower state (i.e. $n_u > n_l$), or if a third pumping level is utilized so that the optical pumping frequency and stimulated emission frequencies are different.

A.2 ENERGY LEVELS OF A LINEAR RIGID ROTOR

To evaluate the energy levels of a rigid linear rotor, the Schrödinger equation needs to be solved for a freely rotating molecule in three dimensions.

Provided that the rotor is rigid, no potential energy is associated with rotation since the motion is that of the molecule as a whole. Hence the potential energy V is zero, and if E_r is the rotational energy of the rotor, then the Schrödinger equation becomes:

$$\frac{\partial^2\psi}{\partial x^2}+\frac{\partial^2\psi}{\partial y^2}+\frac{\partial^2\psi}{\partial z^2}+\frac{8\pi^2\mu}{h^2}E_r\psi = 0 \qquad \text{(A.14)}$$

where μ is the reduced mass of the rotor.

Transforming into spherical polar coordinates:

$$\frac{1}{r^2}\frac{\partial}{\partial r}\left(r^2\frac{\partial\psi}{\partial r}\right)+\frac{1}{r^2\sin^2\theta}\frac{\partial^2\psi}{\partial\phi^2}+\frac{1}{r^2\sin\theta}\frac{\partial}{\partial\theta}\left(\sin\theta\frac{\partial\psi}{\partial\theta}\right)+\frac{8\pi^2\mu E_r\psi}{h^2} = 0 \qquad \text{(A.15)}$$

or

$$\frac{\partial^2\psi}{\partial r^2}+\frac{2}{r}\frac{\partial\psi}{\partial r}+\frac{1}{r^2\sin^2\theta}\frac{\partial^2\psi}{\partial\phi^2}+\frac{1}{r^2\sin\theta}\frac{\partial}{\partial\theta}\left(\sin\theta\frac{\partial\psi}{\partial\theta}\right)+\frac{8\pi^2\mu E_r}{h^2}\psi = 0 \qquad \text{(A.16)}$$

The wavefunction may be written as a product of a radial and an angular function:

$$\psi = R(r)\cdot S(\theta,\phi) \qquad \text{(A.17)}$$

where $R(r)$ is a function of the r alone and $S(\theta,\phi)$ of only θ and ϕ. On differentiation of Equation (A.17) and on substitution for:

$$\frac{\partial\psi}{\partial r},\ \frac{\partial^2\psi}{\partial r^2},\ \frac{\partial^2\psi}{\partial\phi^2},\ \text{and}\ \frac{\partial\psi}{\partial\theta}$$

in Equation (A.16) we obtain, by multiplication throughout by r^2/RS:

$$\frac{r^2}{R}\frac{\partial^2R}{\partial r^2}+\frac{2r}{R}\frac{\partial R}{\partial r}+\frac{1}{S\sin^2\theta}\frac{\partial^2S}{\partial\phi^2}+\frac{1}{S\sin\theta}\frac{\partial}{\partial\theta}\left(\sin\theta\frac{\partial S}{\partial\theta}\right)+\frac{8\pi^2\mu r^2E_r}{h^2} = 0 \qquad \text{(A.18)}$$

where R is an abbreviation for $R(r)$ and S for $S(\theta,\phi)$.

Since for a rigid rotor r is constant, the radial part of the wavefunction must be constant, and the first two terms in Equation (A.16) must vanish. Multiplying through Equation (A.18) by S and making the substitution moment of inertia $= I = \mu r^2$:

$$\frac{1}{\sin^2\theta}\frac{\partial^2 S}{\partial\phi^2}+\frac{1}{\sin\theta}\frac{\partial}{\partial\theta}\left(\sin\theta\frac{\partial S}{\partial\theta}\right)+\frac{8\pi^2 IE_r S}{h^2} = 0 \qquad \text{(A.19)}$$

From mathematical considerations it can be shown that only for certain values of E_r can solutions be obtained which are single-valued and finite. These values of E_r are given by:

$$E_r = J(J+1)h^2/8\pi^2 I \qquad \text{(A.20)}$$

where J is the rotational quantum number which may take only integral values from zero upwards. This is the equation which is employed in the analysis of the rotational structure of a linear molecule, and it is only strictly accurate when the molecule may be regarded as a rigid rotor. In practice the equation holds only for low J values where the molecule behaves nearly as a rigid rotor. In general for linear molecules the rotational energy is described more accurately by:

$$E_r = hc[BJ(J+1) - DJ^2(J+1)^2] \qquad \text{(A.21)}$$

where D is the centrifugal distortion constant. It is important especially for high J values where, as the molecule rotates rapidly, centrifugal forces stretch the bond length in excess of the equilibrium value. The value of D thus depends on the stiffness of the bond (i.e. the force constant, f), and a good approximation is:

$$D = \frac{16\pi^2 B^3 \mu}{f} \qquad \text{(A.22)}$$

Classically, the rotational energy E_r of a rigid body is given by:

$$E_r = \tfrac{1}{2}I\omega^2 \qquad \text{(A.23)}$$

where ω is the angular velocity of rotation and I is the moment of inertia about the axis of rotation. The angular momentum P is:

$$P = I\omega \qquad \text{(A.24)}$$

Substituting Equation (A.24) into Equation (A.23), E_r is given by:

$$E_r = P^2/2I \qquad \text{(A.25)}$$

On comparison of Equations (A.25) and (A.20) it follows that for a linear rigid rotor:

$$P^2/2I = J(J+1)h^2/8\pi^2 I \qquad \text{(A.26)}$$

$$P^2 = J(J+1)h^2/4\pi^2 \qquad \text{(A.27)}$$

or

$$P = \sqrt{[J(J+1)]}\,h/2\pi \qquad \text{(A.28)}$$

Hence, the magnitude of the total angular momentum vector is given by:

$$\sqrt{[J(J+1)]}\,h/2\pi$$

The rotational wavefunctions of importance here are identical to the familiar angular part of the hydrogen atom wavefunctions. Each solution is represented by two quantum numbers J and M_J which take on integral numbers and $|M_J| \leqslant J$. Whereas the total angular momentum vector has a magnitude given by Equation (A.28) the component of angular momentum along a defined axis is given by $M_J h/2\pi$, just as in the atomic case. The angular momentum component in a direction in space is thus quantized since M_J may only adopt certain values.

A.3 SELECTION RULES FOR THE LINEAR RIGID ROTOR

The transition probabilities for the absorption or emission of electromagnetic radiation to bring about a pure rotational energy change of the molecule are governed by the eigenfunctions ψ_m and ψ_n of the rotational energy states between which the transition takes place. In addition, from a knowledge of the eigenfunctions of the two rotational levels it is possible to determine the selection rules for the rotational transitions.

It is convenient to write the probability of absorption as described by Equation (A.1) in terms of the probabilities in three mutually perpendicular directions. Then the probability of a transition between the states m and n is proportional to:

$$|R_x^{nm}|^2 + |R_y^{nm}|^2 + |R_z^{nm}|^2 \qquad \text{(A.29)}$$

where R_x^{nm}, R_y^{nm} and R_z^{nm} are the components of the transition moment in the x, y and z directions:

$$\begin{aligned} R_x^{nm} &= \int \psi_n^* \mu_x \psi_m \, d\tau \\ R_y^{nm} &= \int \psi_n^* \mu_y \psi_m \, d\tau \\ R_z^{nm} &= \int \psi_n^* \mu_z \psi_m \, d\tau \end{aligned} \qquad \text{(A.30)}$$

If one or more of the transition moments, R_x^{nm}, R_y^{nm} and R_z^{nm} are different from zero for the states m and n, there exists a definite probability that radiation may be absorbed or emitted. If, on the other hand, all the transition moments vanish along the x, y and z axes, then the transition is forbidden.

The selection rules can be calculated from these transition moments. The electric dipole moment components of a rotor in terms of polar coordinates are simply:

$$\begin{aligned} \mu_x &= \mu_0 \sin\theta \cos\phi \\ \mu_y &= \mu_0 \sin\theta \sin\phi \\ \mu_z &= \mu_0 \cos\theta \end{aligned} \qquad \text{(A.31)}$$

where θ and ϕ are as defined in Fig. A.1 μ_0 is the permanent dipole moment of the rotor. Substitution of Equation (A.32) into Equation (A.31) gives for the transition moments in the x, y and z directions:

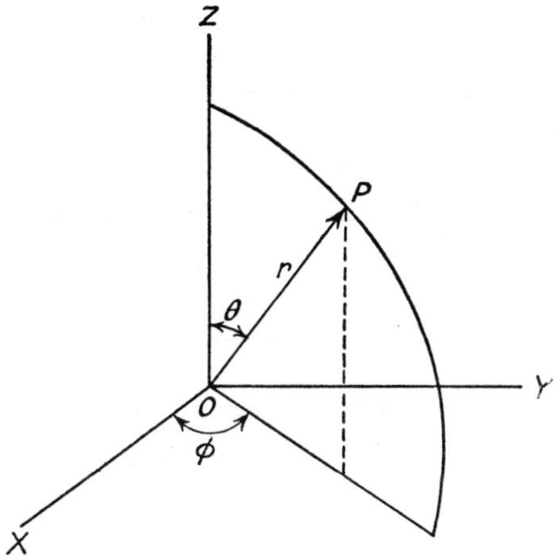

Fig. A.1 Representation of polar co-ordinates.

$$R_x^{J'M'_J J''M''_J} = \mu_0 \int \psi_r^{*J'M'_J} \sin\theta \cos\phi \cdot \psi_r^{J''M''_J} \, d\tau$$

$$R_y^{J'M'_J J''M''_J} = \mu_0 \int \psi_r^{*J'M'_J} \sin\theta \sin\phi \cdot \psi_r^{J''M''_J} \, d\tau \qquad (A.32)$$

$$R_z^{J'M'_J J''M''_J} = \mu_0 \int \psi_r^{*J'M'_J} \cos\theta \cdot \psi_r^{J''M''_J} \, d\tau$$

where ψ_r is the rotor eigenfunction and J' and J'' are the rotational quantum numbers in the upper and lower rotational energy states, while M_J' and M_J'' are the magnetic quantum numbers in the upper and lower energy states. The magnetic quantum number may take the $(2J + 1)$ integral values $M_J = J, (J-1), (J-2), \ldots, -J$.

Obviously if μ_0 is zero, then R_x, R_y, R_z will also be zero, and hence for a linear rotor to absorb or emit radiation it is necessary that μ_0 should not be zero (for transitions between rotational levels only). Furthermore, it can be shown by the detailed evaluation of the transition moment integrals of using the rotational wavefunctions that $R_x^{J'M'_J J''M''_J}$ and $R_y^{J'M'_J J''M''_J}$ vanish except when $J'' = (J' \pm 1)$ and $M_J' = (M_J'' \pm 1)$, and $R_z^{J'M'_J J''M''_J}$ will also vanish except when $J'' = (J' \pm 1)$ and $M_J' = M_J''$. The selection rule for J is then that J can change only by unity, that is $\Delta J = \pm 1$. As regards the additional selection rule $\Delta M_J = 0, \pm 1$ for linear molecules, this is of importance when the $(2J + 1)$ sublevels are not degenerate, e.g. for molecules in electric fields (Stark effect).

Only the interaction of radiation with an electric dipole moment has been considered, but interaction may also take place through a magnetic dipole moment or a quadrupole moment. In the latter two cases the appropriate magnetic dipole or quadrupole moment would have to be substituted for μ_x, μ_y, and

μ_z in Equation (A.18) to find the values of the matrix elements. For example, even if the matrix element of the electric dipole moment is zero, combination of the states m and n could still take place provided that the matrix elements of either the magnetic dipole or quadrupole moment were not equal to zero. (For quadrupole transitions the selection rule $\Delta J = \pm 2$ applies for example). Their intensities, however, are generally much lower than the electric dipole allowed ones.

A.4 ENERGY LEVELS OF A HARMONIC OSCILLATOR

According to classical mechanics the potential energy (V) of a harmonic oscillator consisting of a reduced point mass μ is given by:

$$V = 2\pi^2\mu\omega^2c^2x^2 \tag{A.33}$$

where ω is the frequency of vibration (in cm^{-1}), c is the velocity of light, and x is the displacement of the point mass from its equilibrium position.

The Schrödinger equation for a one-dimensional oscillator is:

$$\frac{\mathrm{d}^2\psi}{\mathrm{d}x^2} + \frac{8\pi^2\mu}{h^2}(E - V)\psi = 0 \tag{A.34}$$

On substitution of Equation (A.33) into Equation (A.34) the wavefunction for a one-dimensional oscillator is obtained:

$$\frac{\mathrm{d}^2\psi}{\mathrm{d}x^2} + \frac{8\pi^2\mu}{h^2}(E - 2\pi^2\mu\omega^2c^2x^2)\psi = 0 \tag{A.35}$$

To simplify Equation (A.35), let:

$$\lambda = 8\pi^2\mu E/h^2 \qquad \text{(A.36)}^\dagger$$

and

$$\alpha = 4\pi^2\mu\omega c/h \tag{A.37}$$

On substitution of Equations (A.36) and (A.37), Equation (A.35) becomes:

$$\frac{\mathrm{d}^2\psi}{\mathrm{d}x^2} + (\lambda - \alpha^2x^2)\psi = 0 \tag{A.38}$$

To obtain satisfactory wavefunctions from Equation (A.38), that is functions $\psi(x)$ which are continuous, single valued, and finite throughout the region $+\infty$ to $-\infty$, the following procedure is adopted.

Initially, the form of ψ is studied in the regions of large positive and negative values of x, then subsequently the general behaviour of ψ is examined. Each case will now be studied in turn.

† This energy parameter λ is not to be confused with wavelength.

A.4.1 Asymptotic solution of the wave equation when x is large.

When x is large then, for any value of the total energy E, λ will be negligibly small compared with $\alpha^2 x^2$, and Equation (A.38) becomes:

$$\frac{d^2\psi}{dx^2} = \alpha^2 x^2 \psi \tag{A.39}$$

Solutions to this equation are:

$$\exp\left(+\frac{\alpha}{2}x^2\right) \tag{A.40}$$

and

$$\exp\left(-\frac{\alpha}{2}x^2\right) \tag{A.41}$$

The first is unacceptable as a wavefunction since it tends rapidly to infinity with increasing values of x. (Wavefunctions must be finite everywhere.)

A.4.2 General solution of the wave equation

In order to obtain an accurate solution of Equation (A.38), we investigate the behaviour of Equation (A.42) where we have multiplied Equation (A.41) by a power series in x, $f(x)$, and see what restrictions must be placed on this function for it to be a solution of Equation (A.38):

$$\psi = \exp\left(-\frac{\alpha}{2}x^2\right) f(x) \tag{A.42}$$

On differentiation of Equation (A.42) twice with respect to x, and writing f for $f(x)$, f' for df/dx, and f'' for d^2f/dx^2:

$$\frac{d^2\psi}{dx^2} = \exp\left(-\frac{\alpha}{2}x^2\right)(\alpha^2x^2f - \alpha f - 2\alpha x f' + f'') \tag{A.43}$$

On substitution for $d^2\psi/dx^2$ from Equation (A.43) into Equation (A.38):

$$\exp\left(-\frac{\alpha}{2}x^2\right)(\alpha^2x^2f - \alpha f - 2\alpha x f' + f'') + (\lambda - \alpha^2x^2)\psi = 0 \tag{A.44}$$

while from Equation (A.42):

$$\exp\left(-\frac{\alpha}{2}x^2\right) = \frac{\psi}{f} \tag{A.45}$$

Thus, Equation (A.44) becomes, on substitution for $\exp[-(\alpha/2)x^2]$ from Equation (A.45):

$$f'' - 2\alpha x f' + (\lambda - \alpha) f = 0 \tag{A.46}$$

For ease in manipulation it is convenient to introduce a new variable s such that:

$$s = x\sqrt{\alpha} \tag{A.47}$$

and to put $f(x) = H(s)$, i.e. expressing $f(x)$ as a power series in s. On differentiation of Equation (A.47):

$$\mathrm{d}s = \sqrt{\alpha}\mathrm{d}x$$

and

$$\frac{\mathrm{d}f}{\mathrm{d}x} = \frac{\mathrm{d}H}{\mathrm{d}s}\frac{\mathrm{d}s}{\mathrm{d}x} = \sqrt{\alpha}\cdot\frac{\mathrm{d}H}{\mathrm{d}s} \tag{A.48}$$

Also:

$$\frac{\mathrm{d}^2 f}{\mathrm{d}x^2} = \sqrt{\alpha}\cdot\frac{\mathrm{d}}{\mathrm{d}s}\left(\sqrt{\alpha}\cdot\frac{\mathrm{d}H}{\mathrm{d}s}\right) = \alpha\frac{\mathrm{d}^2 H}{\mathrm{d}s^2} \tag{A.49}$$

On substitution of Equations (A.48) and (A.49) into Equation (A.46):

$$\frac{\alpha \mathrm{d}^2 H}{\mathrm{d}s^2} - 2\alpha x\sqrt{\alpha}\frac{\mathrm{d}H}{\mathrm{d}s} + (\lambda - \alpha)H = 0 \tag{A.50}$$

or

$$\frac{\mathrm{d}^2 H}{\mathrm{d}s^2} - 2s\frac{\mathrm{d}H}{\mathrm{d}s} + \left(\frac{\lambda}{\alpha} - 1\right)H = 0 \tag{A.51}$$

To solve Equation (8.51), $H(s)$ is represented as the power series:

$$H(s) = \sum a_v s^v = a_0 + a_1 s + a_2 s^2 + a_3 s^3 + \ldots \tag{A.52}$$

$$\frac{\mathrm{d}H}{\mathrm{d}s} = \sum v a_v s^{v-1} = a_1 + 2a_2 s + 3a_3 s^2 + \ldots \tag{A.53}$$

$$\frac{\mathrm{d}^2 H}{\mathrm{d}s^2} = \sum v(v-1) a_v s^{v-2} = 1\cdot 2a_2 + 2\cdot 3a_3 s + \ldots \tag{A.54}$$

On substitution of Equations (A.52)–(A.54) into (A.51) we get:

$$1\cdot 2a_2 + 2\cdot 3a_3 s + 3\cdot 4a_4 s^2 + 4\cdot 5a_5 s^3 + \ldots - 2a_1 s - 2\cdot 2a_2 s^2 - 2\cdot 3a_3 s^3 - \ldots$$
$$+\left(\frac{\lambda}{\alpha} - 1\right)a_0 + \left(\frac{\lambda}{\alpha} - 1\right)a_1 s + \left(\frac{\lambda}{\alpha} - 1\right)a_2 s^2 + \left(\frac{\lambda}{\alpha} - 1\right)a_3 s^3 + \ldots = 0 \tag{A.55}$$

For this series to vanish for all values of s, or in other words, for $H(s)$ to be a solution of Equation (A.51), the coefficients of individual powers of s must vanish separately, that is:

$$
\begin{aligned}
1\cdot 2a_2 + \left(\frac{\lambda}{\alpha} - 1\right) a_0 &= 0 \\
2\cdot 3a_3 - 2a_1 + \left(\frac{\lambda}{\alpha} - 1\right) a_1 &= 0 \\
3\cdot 4a_4 - 2\cdot 2a_2 + \left(\frac{\lambda}{\alpha} - 1\right) a_2 &= 0 \\
4\cdot 5a_5 - 2\cdot 3a_3 + \left(\frac{\lambda}{\alpha} - 1\right) a_3 &= 0
\end{aligned}
\tag{A.56}
$$

In general for the coefficients of s^v Equation (A.57) may be written:

$$(v+1)(v+2)\,a_{v+2} + \left(\frac{\lambda}{\alpha} - 1 - 2v\right) a_v = 0 \tag{A.57}$$

or

$$a_{v+2} = -\frac{(\lambda/\alpha - 2v - 1)}{(v+1)(v+2)} \tag{A.58}$$

Equation (A.58) is called a recursion formula and enables coefficients $a_2, a_3, a_4, \ldots$ to be calculated successively in terms of a_0 and a_1. If $a_0 = 0$, then only odd powers appear, while if $a_1 = 0$, only even powers are present in the series. For arbitrary values of the energy parameter λ the series consists of an infinite number of terms and increases too rapidly to correspond to a satisfactory wavefunction. Values of λ must be chosen such that the series $H(s)$ terminates leaving a polynomial with a finite number of terms. An odd or even polynomial of degree v will be obtained according as $a_0 = 0$ or $a_1 = 0$ respectively; it follows from (A.58) that the λ/α value which causes the series to cease at the vth term is:

$$\frac{\lambda}{\alpha} = (2v + 1) \tag{A.59}$$

On substitution of the values of λ and α from Equations (A.36) and (A.37), respectively, (A.59) becomes:

$$E_v = (v + \tfrac{1}{2})hc\omega \tag{A.60}$$

where $v = 0, 1, 2, 3, \ldots$. This is one of the basic equations in vibrational spectra studies, since the vibrational energy (E_v) is related to the vibrational quantum number which characterizes a particular vibrational mode. For diatomic molecules in infrared and electronic spectra studies, the equation is usually modified, to take account of mechanical anharmonicity, to:

$$E_v = (v + \tfrac{1}{2})hc\omega_e - (v + \tfrac{1}{2})^2 hcx_e\omega_e + (v + \tfrac{1}{2})^3 hcy_e\omega_e + \ldots \tag{A.61}$$

In polyatomic molecules the basic equation is usually retained, and for vibrational modes the vibrational energy in the absence of anharmonicity is given by:

$$E_v = \sum_i (v_i + g_i/2)hc\omega_i \tag{A.62}$$

where g_i is the degeneracy of the vibration.

A.5 CALCULATION OF THE VIBRATIONAL EIGENFUNCTIONS FOR A DIATOMIC MOLECULE

The series whose coefficients are characterized by Equations (A.58) and (A.60) with $a_0 = 0$ (v odd) and $a_1 = 0$ (v even) is the multiple of the Hermite polynomial $H_v(s)$ which can be expressed as:

$$H_v(s) = (-1)^v \exp(s^2) \frac{d^v[\exp(-s^2)]}{ds^v} \tag{A.63}$$

The solution of Equation (A.35) may be written in the form:

$$\psi_v = N_v \exp(-\tfrac{1}{2}s^2) H_v(s) \tag{A.64}$$

in which

$$s = x\sqrt{\alpha}$$

$H_v(s)$ is a polynomial of the vth degree in s. N_v is a constant such that ψ_v is normalized, that is:

$$\int_{-\infty}^{+\infty} \psi_v^* \psi_v \, ds = 1 \tag{A.65}$$

The value of the normalization constant N_v is given by:

$$N_v = \left[\left(\frac{\alpha}{\pi}\right)^{\frac{1}{2}} \frac{1}{2^v v!}\right]^{\frac{1}{2}} \tag{A.66}$$

On substitution of Equation (A.66) into Equation (A.64) the complete wavefunction becomes:

$$\psi_v = \left[\left(\frac{\alpha}{\pi}\right)^{\frac{1}{2}} \frac{1}{2^v v!}\right]^{\frac{1}{2}} \exp(-\tfrac{1}{2}s^2) H_v(s) \tag{A.67}$$

The first four members of the Hermite polynomial are:

$$\begin{aligned} H_0(s) &= 1 \\ H_1(s) &= 2s \\ H_2(s) &= 4s^2 - 2 \\ H_3(s) &= 8s^3 - 12s \end{aligned} \tag{A.68}$$

For the lowest vibrational level ($v = 0$) the value of ψ_0 can be calculated by the use of Equations (A.67) and (A.68) and is:

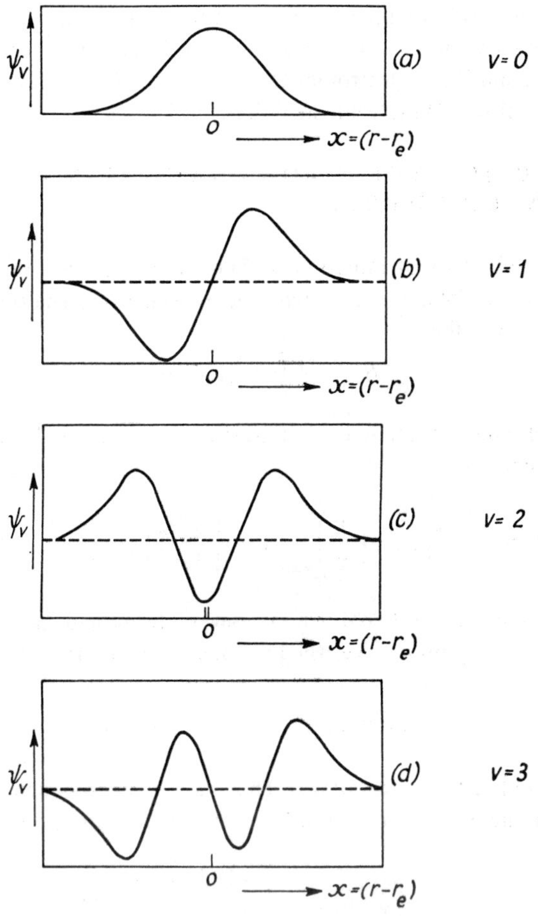

Fig. A.2 Eigenfunctions for the vibrational levels 0, 1, 2, 3, of a harmonic oscillator.

$$\psi_0 = \left(\frac{\alpha}{\pi}\right)^{\frac{1}{4}} \exp(-\tfrac{1}{2}s^2) = \left(\frac{\alpha}{\pi}\right)^{\frac{1}{4}} \exp\left(-\frac{\alpha}{2}x^2\right) \tag{A.69}$$

On substitution of $v = 1, 2, 3, \ldots$ in Equation (A.67) the ψ_1, ψ_2 and ψ_3 values, respectively, can be determined.

From the known reduced mass and fundamental vibrational frequency of a molecule, α be calculated and hence ψ_v written as a function of the displacement (x) of the nuclei in a diatomic molecule from the equilibrium position. The form of the eigenfunctions for values of the vibrational quantum number equal to 0, 1, 2, 3, are shown in Fig. 2 where the eigenfunctions are plotted on the ordinate axis and the displacement x from the equilibrium position along the abscissa. Since H_v is a polynomial of the vth degree with v real zero values,

ψ_v will cross the abscissa v times as shown in the figure. For band intensity considerations the ψ_v values are plotted on the actual potential-energy curves for both the upper and lower electronic states. An example of this procedure is given in Fig. 2.27 for the RbH molecule in Vol. 3.

A.6 SELECTION RULES FOR CHANGES IN VIBRATION QUANTUM NUMBERS

The absorption intensity of Equation (A.7) reduces to a one-dimensional problem for the case of a vibrating diatomic molecule AB. The transition moment may be written:

$$R_{lu}^{x} = \int \psi_l^{*} \mu_x \psi_u \, dx \tag{A.70}$$

where the A—B bond lies along the x-axis. If μ_x is now expanded as in Equation (A.8) above, then:

$$R_{lu}^{x} = \int \psi_l^{*} \left[\mu_0 + \left(\frac{\partial \mu}{\partial x}\right)_{x=0} x + \frac{1}{2}\left(\frac{\partial^2 \mu}{\partial x^2}\right)_{x=0} x^2 + \ldots \right] \psi_u \, dx \tag{A.71}$$

This expression may be broken down into several terms and each one considered separately. The first term involves the permanent dipole moment μ_0 of the molecule:

$$\int \psi_l^{*} \mu_0 \psi_u \, dx = \mu_0 \int \psi_l^{*} \psi_u \, dx \tag{A.72}$$

(A.72) is zero unless $l = u$, i.e. $\Delta v = 0$. This is a direct consequence of the fact that the vibrational eigenfunctions of Section A.5 are orthonormal, i.e.

$$\int \psi_l^{*} \psi_u \, dx = \begin{cases} 1 \text{ if } l = u \\ 0 \text{ otherwise} \end{cases} \tag{A.73}$$

Thus the intensity of absorption due to changes in rotational quantum number with zero change in vibrational quantum number ($\Delta v = 0$) are determined by Equation (A.72). As discussed earlier if $\mu_0 = 0$ (i.e. the molecule has no permanent dipole moment), then no pure rotation spectrum may be observed. A similar comment applies to polyatomic molecules.

The second term in (A.71) involves the change in dipole moment for a unit displacement around the equilibrium position, and may be written:

$$\left(\frac{\partial \mu}{\partial x}\right)_{x=0} \int \psi_l^{*} x \psi_u \, dx \tag{A.74}$$

From the properties of the vibrational eigenfunctions of section A.5, Equation (A.74) will only be non zero if l and u represent vibrational levels which differ in vibrational quantum number, v by 1. Thus the selection rule for a vibrational transition is $\Delta v = \pm 1$ and the intensity of the absorption proportional to $(\partial\mu/\partial x)^2$, i.e. there must be a change in dipole moment on vibration. The term in Equation (8.71) containing $(\partial^2\mu/\partial x^2)$ and x^2 similarly allows transitions between vibrational levels with $\Delta v = \pm 2$. However, this second differential is generally very much smaller than $\partial\mu/\partial x$ so that such overtones ($\Delta v = 2$) are usually considerably weaker in intensity than the corresponding fundamentals ($\Delta v = 1$). The presence of non-zero terms of the type $\partial^2\mu/\partial x^2$ is said to give rise to *electrical anharmonicity* to be compared with the similar terms leading to *mechanical anharmonicity* in the expansion of the vibrational potential energy in similar form to Equation (A.8) as in Equation (6.18) (Chapter 6).

A.7 ABSOLUTE INTENSITIES OF ABSORPTIONS

Using Equation (A.74) the absolute intensity of a vibrational absorption band can be calculated by substitution of algebraic values of ψ_l and ψ_u obtained from Section A.5. Noting that $\mathrm{d}\tau = \mathrm{d}x = \mathrm{d}s/\sqrt{\alpha}$, the transition moment for the vibrational $v = 0 \rightarrow 1$ transition is:

$$\int_{-\infty}^{\infty} \left(\frac{\alpha}{\pi}\right)^{\frac{1}{4}} \exp(-s^2/2) \cdot \left(\frac{\partial\mu}{\partial x}\right)_0 x \left(\frac{\alpha}{\pi}\right)^{\frac{1}{4}} \sqrt{2}s \exp(-s^2/2) \cdot \frac{1}{\sqrt{\alpha}} \mathrm{d}s \qquad \text{(A.75)}$$

which can be simply integrated to give:

$$\mu_{10}{}^x = \frac{1}{\sqrt{(2\alpha)}} \left(\frac{\partial\mu}{\partial x}\right)_0 \qquad \text{(A.76)}$$

On substitution of the value for α [Equation (A.37)], the integrated absorption intensity becomes:

$$A = \frac{\pi N}{3c\mu} \left(\frac{\partial\mu}{\partial x}\right)_0^2 \qquad \text{(A.77)}$$

Experimental determination of A then leads to values for $\partial\mu/\partial x$. For systems other than diatomics the situation is more complex, as noted in Chapter 4, since a particular normal mode is in general a mixture of internal vibrations involving different groups of atoms.

By using a simple model and this result for the vibrational system, the intensity of an electronic absorption band can also be calculated. If it is assumed that the electron is held to the molecule by a Hooke's law (harmonic) type of force, then the harmonic oscillator equations derived in previous sections may be used in the intensity derivation with $\mu = M_e$, the electron mass. For the dipole moment, $\mu^x = ex$, $\mu^y = ey$, and $\mu^z = ez$, and thus the derivatives $\partial\mu/\partial x = e$ etc. From Equation (A.76) and the expression for α (A.37) the transition moment may then be written:

$$R_{01}{}^x = R_{01}{}^y = R_{01}{}^z = \frac{1}{\sqrt{(2\alpha)}}\,e = e\left(\frac{h}{8\pi^2 M_e c\tilde{\nu}}\right)^{\frac{1}{2}} \qquad \text{(A.78)}$$

where $\tilde{\nu}$ is the transition frequency. By adding the contribution from the x, y and z directions:

$$R_{01}{}^2 = \frac{3e^2}{8\pi^2 M_e c\tilde{\nu}} \qquad \text{(A.79)}$$

and on numerical evaluation:

$$A = 2.31 \times 10^{12}\,\text{m}^{-2}\,\text{mol}^{-1}\text{l} \qquad \text{(A.80)}$$

By defining the term *oscillator strength* f, this value is often used as a reference against which to compare observed spectral intensities:

$$f = \frac{\int \epsilon_{\text{obs}}(\nu)\,\text{d}\nu}{2.31 \times 10^{12}} = 4.33 \times 10^{-13} \int \epsilon_{\text{obs}}(\nu)\,\text{d}\nu \qquad \text{(A.81)}$$

Fully allowed electronic transitions have f values close to unity; forbidden transitions which are less intense have f values often several orders of magnitude less.

A.8 ELECTRONIC TRANSITION PROBABILITY AND SPECTRAL INTENSITY

The probability of a transition between two states u and l characterized, respectively, by the total eigenfunctions ψ' and ψ'' is given by the equation:

$$R = \int \psi'^* \mu \psi''\,\text{d}\tau \qquad \text{(A.82)}$$

as seen above.

The total eigenfunction ψ is to a first approximation the product of the electronic ψ_e, vibrational ψ_v, and rotational ψ_r eigenfunctions, respectively, and the reciprocal of the internuclear separation r:

$$\psi = \psi_e \frac{1}{r} \psi_v \psi_r \qquad \text{(A.83)}$$

It can be shown that to a good approximation the rotation of the molecule may be neglected (see Herzberg [A.6]), and Equation (A.83) is modified to:

$$\psi = \psi_e \psi_v \qquad \text{(A.84)}$$

The electric dipole moment operator μ may be divided into two components, the first depending on that for the electrons μ_e and the second on that for the nuclei μ_n. The two dipole moment components are related to μ by:

$$\mu = \mu_e + \mu_n \qquad \text{(A.85)}$$

On substitution of Equations (A.84) and (A.85) into Equation (A.82) the transition moment is given by:

$$R = \int \mu_e \psi_e'^* \psi_v' \psi_e'' \psi_v'' \, d\tau + \int \mu_n \psi_e'^* \psi_v' \psi_e'' \psi_v'' \, d\tau \qquad \text{(A.86)}$$

The volume element $d\tau$ involved in the integrals in Equation (A.86) is the product of two volume elements, namely the volume elements of the nuclear and electron coordinates $d\tau_n$ and $d\tau_e$, respectively. Thus, Equation (A.86) may be written:

$$R = \int \psi_v' \psi_v'' \, d\tau_n \int \mu_e \psi_e'^* \psi_e'' \, d\tau_e + \int \mu_n \psi_v' \psi_v'' \, d\tau_n \int \psi_e'^* \psi_e'' \, d\tau_e \qquad \text{(A.87)}$$

Since $\psi_e'^*$ and ψ_e'' belong to different electronic states it can be shown that they are orthogonal (see Pauling and Wilson (A.1], p. 64) to one another and therefore:

$$\int \psi_e'^* \psi_e'' \, d\tau_e = 0 \qquad \text{(A.88)}$$

Equation (A.87) then becomes:

$$R = \int \psi_v' \psi_v'' \, d\tau_n \int \mu_e \psi_e'^* \psi_e'' \, d\tau_e \qquad \text{(A.89)}$$

Since the only coordinate on which ψ_v depends is the internuclear distance r, $d\tau_n$ may be replaced by dr, and Equation (A.89) becomes:

$$R = \int \psi_v' \psi_v'' \, dr \int \mu_e \psi_e'^* \psi_e'' \, d\tau_e \qquad \text{(A.90)}$$

The matrix element:

$$\int \mu_e \psi_e'^* \psi_e'' \, d\tau_e$$

is called the electronic transition moment R_e, where $|R_e|^2$ is proportional to the electronic transition probability as shown above. As has been shown in previous chapters, symmetry considerations on μ_e, ψ_e', and ψ_e'' will determine whether this integral is zero or non-zero.

Since for different internuclear distances the electron potential energy is different, it follows that the electron eigenfunction ψ_e must depend to some extent on the internuclear separation. Hence, R_e is also dependent on r, but since the variation of ψ_e with r is slow, this variation is often neglected, and R_e is replaced by an average value $\overline{R}_e$. For an electronic transition between the vibrational levels v' and v'' Equation (A.90) becomes on substituting for $\overline{R}_e$:

$$R^{v'v''} = \overline{R}_e \int \psi_v' \psi_v'' \, dr \qquad \text{(A.91)}$$

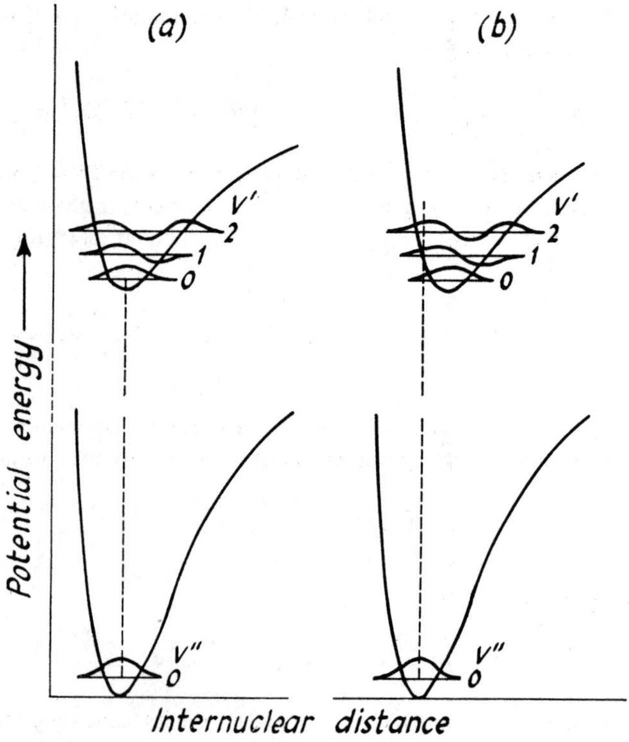

Fig. A.3 Electronic transition where (a) the r_e values for the upper and lower states are the same, (b) the r_e values differ.

The integral over the products of the vibrational eigenfunctions of the two states in Equation (A.91) is known as the overlap integral.

The emission intensity of an electronic transition can simply be derived from Equation (A.9) by noting that the intensity of light is equal to the product (the number of molecules present $N_{v'}$) × (the photon energy) × (the rate of spontaneous emission), i.e.

$$\text{Intensity} = \tfrac{64}{3}\pi^4 N_{v'}\nu^4 \bar{R}_e^{\,2} \left| \int \psi_{v'}\psi_{v''}\, \mathrm{d}r \right|^2 \tag{A.92}$$

In absorption the intensity is given by substitution into Equation (A.7).

In Fig. A.3(a) and (b) potential curves for an upper and the lower electronic states may be observed. Superimposed on these curves are the vibrational eigenfunctions. In Fig. A.3(a) the minima of the potential energy curves lie one above the other while in Fig. A.3(b) the minima are displaced relative to one another. In Fig. A.3(a) the eigenfunctions for the vibrational levels $v' = 0$ and $v'' = 0$ having a maximum value of the integral $\int \psi_{v'}\psi_{v''}\,\mathrm{d}r$ will have a maximum value for this (0, 0) band, which in consequence will be a most intense band. As the minima of the potential energy curves are displaced relative to one another, then the overlap integral value becomes smaller, and the intensity of the (0, 0) band is diminished. In Fig. A.3(b) the best overlap of the vibrational eigenfunctions is seen to be for the (2, 0) band, and therefore will be the most intense band.

Since for the higher vibrational levels in both the electronic states the eigenfunctions have broad maxima or minima near the turning points of the vibrations, maximum values of the overlap integral are obtained when the maximum of the eigenfunction of the lower state lies vertically below the broad maximum or minimum of the upper state. These facts are in accordance with the elementary treatment of the Franck–Condon principle given in Vol. 3. Whether such transitions are observed in practice depends on there being a sufficient number of molecules present in the vibrational levels from which the transition takes place. For example, in absorption only the lowest vibrational levels (v'') are sufficiently populated for the transition to be detected.

REFERENCES

A.1 Pauling, L. and Wilson, E.B., *Introduction to Quantum Mechanics*, McGraw-Hill, New York (1935).

A.2 Atkins, P.W., *Molecular Quantum Mechanics,* Oxford University Press (1970).

A.3 Eyring, W.J., Walter, J. and Kimball, G.E., *Quantum Chemicstry*, John Wiley, New York (1944).

A.4 Wilson, E.B., JR., Decius, J.C. and Cross, P.C., *Molecular Vibrations*, McGraw-Hill, New York (1955).

A.5 Schutte, C.J.H., *Wave Mechanics of Atoms, Molecules and Ions*, Arnold, London (1968).

A.6 Herzberg, G., *Spectra of Diatomic Molecules*, Van Nostrand, New York (1950).

Index